T0258158

# Encyclopedia of Hydrodynamics: Natural Water Sources

# Volume II

# Encyclopedia of Hydrodynamics: Natural Water Sources
# Volume II

Edited by **Fay McGuire**

New York

Published by NY Research Press,
23 West, 55th Street, Suite 816,
New York, NY 10019, USA
www.nyresearchpress.com

**Encyclopedia of Hydrodynamics: Natural Water Sources**
**Volume II**
Edited by Fay McGuire

International Standard Book Number: 978-1-63238-134-7 (Hardback)

Printed in the United States of America.

# Contents

# Preface

The main aim of this book is to educate learners and enhance their research focus by presenting diverse topics covering this vast field. This is an advanced book which compiles significant studies by distinguished experts in the area of analysis. This book addresses successive solutions to the challenges arising in the area of application, along with it; the book provides scope for future developments.

In-depth information about the concepts of hydrodynamics and natural water sources has been described in this all-inclusive book. Information regarding the properties of fluids and their capability to transport materials and physical features is extremely significant. The quantification of the activity of fluids is a complicated task. This complexity is demonstrated primarily when natural flows occurring in large scales (rivers, lakes, oceans) are considered. This book discusses various features of flows in natural water bodies, like the evolution of plumes, the transport of sediments, air-water mixtures, etc. The major topics covered in the book include tidal and wave dynamics of various water bodies, and multiphase phenomena such as air-water flows and sediments. The book provides conceptual arguments, experimental and numerical results, and demonstrates practical functions of the procedures and tools of hydrodynamics.

It was a great honour to edit this book, though there were challenges, as it involved a lot of communication and networking between me and the editorial team. However, the end result was this all-inclusive book covering diverse themes in the field.

Finally, it is important to acknowledge the efforts of the contributors for their excellent chapters, through which a wide variety of issues have been addressed. I would also like to thank my colleagues for their valuable feedback during the making of this book.

**Editor**

# Part 1

# Tidal and Wave Dynamics:
# Rivers, Lakes and Reservoirs

# A Hydroinformatic Tool for Sustainable Estuarine Management

António A.L.S. Duarte
*University of Minho*
*Portugal*

## 1. Introduction

Hydrodynamics and pollutant loads dispersion characteristics are determinant factors for an integrated river basin management, where different waters uses and aquatic ecosystems protection must be considered. Strategic Environmental Assessment (SEA) of river basin planning process is crucial to promote a sustainable development. Towards this purpose, the European Water Framework Directive (WFD) establishes a scheduled strategy to reach good ecological status and chemical quality for all European water bodies.

As transitional aquatic environments, where fresh and marine waters meet, estuaries are generally characterized by complex interactions, with strong gradients and discontinuities, between physical, chemical and biological processes. This complexity is often increased by intensive anthropogenic inputs (nutrients and pollutants) from urban, agricultural and industrial effluents, leading to sensitive structural changes (Paerl, 2006) that modify both the trophic state and the health of the whole estuarine ecosystem. As a response to this, there has been an enormous increase in restoration plans for reversing habitat degradation, based on knowledge of the processes which led to the observed ecological changes (Valiela et. al., 1997).

Estuaries are recognised worldwide for providing essential ecological functions (fish nursery, decomposition, nutrient cycling, and shoreline protection) and support multiple human activities (fisheries resources, harbours, and recreational purposes). Each estuary is unique, because of its specific geological structure, morphology, hydrodynamics, land use, and the inflowing freshwater's characteristics (amount and quality).

Estuarine waters are generally characterized by intense biogeochemical processes that can renew the aquatic compartment, but their flushing capacity is mainly dependent on the hydrodynamic processes. The major driving forces of estuarine circulation are tides, wind, freshwater inflow, and general morphology (bathymetry, intertidal areas extension, roughness). The mixing and dispersion processes are critically dependent upon the salinity intrusion type (concerning it spatial distribution), which defines estuaries ranging from those with a highly stratified salt-wedge and a sharp halocline in the vertical structure to well-mixed systems.

The description of the estuarine transport process can be expressed by the definition of a transport time scale. This time scale is generally shorter than the time scale of the biogeochemical renewal processes and gives an estimate of the water-mass retention within the river basin system. So, the influence of hydrodynamics must not be neglected on estuarine eutrophication vulnerability assessment, because flushing time is determinant for the transport capacity and the permanence of substances, like pollutants or nutrients, inside an estuary (Duarte, 2005).

Excessive nutrient input, associated with high residence times, leads to eutrophication of estuarine waters and habitat degradation. It is widely recognized as a major worldwide threat, originating sensitive structural changes in estuarine ecosystems due to strong stimulation of opportunistic macroalgae growth, with the consequent occurrence of algal blooms (Pardal et al., 2004).

Much progress has been made in understanding eutrophication processes and in constructing modelling frameworks useful for predicting the effectiveness of nutrient reduction strategies (Thomann & Linker, 1998) and the increase of the estuarine flushing capacity in order to reverse habitat degradation, based on knowledge of the major processes that drive the observed ecological changes (Duarte et. al., 2001).

Residence time (RT) is a concept related with the water constituents (conservatives or not) permanence inside an aquatic system. Therefore, it could be a key-parameter towards the sustainable management of estuarine systems, because its values can represent the time scale of physical transport and processes, and are often used for comparison with time scales of biogeochemical processes, like primary production rate (Dettmann, 2001). In fact, estuaries with nutrients residence time values shorter than the algal cells doubling time will inhibit algae blooms occurrence (Duarte & Vieira, 2009a).

Estuarine water retention (or residence) time (WRT) has a strong spatial and temporal variability, which is accentuated by exchanges between the estuary and the coastal ocean due to chaotic stirring at the mouth (Duarte et. al., 2002). So, the concept of a single WRT value per estuary, while convenient from both ecological and engineering viewpoints, is shown to be an oversimplification (Oliveira & Baptista, 1997). The WRT (so called as transport time scale) has been assessed by many authors to be a fundamental parameter for the understanding of the ecological dynamics that interest estuarine and lagoon environments (Monsen et al., 2002).

The WRT variability within the basin has been related, in many research works, with the variability of some important environmental variables (dissolved nutrient concentrations, mineralization rate of organic matter, primary production rate, and dissolved organic carbon concentration). In literature, the WRT is defined through many different concepts: age, flushing time, residence time, transit time and turn-over time. Nevertheless, the definitions of these concepts are often not uniquely defined and generally confusing.

WRT estimation can be done considering an Eulerian or a Lagrangian approach. In the first option, WRT is identified as the time required for the total mass of a conservative tracer originally within the whole or a segment of the water body to be reduce to a factor "1/e" (Sanford et al., 1992; Luketina, 1998, Wang et al., 2004; Rueda & Moreno-Ostos, 2006; Cucco & Umgiesser, 2006), being a property of a specific location within the water body that is flushed by the hydrodynamic processes. In the second one, it is identified as the water transit time that corresponds to the time it takes for any water particles of the sample to leave the lagoon through its outlet (Dronkers & Zimmerman, 1982; Marinov & Norro, 2006; Bendoricchio, 2006), being a property of the water parcel that is carried within and out of the basin by the hydrodynamic processes.

The two methods give similar results for transport time scales calculation only when applied to simple cases, such as regular basins or artificial channels (Takeoka, 1984). However, sensitive differences arise in applications to basins characterized by complex morphology and hydrodynamics, mostly induced by the tidal range variability. It should be noted that the Lagrangian technique used for water transport time computation neglects the return flow effect at the estuarine mouth, which does not happen with the Eulerian approach. So, from a hydrological analysis in order to understand the flushing capacity of a tidal embayment, the Eulerian transport time scale seems to be the most representative parameter of all the processes occurring in the basin (Cucco et. al., 2009) and, being less dependent of tide variability, is able to describe the long term flushing dynamics of an estuarine system.

A numerical modelling study applied to Tampa Bay (Florida) was performed comparing the residence times by this two different methods: Eulerian concentration based, and Lagrangian particle tracking. The results obtained with the Lagrangian approach showed a doubling of overall residence time and strong spatial gradients in residence time values (Burwell, 2001).

Since the lower WRT values can increase the estuarine eutrophication processes, an enhanced Eulerian approach was adopted in this research study, conceptualising the residence time (RT) as a characteristic of water constituents, also including the no conservative substances. Thus, RT values were calculated, for each location and instant, as an interval of time that is necessary for that corresponding initial mass to reduce to a pre-defined percentage of that value, using the developed *TemResid* module (Duarte, 2005). In this work, a value of 10% was defined for the residual concentration of the substance, attending to the fact that the effect of the re-entry of the mass in the estuary during tidal flooding is considered (a significant effect for dry-weather river flow rates).

Mathematical models are well known as useful tools for water management practices. They can be applied to solve or understand either simple water quality problems or complex water management problems of estuaries, trans-boundary rivers or multiple-purpose and stratified reservoirs. Accidental spills of pollutants are of general concern and could be harmful to water users along the river basins, becoming crucial to get knowledge of the dispersive behaviour of such pollutants.

In this context, the mathematical modelling of dispersion phenomena can play an important role. Additionally, a craterous selection of mathematical models for application in a specific river basin management plan can mitigate prediction uncertainty. Therefore, intervention measures and times will be established with better reliability and alarm systems could efficiently protect the aquatic ecosystems, the water uses and the public health (Duarte & Boaventura, 2008). The benefits of the synergy between modelling and monitoring are often mentioned by several authors and the linkage of both approaches makes possible to apply cost-benefit measures (Harremoës & Madsen, 1999). Therefore, it is essential to correlate monitoring and modelling information with a continuous feedback, in order to optimize both processes, the monitoring network and the simulation scenarios formulation.

An integrated approach (hydrodynamics and water quality issues) is fundamental to prioritise risk reduction options in order to protect water sources and to get a high quality of the raw material for the water supply systems (Vieira et. al., 1999). Moreover, integrated models allow the optimization of the designed monitoring network (Fig. 1, adopted from Stamou et. al., 2007), based on hydrodynamic and water quality parameters calculation at any section using data from a monitoring programme (necessarily applied to limited number of sampling or measuring stations).

The analysis of water column and benthos field data observed in the Mondego estuary (Portugal), over the last two decades, allowed us to conclude that hydrodynamics was a major factor controlling the occurrence of macroalgae blooms, as determinant of nutrients availability and uptake conditions (Martins et al., 2001). Thus, the development of hydrodynamic (transport) processes characterization was obviously pertinent and useful.

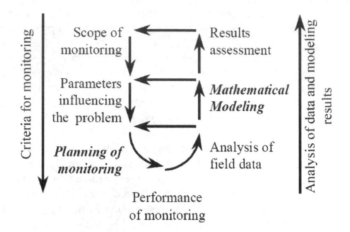

Fig. 1. Interaction between monitoring and modelling for monitoring network optimization

The aims of this chapter are to present the structure of a hydroinformatic tool developed for the Mondego estuary – named *MONDEST model* – linking hydrodynamics, water quality and residence time calculation modules, in order to simulate estuarine hydrodynamic behaviour, salinity and residence times spatial distributions, at different simulated management scenarios. Model calibration and validation was performed using field data obtained from the sampling carried out over the past two decades (Duarte, 2005).The results of the model simulations, considering different river water flow scenarios, illustrate the strong asymmetry of flood and ebb duration time at the inner sections of this estuary, a key-parameter for a correct tidal flow estimation, as the major driving force of the southern arm flushing capacity. The saline wedge propagation into the estuary and the spatial variation of residence time values are also assessed under different management scenarios. The RT values obtained show a strong spatial and temporal variability, as expected in complex aquatic ecosystems with extensive intertidal areas (Duarte & Vieira, 2009b)

The conclusion of this chapter will confirm the crucial influence of hydrodynamics on estuarine water quality status (chemical and ecological) and the usefulness of this hydroinformatic tool as contribution to support better management practices and measures of this complex aquatic ecosystem, like nutrient loads reduction or dislocation and hydrodynamic circulation improvement, in order to contribute for a true sustainable development.

## 2. Methods

### 2.1 Study site

The Mondego river basin is located in the central region of Portugal. The drainage area is about 6670 $km^2$ and the annual mean rainfall is between 1000 and 1200 mm. The area covered in this study refers to the whole Mondego estuary (Fig. 2), 32 km in length from its ocean boundary defined approximately 3 km outward from the mouth to Pereira bridge.

Fig. 2. Location and layout of river Mondego estuary

This complex and sensitive ecosystem was under severe environmental stress due to human activities: industries, aquaculture farms and nutrients discharge from agricultural lands of low river Mondego valley.
The Mondego estuary main zone (40°08'N 8°50'W), with only about 10 km long, is divided into two arms (north and south) with very different hydrological characteristics, separated by the Murraceira Island (Fig. 3).

Fig. 3. Aerial views of Mondego estuarine main zone

The north arm is deeper and receives the majority of freshwater input (from Mondego River), while the south arm of this estuary is shallower (2 to 4 m deep, during high tide) and presents an extensive intertidal zone covering almost 75% of its total area during the ebb tide. The irregularity of its morphology and bathymetry is depicted in Fig. 4 (Duarte, 2005).

Fig. 4. The Mondego estuary (main zone) bathymetry

For some decades, the river Mondego estuary was under severe ecological stress, mainly caused by eutrophication of its south arm due to the combination of the nutrient surplus with low hydrodynamics and high salinity, because, until the end of 1998, this sub-system was almost silted up in the upstream areas (Fig. 5), drastically reducing the Mondego river water inflow. Hence, the south arm estuary water circulation was mainly driven by tide and wind, originating, in dry-weather conditions, a coastal lagoon-like behaviour. The freshwater inflow was seasonal and only provided by the (small) discharges of the Pranto River, a tributary artificially controlled by the *Alvo* sluices, located 1 km upstream from its mouth.

The most visible effect of this important hydrodynamic constrain was the occurrence of episodic macroalgae blooms and the concomitant severe decrease of the area occupied by *Zostera noltii* beds. So, for the control of this eutrophication process, it became crucial to obtain field data to characterize the real trophic status of this aquatic ecosystem, as well as to better understand the major mechanisms that regulate the abundance of opportunistic macroalgae in order to eradicate its periodic early spring algal blooms.

Fig. 5. Silting up process occurred in the upstream areas of the estuary south arm

Figure 6 shows the size-grain distribution of the sediments in the Mondego estuary main zone (Cunha & Dinis, 2002). A strong correlation was found with the flow channels configuration that occurs during low tide. This information could be very useful for the roughness coefficient definition along the estuarine system, considering or not the variability of the bottom shear stress.

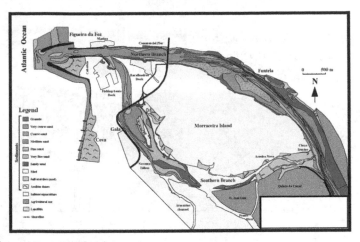

Fig. 6. Mondego estuary grain-size map

The Mondego River monthly inflows were calculated based on the analysis conducted for the daily average values measured at the Coimbra dam-bridge in the period 1990-2004 (Fig. 7).

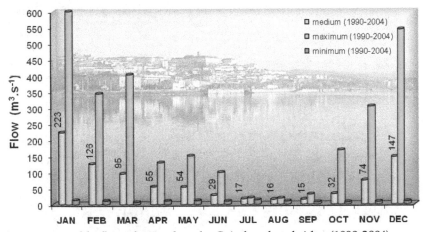

Fig. 7. Average monthly flow observed at the Coimbra dam-bridge (1990-2004).

Based on this available data, the typical dry-weather flow (corresponding to the 90% percentile on the cumulative flow rate curve) is about 15 $m^3.s^{-1}$, while the annual average flow value was 75 $m^3.s^{-1}$. The maximum flow value for sizing the minor bed of the main channel was estimated about 340 $m^3.s^{-1}$.

The values that were estimated for the Pranto River inflow to the Mondego estuary south arm correspond to those observed during field work, considering the flow discharge curves of the three Alvo sluices (Fig. 8).

So, average daily values of 0 (closed sluices), 15 and 30 $m^3.s^{-1}$ were considered. They correspond, respectively, to discharges carried out during part of the tidal cycle and continuous discharges that are usual in periods of greater rainfall, considering the water demand for existing intensive oriziculture activity in the Pranto river catchment.

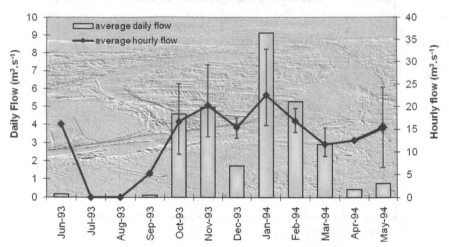

Fig. 8. Pranto river annual (1993-94) flow discharge into the Mondego estuary south arm

In this study, the tidal harmonic signal at Figueira da Foz harbour was generated, for each simulated period, using the programme SR95 (JPL, 1996). Fig. 9 presents an example of a monthly tidal signal used in the *Mondest* model as a downstream boundary condition, during its calibration procedure (Duarte, 2005).

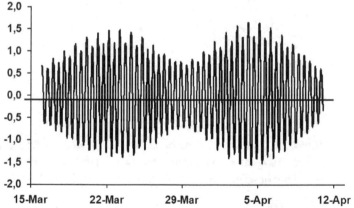

Fig. 9. Monthly tidal harmonic signal at Figueira da Foz harbour using the SR95 programme

## 2.2 Sampling programme
An extensive sampling programme was carried out during last two decades at three benthic stations. The choice of benthic stations was related with the observation of an eutrophication gradient in the south arm of the estuary, involving the replacement of eelgrass, *Zostera noltii* by opportunistic green macroalgae such *as Enteromorpha spp.* and *Ulva spp.*

Water column monitoring was performed by specific sampling campaigns, some of them in simultaneous with the benthic ones, at three other sites: Pranto river mouth (S3); *Armazéns* channel mouth (S2); and Lota (S1), downstream the *Gala* bridge). The location of water monitoring stations at Mondego estuary south arm were selected in order to represent the different flow regimes observed in this system. Water level, velocity, salinity, temperature and dissolved oxygen were measured in situ and water samples were collected for physical and chemical system characterization.

Dissolved fraction seems to be the most representative of nutrients transport inside the south arm of this estuary, followed by the suspended particulate matter fraction. This finding was very relevant to understand the high eutrophication vulnerability of this sub-system, since these fractions represent the nutrients immediately accessible to the macroalgae tissues incorporation on the growing process.

An example of the sampling programme results is depicted in Figure 10 showing the average monthly values of salinity obtained (in 2000-2001) at Lota station (S1) and Pranto river mouth station (S3), as well as its variation over a medium tidal cycle.

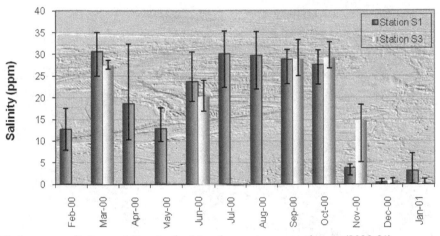

Fig. 10. Average salinity variation in the Mondego estuary south arm (2000-01)

The sampling data analysis was crucial to better understand eutrophication mechanisms and allowed us to conclude that the occurrence of green macroalgae blooms is strongly dependent on the estuarine flushing conditions, salinity gradients and nutrient loading characteristics, availability and residence time (Martins et al., 2001; Duarte et al., 2002).

## 2.3 Dye tracer experiments

Hydrodynamics and pollutant discharge dispersion characteristics are determinant factors in river basin planning and management, where different waters uses and aquatic ecosystems protection must be considered.

Net advection and longitudinal dispersion play important roles in determining transport and mixing of substances and pollutants discharged into the aquatic systems. In order to enhance water sources protection, the knowledge of transport processes is of increasing importance concerning the prediction of the pollutant concentration distribution, particularly when resulting from a continuous or accidental spill event caused by industrial and mining activities or road-river accidents.

Generally, there are two approaches to calculate the transport of solutes in water bodies. One is the more classical calculation based on exact river morphological and hydraulic input data and the other is the calculation based on estimation of transport parameters such as travel time and dispersion coefficients. Since exact morphological data are often unavailable, the parameter estimation technique is more promising.

In both approaches, tracer experiments are needed to provide field data for water quality models calibration and validation procedures. Indeed, model calibration is often a weak step in its development and using experimental tracer techniques, the calibration and validation problems can be solved satisfactorily, improving the needed feasibility of the early warning systems used by many water supply utilities.

Tracer experiments are typically conducted with artificial fluorescent dyes (like rhodamine WT) (Fig. 11), whose concentrations are easily measured with a fluorometre. These tracers should be easily detected, non toxic and non-reactive, as well as, have high diffusivity, low acidity and sorption for a quasi-conservative behaviour.

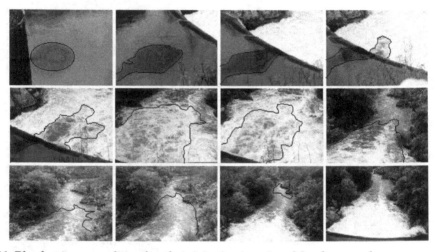

Fig. 11. Rhodamine spreading after their injection in a river Mondego reach

Based on field experiments data, many investigators have derived semi-empirical equations (Hubbard et al., 1982; Chapra, 1997; Addler et al., 1999) or applied one-dimensional models (Duarte & Boaventura, 2008) to calculate experimental longitudinal dispersion coefficients from concentration time curves at consecutive sampling sites, using the analytical solution of first order decay kinetics (Table 1).

The injected tracer dye mass must be calculated considering the water volume estimated in the river reach or reservoir system and the fluorometre detection limit. Specific problems of the application of tracers to surface water researches include the photosensitivity of dyes, such as fluorescence tracers, and recovery efficiency, which may imply the use of correction techniques for tracer losses. The tracer mass recovered at each site allowed the assessment of the importance of physical and biochemical river processes by quantifying precipitation, sorption, retention and assimilation losses. Usually, total tracer mass losses resulting from all these sinks can reach 40 to 50% of the injected mass (Duarte & Boaventura, 2008; Addler et al., 1999).

In some recent experiments, a gas tracer (SF$_6$) has been shown to be a powerful tool for examining mixing, dispersion, and residence time on large scales in rivers and estuaries

(Caplow et al. 2004) as an alternative method to dye tracer experiments used for advection and dispersion characterisation.

| MONITORING PROGRAM | REACH | AVERAGE VELOCITY (ms⁻¹) | | TRAVEL TIME (h) | | DISPERSION COEFFICIENT (m²s⁻¹) | | RECOVERED MASS |
|---|---|---|---|---|---|---|---|---|
| | | EXPER. | DUFLOW | EXPER. | DUFLOW | EXPER. | DUFLOW | (%) |
| 3 rd.<br>(Nov.-90) | S1 – S2 | 0.526 | Var. | 2:37 | 2:35 | 14 | 10 | 57 |
| | S2 – S3 | 0.497 | Var. | 2:41 | 2:41 | 51 | 45 | 56 |
| | S3 – S5 | 0.473 | Var. | 3:21 | 3:19 | 37 | 35 | 55 |
| | S1 – S3 | 0.511 | Var. | 5:18 | 5:16 | 34 | - | - |
| | S1 – S5 | 0.497 | Var. | 8:38 | 8:35 | 35 | - | - |
| 1 st.<br>(Dec.-89) | S1 – S2 | 1.105 | Var. | 1:14 | 1:14 | 52 | 40 | 62 |
| | S2 – S3 | 0.949 | Var. | 1:24 | 1:24 | 61 | 70 | 62 |
| | S1 – S3 | 1.023 | Var. | 2:38 | 2:38 | 58 | - | - |

Table 1. Hydraulic and dispersion parameters estimation using tracer dye experiments in a non-tidal reach of river Mondego

The dispersion processes in rivers are combined with a specific dynamic characterized by a decrease in maximum dye concentration (Fig. 12). The distribution of the tracer in all directions follows the sluggish injection into the channel. In non-tidal rivers, the lateral and vertical dispersion processes are almost always faster than the continuing longitudinal dispersion process.

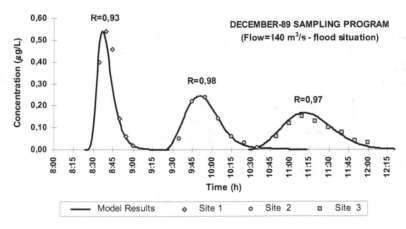

Fig. 12. River Mondego model calibration: correlation between field tracer experiment data and model results.

One-dimensional modelling is a reasonably reliable tool to be considered for estimating the distribution of solutes in large rivers. Complex processes, for example in dead zones or downstream from the confluence of two rivers, have to be investigated by direct measurements and should be described by two-dimensional transport models. Calculation of net advection in tidal rivers is fairly straightforward, but longitudinal dispersion is difficult to determine *a priori*, and the application of two or three-dimensional transport models are often required.

Ever increasing computational capacities provide the development of powerful and user-friendly mathematical models for the simulation and forecast of quality changes in receiving waters after land runoff, mining and wastewater discharges.

The results of several research works have showed that the linkage of tracer experimental approach with mathematical modelling can constitute a power and useful operational tool to establish better warning systems and to improve management practices for the efficiently protection of water supply sources and, consequently, public health.

## 2.4 Mathematical modelling

Numerical modelling is a multifaceted tool that enables a better understanding of physical, chemical and biological processes in the water bodies, based on a "simplified version of the real" described by a set of equations, which are usually solved by numerical methods.

The models to be used for the implementation of the WFD management strategies should ideally have the highest possible degree of integration to comply with the integrated river basin approach, coupling hydrological, hydrodynamic, water quality and ecological modules as a function of the specific environmental issues to analyse.

The *Mondego Estuary (MONDEST)* model was conceptualized (Fig. 13) as an integrated hydroinformatic tool, linking hydrodynamics, water quality and residence time (*TempResid*) modules (Duarte, 2005).

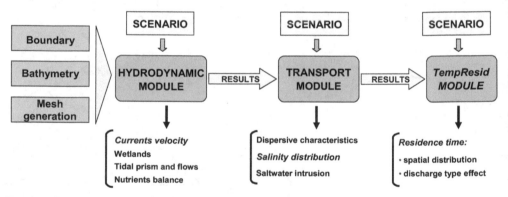

Fig. 13. The *MONDEST* model conceptualization

The formulation of an accurate model requires the best possible definition of the geometry and bathymetry of the water body and the interactions with the boundary conditions, as stated in previous items.

This model is based on generalized computer programmes RMA2 and RMA4 (WES-HL, 1996; 2000), which were applied and adapted to this specific estuarine ecosystem. The CEWES version of RMA4 is a revised version of RMA4 as developed by King & Rachiele (1989).

The RMA2 programme solves depth-integrated equations of fluid mass and momentum conservation in two horizontal directions by the finite element method (FEM) using the Galerkin Method of weighted residuals. The shape (or basis) functions are quadratic for velocity and linear for depth. Integration in space is performed by Gaussian integration. Derivatives in time are replaced by a nonlinear finite difference approximation.

The RMA4 programme solves depth-integrated equations of the transport and mixing process using the Galerkin Method of weighted residuals. The form of the depth averaged transport equation is given by equation (1)

$$h\left(\frac{\partial c}{\partial t}+u\frac{\partial c}{\partial x}+v\frac{\partial c}{\partial y}-\frac{\partial}{\partial x}D_x\frac{\partial c}{\partial x}-\frac{\partial}{\partial y}D_y\frac{\partial c}{\partial y}-\sigma+kc+\frac{R(c)}{h}\right)=0 \qquad (1)$$

Where
h = water depth;
c = concentration of pollutant for a given constituent;
t = time;
u, v = velocity in x direction and y direction;
Dx, Dy, = turbulent mixing (dispersion) coefficient;
k = first order decay of pollutant;
σ = source/sink of constituent;
R(c) = rainfall/evaporation rate.

As with the hydrodynamic model RMA2, the transport model RMA4 handles one-dimensional segments or two-dimensional quadrilaterals, triangles or curved element edges. Spatial integration of the equations is performed by Gaussian techniques and the temporal variations are handled by nonlinear finite differences consistent with the method described for RMA2.

The numerical computation was carried out for all Mondego estuary spatial domains. Several sections were carefully selected and used for calibrating and analysis of the simulation results (Duarte, 2005). The legend includes the designation, section code and their distance to the mouth of the estuary (Fig. 14).

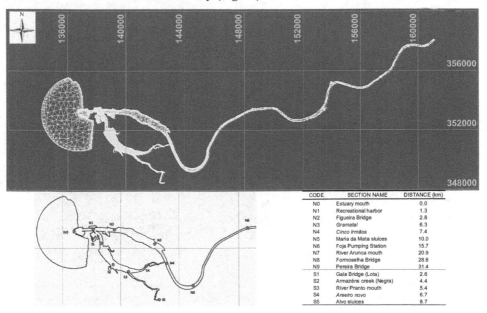

| CODE | SECTION NAME | DISTANCE (km) |
|---|---|---|
| N0 | Estuary mouth | 0.0 |
| N1 | Recreational harbor | 1.3 |
| N2 | Figueira Bridge | 2.8 |
| N3 | Gramatal | 6.3 |
| N4 | Cinco Irmãos | 7.4 |
| N5 | Maria da Mata sluices | 10.0 |
| N6 | Foja Pumping Station | 15.7 |
| N7 | River Arunca mouth | 20.9 |
| N8 | Formoselha Bridge | 28.6 |
| N9 | Pereira Bridge | 31.4 |
| S1 | Gala Bridge (Lota) | 2.6 |
| S2 | Armazéns creek (Negra) | 4.4 |
| S3 | River Pranto mouth | 5.4 |
| S4 | Areeiro novo | 6.7 |
| S5 | Alvo sluices | 8.7 |

Fig. 14. The *MONDEST* model finite elements mesh and outline of the control sections

The size of the elements to consider in the spatial discrimination of the simulated domain of numerical models must be established as a function of larger or smaller spatial gradients than those displayed by the variables (water level and velocity) in that domain. In the case of the Mondego estuary, since the south arm was the preferred object for studying, the network of finite elements was refined in that sub-domain, thereby reducing the maximum area of its (triangular) elements to 500 m$^2$ (Duarte, 2005).

In the *MONDEST model*, the hydrodynamic module provides flow velocities and water levels for the water quality module, whose results acts as input on the *TempResid* module, feeding the constituents concentration over the aquatic system. The post-processing and mapping of model results was performed using SMS package (Boss SMS, 1996).

The *TempResid* module was integrally developed in this research work aiming to compute RT values of each water constituent (conservative or not) and allowing to map its spatial distribution over all the estuarine system, considering different simulated management scenarios.

RT value of a substance was calculated for each location and instant, as an interval of time that is necessary for that corresponding initial mass to reduce to a pre-defined percentage of that value. In this work, a value of 10% was adopted for the residual concentration of the substance, attending to the fact that the effect of the re-entry of the mass in the estuary during tidal flooding is considered (a significant effect for dry-weather river flow rates).

The determination of the RT in several stations along the estuary, where the eutrophication gradient occurred, was carried out by applying the *TempResid* programme to the results of the simulations that were performed with the transport module of the MONDEST model. Figure 15 shows an example of the MONDEST model transport module results for the management scenario considered as the most favourable to macroalgae blooms occurrence (Duarte, 2005), due to low freshwater inputs and consequent reduction of estuarine waters renovation (scenario RT1).

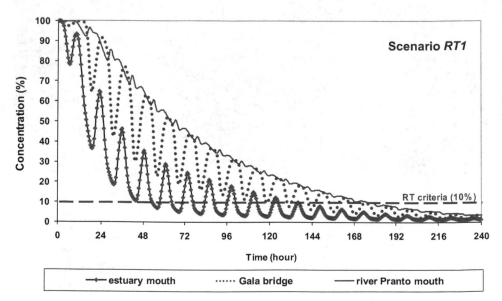

Fig. 15. Residence time computation using *TempResid* module

This graph presents the concentration decrease of a conservative constituent, in three control points (N0 - estuary mouth; S1 - Gala bridge/Lota; and S3- Pranto river mouth), due to estuarine flushing currents, considering the well known re-entrance phenomena at the estuary mouth.

## 2.5 Simulated management scenarios

For hydrodynamic modelling purpose, a wide range (sixteen) of management scenarios were judiciously selected covering a representative set of hydraulic conditions (Table 2), resulting from the combination of typical tidal amplitudes (0.60, 1.15, and 1.60 m) and freshwater flow inputs (from Mondego and Pranto).

| Freshwater flow (m$^3$.s$^{-1}$) | | TIDE | | |
|---|---|---|---|---|
| Mondego | Pranto | Medium | Spring | Neap |
| | 0 | H 1 | H 2 | H3 |
| 15 | 15 | H 4 | - | - |
| | 30 | H 5 | - | - |
| 75 | 0 | H 6 | H 7 | H 8 |
| | 0 | H 9 | H 10 | H 11 |
| 340 | 15 | H 12 | - | - |
| | 30 | H 13 | - | - |
| 500 | 30 | - | H 14 | - |
| 800 | 30 | - | H 15 | H 16 |

Table 2. Simulated management scenarios for the hydrodynamic modelling

For the *Mondest* transport model calibration and validation, the salinity was adopted as a natural tracer. Several management scenarios (nine) were also carefully selected (Table 3) considering the most representative hydrodynamic conditions in order to estimate salt wedge propagation into the estuary and to identify the areas (in both arms) where favourable salinity values for macroalgae growth can potentiate the estuarine eutrophication vulnerability.

| Freshwater flow (m$^3$.s$^{-1}$) | | TIDE | | |
|---|---|---|---|---|
| Mondego | Pranto | Medium | Spring | Neap |
| | 0 | SL 1 | SL 6 | SL 9 |
| 15 | 15 | SL 2 | - | - |
| | 30 | SL 3 | - | - |
| 75 | 0 | SL 4 | SL 7 | - |
| 340 | 15 | SL 5 | SL 8 | - |

Table 3. Simulated management scenarios for the hydrodynamic modelling

For the RT values calculation using the *TempResid* module, the simulated management scenarios (fourteen) were defined considering not only the most critical hydrodynamic conditions, but also by carefully selecting distinct pollutant load characteristics (e.g. location, duration and type of the discharge event, instant of tidal cycle when the release occurs) and constituent decay rates (Table 4) in order to assess and confirm the highest eutrophication vulnerability of the inner areas of the Mondego estuary south arm, due to the expected occurrence of higher RT values.

| SCENARIO | RIVER FLOW $(m^3.s^{-1})$ | | TIDE | LOAD | DECAY RATE $(day^{-1})$ |
|---|---|---|---|---|---|
| | Mondego | Pranto | | | |
| RT 1 | 15 | 0 | medium | point | 0 |
| RT 2 | | | spring | | |
| RT 3 | | | neap | | |
| RT 4 | | | medium | | 1 |
| RT 5 | | | | | 10 |
| RT 6 | | 15 | | | 0 |
| RT 7 | 1 | | | | |
| RT 8 | 75 | | | | |
| RT 9 | 340 | | | | |
| RT 10 | 15 | 0 | | diffuse | |
| RT 11 | | | | | 1 |
| RT12 | 75 | | | | 0 |
| RT 13 | | | | | 1 |
| RT 14 | | | | | 0,5 |

Table 4. Simulated management scenarios for estuarine residence time calculation

In this work only a few examples of the very large amount of MONDEST model results obtained for those different simulated scenarios can be presented. The main aim of the following item will be to highlight the evident influence of hydrodynamics (tidal regime and freshwater inflows) on estuarine residence time spatial variation, which can play a special role in estuarine eutrophication vulnerability assessment.

## 3. Results and discussion

### 3.1 Hydrodynamic modelling

Hydrodynamic modelling results allowed to evaluate the water level and magnitude of currents velocity in both arms during tidal ebbing and flooding situations, and to assess the influence of tidal and freshwater inflows regimes on its variability.

For dry weather conditions, the higher velocity values were obtained in the southern arm, near Gala Bridge, reaching 0.35 (neap tide, scenario H3) to 0.70 m.s⁻¹ (spring tide, scenario H2) while in the northern arm these maximum values (which occur in the section N4) are lower, reaching 0.33 (neap tide) to 0.60 m.s⁻¹ (spring tide), at 1km upstream the Figueira da Foz bridge. These results are depicted on Figure 16 mapping the effect of extreme tidal regimes on maximum currents velocity magnitude during the flooding period and considering dry-weather conditions.

In the southern arm, the flooding time, which decreases at the inner zones, is much shorter than the ebbing time, due to shallow waters and to large intertidal mudflats areas. This asymmetry is influenced by the tidal regime and has a fast increase into the inner areas of this arm reaching 2.5 hours: 5 hours for flooding and 7.5 hours for ebbing time. In the northern arm, between the sections N1 and N4, there is a little delay of fifteen minutes in the high tide occurrence and a bigger delay in ebb tide (about two hours).

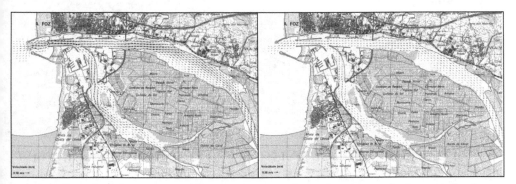

Fig. 16. Effect of tidal regime on ebbing maximum values of currents velocity magnitude (scenarios H2 and H3)

Figure 17(a) shows an example of the tidal regime effect in the mean velocity magnitude (MVM) variation, at section N4 (where maximum values of this parameter occurred). It should be noted that for a neap tide, the VMM during the tidal flooding period is almost an half of the value reached for a typical sprig tide.

For upstream estuarine sections, water surface levels in high tide are similar, but, in ebb tide, water surface level increases in the inner section due to the effect of the estuarine bathimetry (elevation of bottom level) (Fig. 17b).

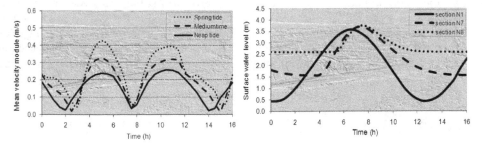

Fig. 17. (a) Effect of tidal regime on ebbing maximum values of currents velocity magnitude (section N4); (b) Surface water level variation along the estuarine system (N1, N7, N8)

### 3.2 Model calibration and validation

The velocities and water levels field data obtained from the sampling programme were used for model calibration and validation. Figure 18 shows an example of a specific procedure performed in section S1 (Gala bridge/Lota) for the parameter "surface water level (SWL)".

Two different sensitivity analyses were carried out to define the accurate values to adopt for the main calibration parameters used in both (hydrodynamic and water transport) modules of *Mondest* model: one for the Manning bottom friction coefficient (n) and horizontal Eddy viscosity coefficient ($E_h$); and the other for the horizontal dispersion coefficient ($D_h$). For each calibration parameter, three different values were tested comparing field data with the corresponding model results.

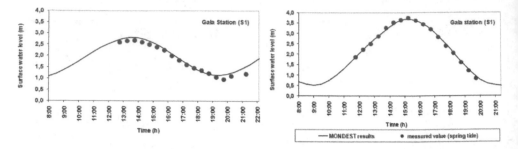

Fig. 18. Hydrodynamic module calibration (spring tide) and validation (neap tide) (station S1)

For the simulated management scenarios and based on calculated correlation coefficients, the best agreements were obtained considering the following parameters values: the ordered pair (n=0.02 m$^{-1/3}$.s; $E_h$= 20 m$^2$.s$^{-1}$), for the hydrodynamic module; and $D_h$= 30 m$^2$.s$^{-1}$, for the water transport module.

A more detailed description of these sensitivity analyses (scenarios, results and discussion) can be found in Duarte (2005).

## 3.3 Tidal prism and flow estimation

In this work a new approach was developed for tidal flow estimation, based on the previous tidal prism calculation using mathematical modelling. The adopted approach allows to consider the temporal variation of the cross section area during the tidal cycle and, mainly, the real asymmetry of tidal flooding and ebbing periods verified in the inner estuarine areas. Tidal prisms were calculated as the difference between the water volume in a specific high tide and the correspondent previous ebb tide, which can be automatically given by the query tools of the post-processor module (SMS). Figure 19 shows the spatial variation of tidal

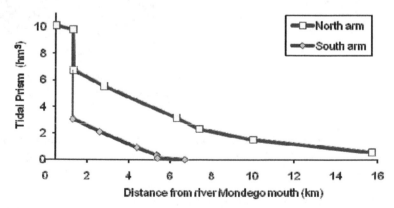

Fig. 19. Tidal prism spatial variation in both estuary arms (flooding of scenario H1)

prism for the both estuary arms (north and south) based on this procedure calculation for each control sections along the Mondego estuary, considering the flooding period of the scenario H1.

The mean tidal flow estimation in each estuarine section can be performed using the correspondents' tidal prism values and the real duration of the ebbing and flood events. The mean tidal flow values obtained for several hydrodynamic scenarios in the sections N0 and S1 are summarized in Table 5.

| Section | Scenario | Tidal prism (hm³) | | Duration (h) | | Mean tidal flow (m³.s⁻¹) | |
|---------|----------|---------|--------|----------|--------|----------|--------|
| | | flooding | ebbing | flooding | ebbing | flooding | ebbing |
| | H 1 | 9.178 | 9.894 | 6.25 | 6.25 | 408 | 440 |
| | H 2 | 12.02 | 13.063 | 6.25 | 6.25 | 534 | 581 |
| N0 | H 3 | 5.818 | 5.692 | 6.25 | 6.25 | 259 | 253 |
| | H 7 | 14.792 | 15.386 | 6,25 | 6,25 | 657 | 684 |
| | H 10 | 11.387 | 12.089 | 6.00 | 6.50 | 527 | 517 |
| | H 1 | 2.334 | 2.341 | 5.50 | 7.00 | 118 | 93 |
| | H 2 | 3.265 | 3.276 | 5.50 | 7.00 | 165 | 130 |
| S1 | H 3 | 1.269 | 1.266 | 6.00 | 6.50 | 59 | 54 |
| | H 7 | 3.449 | 345 | 5.50 | 7.00 | 174 | 137 |
| | H 10 | 3.325 | 3.337 | 5.50 | 7.00 | 168 | 132 |

Table 5. Synthesis of mean tidal flow calculation (sections N0 and S1)

### 3.4 Hydrodynamic influence on estuarine salinity distribution

The analysis of the salinity distribution in the estuary had, as a primary goal, the identification of the areas that, throughout the tidal cycle, present salinity values within the range of 17 to 22‰, defined by Martins et al. (2001) as the most favourable for algal growth in this specific aquatic ecosystem.

The Pranto river inflow in estuary southern arm has shown a strong influence on salinity distribution decreasing drastically its values to a range far from the one defined as the most favourable for this estuarine eutrophication process. Figure 20 shows the opening Alvo sluices effect on southern arm salinity gradients caused by Pranto river flow discharge of 30 m³.s⁻¹, during the ending of ebbing and the beginning of tidal flooding periods (scenarios SL 3 and SL1) (Duarte & Vieira, 2009a).

Fig. 20. Effect of Pranto river flow discharge on estuarine salinity distribution (high tide)

The effect of tidal regime on saline wedge propagation into the Mondego estuary can be assessed by comparing the saline front position at high or ebb tide achieved for the extreme tidal amplitudes (spring and neap tides). For the simulated conditions (scenarios SL2 and SL3) a difference of about 4 km in the estuarine saline wedge intrusion was observed: 12.5 km for a spring tide and only 8.5 km for a neap tide. Figure 21 depicts the differences on the saline wedge return (ebb tide) for these two extreme tidal regimes.

Fig. 21. Effect of tidal regime on saline wedge reflux (ebb tide) (scenarios SL2 and SL3)

### 3.5 Hydrodynamic influence on estuarine residence time distribution

During the warm season (late spring and summer), the Alvo sluices are almost closed (scenario RT1). For this operational condition, the RT values near Pranto mouth station can quintuplicate when compared with those resulting from a Pranto river flow discharge of 15 $m^3.s^{-1}$ (scenario RT6), both under dry-weather conditions (low river Mondego inflows). Figure 22 shows this sensitive increase on flushing capacity of the Mondego estuary south arm due to Pranto river discharges from Alvo sluices opening.

Fig. 22. Effect of Pranto river discharge on RT values distribution (scenarios RT1 and RT6)

For the other hand, when the Alvo sluices remain closed the salinity and the RT values inside the southern arm are strongly influenced by tidal regime. Figure 23 illustrates the gradient of RT spatial distribution, which was mapped applying the *TemResid* module computing availability for the simulation of management scenarios RT2 and RT3.

Simulation results for these two tidal scenarios showed a RT values increase of 50% for a neap tide, when compared with a spring tide, both in the south arm and in the north arm reach, between N1 and N2 control points. This increase is smoothed in the northern arm inner areas, with the lowest increase (only 17%) at the Mondego estuary mouth. The minimum RT values (3.2 days) occurred in the Mondego estuary mouth (N0) and in the mesotrophic wetland zone of the south arm (near station S2). The maximum RT values (9.5 days) were obtained for the zone (near station S3) with higher eutrophication vulnerability. Concerning the periodicity of tidal regime recurrence, its effect could be very relevant for estuarine biochemical processes with a time scale lower than 6 days.

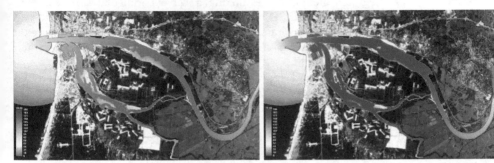

Fig. 23. Effect of tidal regime on RT values distribution (scenarios RT2 and RT3)

## 4. Conclusion

The analysis of the results obtained in the performed simulations allows the confirmation that there is a significant influence of bathymetry in the spatial variation of the RT along the Mondego estuary and consequently, the definition of typical (unique) values for each one of its arms becomes inadequate if they are not associated to local and specific hydrodynamic scenarios.

The results obtained from hydrodynamic modelling have shown a strong asymmetry of ebbing and flooding times in the inner estuary south arm areas due to their complex geo-morphology (extensive wetlands and salt marsh zones, over 75% of its total area). This information allows a better understand of the estuarine circulation pattern, since tide is the major driving force of the southern arm flushing capacity, when the Alvo sluices remain closed. Indeed, the absence of the Pranto river discharge (a typical dry-weather condition) drastically increases salinity and RT values in the inner estuary southern arm and, consequently, the nutrients availability for algae uptake is higher, enhancing estuarine vulnerability to eutrophication.

From the analysis of the results obtained, it is possible to conclude that in both arms of this estuary, the tidal prism volumes are influenced by the bathymetry (extensive wetland areas), tidal regime and freshwater inputs. However, the influence of the tidal regime on the tide prism values is much greater than that of the freshwater inflows, and it is possible to verify that those values do not increase proportionally to the incremental values of the Mondego River flow rate.

The knowledge of the ebbing and flooding duration asymmetry is crucial for a more accurate tidal flow calculation, based on previous tidal prim estimation using mathematical modelling tools. With this new approach for mean tidal flow estimation the variation of cross section area can also be computed increasing the feasibility of the obtained results.

For the simulated conditions a difference of about 4 km in the estuarine saline wedge intrusion was observed: 12.5 km for a spring tide and only 8.5 km for a neap tide. However, a sensitive surface water elevation was monitored in the upper control section (N8), near the

Formoselha bridge (located 30 km upstream the estuary mouth), during a spring tide propagation.

For medium typical tide, drought conditions and conservative constituents, simulation results showed that estuarine RT values range between 6 days (at both arms) and 4 days in the downstream reach of its two arms confluence (control point N1).

The development of integrated methodologies linking tracer experimental approach with hydroinformatic tools (based on 2D and 3D mathematical models) is of paramount interest because they can constitute a accurate and useful operational tool to establish better warning systems and to improve management practices for efficiently protecting water sources and, consequently, public health.

The MONDEST model developed and applied in this work allowed the evaluation and ranking of potential mitigation measures (like nutrient loads reduction or dredging works for hydrodynamic circulation improvement). So, the proposed methodology, integrating hydrodynamics and water quality, constitutes a powerful hydroinformatic tool for enhancing estuarine eutrophication vulnerability assessment, in order to contribute for better water quality management practices and to achieve a true sustainable development.

## 5. References

Addler, M.J.; Stancalie, G. & Raducu, C. (1999). Integrating tracer with remote sensing techniques for determining dispersion coefficients of the Dâmbovita River, Romania. In: *Integrated Methods in Catchment Hydrology – Tracer, Remote Sensing and New Hydrometric Techniques (Proceedings of IUGG 99-Symposium HS4)*, IAHS Publ. No. 258, pp. 75-81, Birmingham, July, 1999.

Bendoricchio, G.D.B. (2006). A water-quality model for the lagoon of Venice, Italy. *Ecological Modelling*, 184, pp. 69–81, ISSN 0304-3800

Boss SMS (1996). *Boss Surface Modeling System-User's Manual*, Brigham Young University Press, USA.

Burwell, D.C. (2001). *Modelling the spatial structure of estuarine residence time : eulerian and lagrangian approaches*. PhD. Thesis, College of Marine Science, University of South Florida, USA.

Caplow, T.; Schlosser, P.; Ho, D. T. & Enriquez, R. C. (2004). Effect of tides on solute flushing from a strait: imaging flow and transport in the East River with $SF_6$. *Environ. Sci. Technol.*, Vol.38, No.17, pp. 4562–4571, ISSN 1520-5851

Chapra, S. C. (1997). *Surface Water Quality Modelling*, McGraw-Hill, New York, USA.

Cucco, A., Umgiesser, G.; Ferrarinb, C.; Perilli A., Canuc, D.M. & Solidoroc, C. (2009). Eulerian and lagrangian transport time scales of a tidal active coastal basin, *Ecological Modelling*, Vol.220, No.7, pp. 913–922, ISSN 0304-3800

Cucco, A. & Umgiesser, G. (2006). Modelling the Venice lagoon water residence time. *Ecological Modelling*, Vol.193, pp. 34–51, ISSN 0304-3800

Cunha, P.P. & Dinis, J. (2002). Sedimentary dynamics of the Mondego estuary. In: *Aquatic ecology of the Mondego river basin. Global importance of local experience*, Pardal M.A., Marques J.C. & Graça M.S. (eds.), pp. 43-62, Coimbra University Press, IBSN 972-8704-04-6, Coimbra, Portugal.

Dronkers, J. & Zimmerman, J.T.F. (1982). Some principles of mixing in tidal lagoons. In: *Oceanologica Acta. Procedings of the International Symposium on Coastal Lagoons*, pp. 107–117, Bordeaux, France, September 9–14, 1981

Dettmann, E. (2001). Effect of water residence time on annual export and denitrification of nutrient in estuaries: a model analysis. *Estuaries*, Vol.24, No.4, pp. 481–490, ISSN 1559-2723

Duarte, A.A.L.S. & Vieira, J.M.P. (2009a). Mitigation of estuarine eutrophication proceses by controlling freshwater inflows. In: *River Basin Management V*, ISBN 978-1-84564-198-6, and *WIT Transactions on Ecology and the Environment*, pp. 339-350, ISSN: 1743-3541, WIT Press, Ashusrt, Reino Unido.

Duarte, A.A.L.S. & Vieira, J.M.P. (2009b). Estuarine hydrodynamic as a key-parameter to control eutrophication processes. *WSEAS Transactions on Fluid Mechanics*, Vol.4, No.4, (October 2009), pp. 137-147, ISSN 1790-5087

Duarte, A.A.L.S. & Boaventura, R.A.R (2008). Pollutant dispersion modelling for Portuguese river water uses protection linked to tracer dye experimental data. *WSEAS Transactions on Environment and Development*, Vol.4, No.12, (December 2008), pp. 1047-1056, ISSN 1790-5079

Duarte, A.A.L.S. (2005). *Hydrodynamics influence on estuarine eutrophication processes*. PhD. Thesis, Civil Engineering Dept., University of Minho, Braga, Portugal (in Portuguese).

Duarte, A.A.L.S.; Pinho, J.L.S.; Vieira, J.M.P. & Seabra-Santos, F. (2002). Hydrodynamic modelling for Mondego estuary water quality management. In: *Aquatic ecology of the Mondego river basin. Global importance of local experience*, Pardal M.A., Marques J.C. & Graça M.S. (eds.), pp. 29-42, Coimbra University Press, IBSN 972-8704-04-6, Coimbra, Portugal.

Duarte, A.A.L.S.; Pinho, J.L.S.; Pardal, M.A.; Neto, J.M.; Vieira, J.M.P. & Seabra-Santos, F. (2001). Effect of Residence Times on River Mondego Estuary Eutrophication Vulnerability. *Water Science and Technology*, Vol.44, No.2/3, pp. 329-336, ISSN 0273-1223.

Harremoës, P. & Madsen, H. (1999). Fiction and reality in the modelling world – Balance between simplicity and complexity, calibration and identifiably, verification and falsification. *Water Science and Technology*, Vol.39, No.9, pp. 47–54, ISSN 0273-1223.

Hubbard, E.F.; Kilpatrick, F.A.; Martens, C.A. & Wilson, J.F. (1982). *Measurement of Time of Travel and Dispersion in Streams by Dye Tracing*, Geological Survey, U.S. Dept. of the Interior, Washington, EUA

JPL (1996). *A Collection of Global Ocean Tide Models*. Jet Propulsion Laboratory, Physical Oceanography Distributed Active Archive Center, Pasadena, CA.

King, I.P. & Rachiele, R.R. (1989). *Program Documentation: RMA4 - A two Dimensional Finite Element Water Quality Model, Version 3.0*, ed. Resource Management Associates, January, 1989.

Luketina, D. (1998). Simple tidal prism model revisited. *Estuarine Coastal and Shelf Science*, Vol.46, pp.77–84, ISSN 0272-7714

Marinov, D. & Norro, A.J.M.Z. (2006). Application of COHERENS model for hydrodynamic investigation of Sacca di Goro coastal lagoon (Italian Adriatic Sea shore). *Ecological Modelling*, Vol.193, No.1, pp. 52–68, ISSN 0304-3800

Monsen, N.E.; Cloern, J.E. & Lucas, L.V. (2002). A comment on the use of flushing time, residence time, and age as transport time scales. *Limnology & Oceanography*, Vol.47, No.5, (May 2002), pp. 1545–1553, ISSN 0024-3590

Oliveira, A. P. & Baptista, A.M. (1997). Diagnostic modelling of residence times in estuaries. *Water Resources Reseach*, Vol.33, pp.1935–1946, ISSN 0024-3590

Paerl, H.W (2006). Assessing and managing nutrient-enhanced eutrophication in estuarine and coastal waters: interactive effects of human and climatic perturbations. *Ecological Engineering*, Vol.26, No.1, (January 2006), pp. 40-54, ISSN 0925-8574

Pardal, M.A.; Cardoso, P.G.; Sousa, J.P.; Marques, J.C. & Raffaelli, D.G. (2004). Assessing environmental quality: a novel approach. *Marine Ecology Progress Series*, Vol.267, pp. 1-8, ISSN 0171-8630.

Sanford, L.; Boicourt, W. & Rives, S. (1992). Model for estimating tidal flushing of small embayments. *Journal of Waterway, Port, Coastal and Ocean Engineering*, Vol.118, No.6, pp. 913–935, ISSN 1943-5460

Stamou, A.I.; Nanou-Giannarou, K. & Spanoudaki, K. (2007). Best modelling practices in the application of the Directive 2000/60 in Greece. *Proceedings of the 3rd IASME/WSEAS Int. Conference on Energy, Environment, Ecosystems and Sustainable Development*, pp. 388-397, Agios Nikolaos, Crete Island, Greece, July 24-26, 2007.

Takeoka, H. (1984). Fundamental concepts of exchange and transport time scales in a coastal sea. *Continental Shelf Research*, Vol.3, No.3, pp. 311–326, ISSN 0278-4343

Thomann, R.V. & Linker, L.C. (1998). Contemporary issues in watershed and water quality modelling for eutrophication control. *Water Science & Technology*, Vol.37, pp. 93-102, ISSN 0273-1223

Valiela, I., McClelland, J., Hauxwell, J., Behr, P.J., Hersh, D. & Foreman, K. (1997) Macroalgae blooms in shallow estuaries: controls, ecophysiological and ecosystem consequences. *Limnology & Oceanography*, Vol.42, No.5, (January 1997), pp. 1105–1118, ISSN 0024-3590

Vieira, J.M.P; Duarte, A.A.L.S.; Pinho, J.L.S. & Boaventura, R.A. (1999). A Contribution to Drinking Water Sources Protection Strategies in a Portuguese River Basin, *Proceedings of the XXII World Water Congress*, CD-Rom, Buenos Aires, Argentina, September, 5-7, 1999.

Wang, C.F.; Hsu,M. & Kuo, A.Y.(2004). Residence time of the Danshuei Estuary, Taiwan. *Estuarine, Coastal and Shelf. Science*, Vol.60, pp. 381-393, ISSN 1906-0015

WES-HL (1996). *Users Guide to RMA2 Version 4.3*. US Army Corps of Engineers, Waterways Experiment Station Hydraulics Laboratory, Vicksburg, USA.

WES-HL (2000). *Users Guide to RMA4 WES Version 4.5*. US Army Corps of Engineers, Waterways Experiment Station Hydraulics Laboratory, Vicksburg, USA.

# 2

# Hydrodynamic Pressure Evaluation of Reservoir Subjected to Ground Excitation Based on SBFEM

Shangming Li

*Institute of Structural Mechanics, China Academy of Engineering Physics*
*Mianyang City, Sichuan Province*
*China*

## 1. Introduction

Dynamic responses of dam-reservoir systems subjected to ground motions are often a major concern in the design. To ensure that dams are adequately designed for, the hydrodynamic pressure distribution along the dam-reservoir interface must be determined for assessment of safety.

Due to the fact that analytical methods are not readily available for dam-reservoir systems with arbitrary geometry shape, numerical methods are often used to analyze responses of dam-reservoir systems. In numerical methods, dams are often discretized into solid finite elements through Finite Element Method (FEM), while the reservoir is either directly modeled by Boundary Element Method (BEM) or is divided into two parts: a near field with arbitrary geometry shape and a far field with a uniform cross section. The near field is discretized into acoustic fluid finite elements by using FEM or boundary elements by BEM, while the far field is modeled by BEM or a Transmitting Boundary Condition (TBC). Based on these numerical methods, several coupling procedures were developed.

A FEM-BEM coupling procedure was used to implement the linear and non-linear analysis of dam-reservoir interaction problems (Tsai & Lee, 1987; Czygan & Von Estorff, 2002), respectively, in which the dam was modeled by FEM, while the reservoir was modeled by BEM. A BEM-TBC coupling method was adopted to solve dam-water-foundation interaction problems and dam-reservoir-sediment-foundation interaction problems (Dominguez & Maeso, 1993; Dominguez et al., 1997). The dam and the near field of the reservoir were discretized by using BEM, while the far field of the reservoir was represented by a TBC. As a traditional numerical method, BEM has been popular in simulating unbounded medium, but it needs a fundamental solution and includes a singular integral, which affect its application. In order to avoid deriving a fundamental solution required in BEM, the TBC attracted some researchers' interests. A Sommerfeld-type TBC was used to represent the far field (Kucukarslan et al., 2005), while a Sharan-type TBC was proposed for infinite fluid (Sharan, 1987). The Sommerfeld-type and Sharan-type TBCs are readily implemented in FEM due to their conciseness, but a sufficiently large near field is required to model accurately the damping effect of semi-infinite reservoir. Except for the aforementioned TBCs, an exact TBC (Tsai & Lee, 1991), a novel TBC (Maity &

Bhattacharyya, 1999) and a non-reflecting TBC (Gogoi & Maity, 2006) were proposed, respectively. These complicated TBCs gave better results even when a small near field was chosen, but their implementations in a finite element code became complex and tedious.

In this chapter, the scaled boundary finite element method (SBFEM) was chosen to model the far field. The SBFEM does not require fundamental solutions and is able to model accurately the damping effect of semi-infinite reservoir and incorporate with FEM readily, but the SBFEM requires the geometry of far field is layered (or tapered). Although BEM and some of TBCs can handle far fields with arbitrary geometry, far fields in most dam-reservoir systems are always chosen to be layered with a uniform cross section, which ensures the SBFEM can be used in dam-reservoir interaction problems.

Based on a mechanically-based derivation, the SBFEM was proposed for infinite medium (Wolf & Song, 1996a; Song & Wolf, 1996) which was governed by a three-dimensional scalar wave equation and a three-dimensional vector wave equation, respectively. A dynamic stiffness matrix and a dynamic mass matrix were introduced to represent infinite medium in the frequency domain and the time domain, respectively. The dynamic stiffness matrix satisfies a non-linear ordinary differential equation of first order, while the dynamic mass matrix is governed by an integral convolution equation. The SBFEM reduces spatial dimensions by one. Only boundaries need discretization and its solutions in the radial direction are analytical. Therefore, it can handle well bounded domain problems with cracks and stress singularities and unbounded domain problems including infinite soil or unbounded acoustic fluid medium. In analyzing crack and stress singularities problems, the SBFEM placed the scaling center on the crack tip and only discretized the boundary of bounded domain using supper elements except the straight traction free crack faces, which permitted a rigorous representation of the stress singularities around the crack tip (Song, 2004; Song & Wolf, 2002; Yang & Deeks, 2007). The response of unbounded domain problems was obtained by using the SBFEM alone or coupling FEM and the SBFEM. A FEM-SBFEM coupling procedure was used to analyze unbounded soil-structure interaction problems in the time domain (Ekevid & Wiberg, 2002; Bazyar & Song, 2008). For unbounded acoustic fluid medium problems, a FEM-SBFEM coupling procedure combined with acoustic approximations was proposed to evaluate the responses of submerged structures subjected to underwater shock waves in the time domain (Fan et al., 2005; Li & Fan, 2007). Results showed that the SBFEM was able to model accurately the damping behavior of the unbounded soil and infinite acoustic fluid medium, but it was computationally expensive because the evaluations of the dynamic mass matrix and dynamic responses need solving integral convolution equations. In the frequency domain, dynamic condensation and substructure deletion methods were used to evaluate the dynamic stiffness matrix, which avoid evaluating integral convolution equations, but evaluation errors increased with frequency increasing so that results at high frequencies were not acceptable (Wolf & Song, 1996b). To evaluate accurately high frequencies behaviors of the dynamic stiffness matrix, a Pade series was presented to analyze out-of-plane motion of circular cavity embedded in full-plane through using the SBFEM alone (Song & Bazyar, 2007). Good results were obtained at high frequencies, but results at low frequencies were inferior even if a high order Pade series was used. The high order Pade series was not only complex, and also increased computational cost. A simplified SBFEM formulation was presented through discovering a zero matrix and a FEM-SBFEM coupling procedure was used to analyze dam-reservoir interaction problems subjected to ground motions (Fan & Li, 2008). The simplified SBFEM was well suitable for all frequencies and no additional computational costs were increased for low frequency analysis in comparison with for high frequency analysis. Its advantages were exhibited by analyzing the harmonic responses of dam-reservoir systems in the frequency domain. However in the time domain, its advantages are not as obvious as those in the frequency domain because integral convolutions still need evaluating. Although a Riccati equation and Lyapunov equations were presented to solve the integral convolutions (Wolf & Song, 1996b), solving them needed great computational costs, especially for large-scale systems, which limited the SBFEM applications in the time domain. To simplify the integral convolutions and save computational costs, some recursive formulations were proposed (Paronesso & Wolf, 1998; Yan et al., 2004), based on a diagonalization procedure and the linear system theory (Paronesso & Wolf, 1995). The integral convolution was transformed into an equivalent system of linear equations, named state-variable description which was represented by finite-difference equations. However, the coefficient matrix quaternion of finite-difference equations was calculated by using Hankel matrix realization algorithms, which complicated the analysis. Furthermore, the diagonalization procedure increased the order of the dynamic mass matrix, and some global lumped parameters, such as springs, dashpots and masses, used in the diagonalization procedure must be introduced at additional internal nodes corresponding to inner variables in the state-variable description, besides the nodes on the structure-medium interface. The number of global lumped parameters would become very large for large-scale systems. This weakened the feasibility of the diagonalization procedure. A new diagonalization procedure of the SBFEM for semi-infinite reservoir was proposed (Li, 2009), whose calculation efficiency was proven to be high, although it still included convolution integrals. With the improvement of the SBFEM evaluation efficiency in the time domain analysis, the SBFEM will show gradually its advantages and potential to solve problems including unbounded soil or unbounded acoustic fluid medium, such as the dam-reservoir interaction problems.

## 2. Problem statement

Consider dam-reservoir interaction problems subjected to horizontal ground accelerations. The dam-reservoir system and its Cartesian coordinate system were shown in Fig.1. The

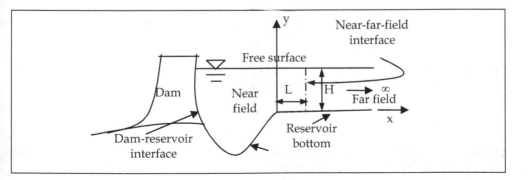

Fig. 1. Dam-reservoir system

dam was subjected to a horizontal ground acceleration $a_x$ and the semi-infinite reservoir was filled with an inviscid isentropic fluid. To evaluate the response of the dam-reservoir system under a horizontal ground acceleration $a_x$ excitation, the semi-infinite reservoir was divided into two parts: a near field and a far field. The near field was located between the dam-reservoir interface and the radiation boundary (the near-far-field interface at $x = L$), while the far field was from $x = L$ to $\infty$. Note that the geometry of the reservoir was chosen to be arbitrary for $x < 0$ and flat for $x \geq 0$.

For an inviscid isentropic fluid (acoustic fluid) with the fluid particles undergoing only small displacements and not including body force effects, the governing equations is expressed as

$$\nabla^2 \phi = \frac{1}{c^2} \ddot{\phi} \tag{1}$$

where $\phi$ denotes velocity potential and $c$ denotes the sound speed in fluid. Reservoir pressure $p$, the velocity vector $\mathbf{v}$ and the velocity potential $\phi$ have a relationship as follows:

$$\mathbf{v} = \nabla \phi \tag{2a}$$

$$p = -\rho \dot{\phi} \tag{2b}$$

where $\rho$ denotes fluid density. Boundary conditions of the near field for Eq.(1) are following. Along the dam-reservoir interface, one has

$$\mathbf{v} \bullet \mathbf{n} = \frac{\partial \phi}{\partial n} = v_n \tag{3}$$

where the unit vector $\mathbf{n}$ is perpendicular to the dam-reservoir interface and points outward of fluid; $v_n$ is the normal velocity of the dam-reservoir interface. The boundary condition along the reservoir bottom is

$$\frac{\partial \phi}{\partial n} + q\dot{\phi} = v_n \tag{4}$$

where $q$ is defined as

$$q = \frac{1}{c}\left(\frac{1 - \alpha_r}{1 + \alpha_r}\right) \tag{5}$$

in which $\alpha_r$ denotes a reflection coefficient of pressure striking the bottom of the reservoir. By ignoring effects of surface waves of fluid, the boundary condition of the free surface is taken as

$$\phi = 0 \tag{6}$$

The boundary condition on the radiation boundary (near-far-filed interface) should include effects of the radiation damping of infinite reservoir and those of energy dissipation in the reservoir due to the absorptive reservoir bottom. To model these effects accurately, the SBFEM was adopted in this chapter.

## 3. SBFEM formulation

Fig.2 showed the SBFEM discretization model of the far field shown in Fig.1, which was a layered semi-infinite fluid medium whose scaling center was located at minus infinity. The whole semi-infinite layered far field was divided into some layered sub-fields. Each layered sub-field was represented by one element on the near-far-field interface, so the whole far field was discretized into some elements on the near-far-field interface. Based on the discretization, a dynamic stiffness or mass matrix was introduced to describe the characteristics of the far field in the SBFEM. The interaction between the near field and the far field was expressed as the following SBFEM formulation.

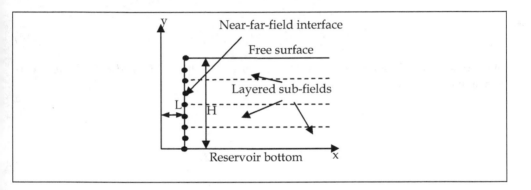

Fig. 2. SBFEM discretization model of layered far field

### 3.1 SBFEM formulation in the frequency domain

On the discretized near-far-field interface, the SBFEM formulation in the frequency domain (Fan & Li, 2008; Li et al., 2008) for the far field filled with unbounded acoustic fluid medium is written as

$$\mathbf{V}_n(\omega) = \mathbf{S}^\infty(\omega)\mathbf{\Phi}(\omega) \tag{7}$$

where $\mathbf{\Phi}(\omega)$ denotes the column vector composed of nodal velocity potentials $\phi$ ; $\mathbf{S}^\infty(\omega)$ is the dynamic stiffness matrix of the far field and $\mathbf{V}_n(\omega)$ satisfies

$$\mathbf{V}_n(\omega) = \sum_e \int_{\Gamma_w^e} \mathbf{N}_f^T v_n(\omega) d\Gamma_w^e \tag{8}$$

in which $v_n$ is the normal velocity; $\Gamma_w$ denotes the near-far-field interface; $\mathbf{N}_f$ is the shape function for a typical discretized acoustic fluid finite element; and $\Sigma_e$ denotes an assemblage of all fluid elements on the near-far-field interface. The dynamic stiffness matrix $\mathbf{S}^\infty(\omega)$ (Li, et al., 2008) satisfies

$$\left(\mathbf{S}^\infty(\omega) + \mathbf{E}^1\right)\mathbf{E}^{0-1}\left(\mathbf{S}^\infty(\omega) + \mathbf{E}^{1T}\right) - \mathbf{E}^2 - i\omega\mathbf{C}^0 - (i\omega)^2\mathbf{M}^0 = 0 \tag{9}$$

where global coefficient matrices $\mathbf{E}^0$, $\mathbf{E}^1$, $\mathbf{E}^2$, $\mathbf{C}^0$ and $\mathbf{M}^0$ only depend on the geometry of the near-far-field interface and the reflection coefficient $\alpha_r$. They are obtained through assembling all elements' $\mathbf{E}_e^0$, $\mathbf{E}_e^1$, $\mathbf{E}_e^2$, $\mathbf{C}_e^0$ and $\mathbf{M}_e^0$ on the near-far-field interface. The matrices $\mathbf{E}_e^0$, $\mathbf{E}_e^1$, $\mathbf{E}_e^2$, $\mathbf{C}_e^0$ and $\mathbf{M}_e^0$ corresponding to each element can be evaluated numerically or analytically using the following equations.

$$E_e^0 = \int_{-1}^{1}\int_{-1}^{1} \mathbf{B}^{1T}\mathbf{B}^1 |\mathbf{J}|\, d\eta\, d\varsigma \tag{10a}$$

$$E_e^1 = \int_{-1}^{1}\int_{-1}^{1} \mathbf{B}^{2T}\mathbf{B}^1 |\mathbf{J}|\, d\eta\, d\varsigma \tag{10b}$$

$$E_e^2 = \int_{-1}^{1}\int_{-1}^{1} \mathbf{B}^{2T}\mathbf{B}^2 |\mathbf{J}|\, d\eta\, d\varsigma \tag{10c}$$

$$\mathbf{M}_e^0 = \int_{-1}^{1}\int_{-1}^{1} \frac{1}{c^2}\mathbf{N}_f^T\mathbf{N}_f |\mathbf{J}|\, d\eta\, d\varsigma \tag{10d}$$

where the $\mathbf{N}_f$ is defined in Eq.(8) and the others $\mathbf{J}$, $\mathbf{B}^1$, $\mathbf{B}^2$ are defined below. The matrix $\mathbf{J}$ is defined as

$$\mathbf{J} = \begin{bmatrix} H & 0 & 0 \\ \dfrac{d\mathbf{N}_f}{d\eta}\mathbf{x} & \dfrac{d\mathbf{N}_f}{d\eta}\mathbf{y} & \dfrac{d\mathbf{N}_f}{d\eta}\mathbf{z} \\ \dfrac{d\mathbf{N}_f}{d\varsigma}\mathbf{x} & \dfrac{d\mathbf{N}_f}{d\varsigma}\mathbf{y} & \dfrac{d\mathbf{N}_f}{d\varsigma}\mathbf{z} \end{bmatrix} \tag{11a}$$

where the symbol $H$ denotes the water depth in the far field and $\mathbf{x}$, $\mathbf{y}$ and $\mathbf{z}$ are element nodal coordinates column vectors. Due to the fact that $x$ coordinates of all nodes inside the near-far-field interface (vertical surface) are same, the matrix $\mathbf{J}$ becomes

$$\mathbf{J} = \begin{bmatrix} H & 0 & 0 \\ 0 & \dfrac{d\mathbf{N}_f}{d\eta}\mathbf{y} & \dfrac{d\mathbf{N}_f}{d\eta}\mathbf{z} \\ 0 & \dfrac{d\mathbf{N}_f}{d\varsigma}\mathbf{y} & \dfrac{d\mathbf{N}_f}{d\varsigma}\mathbf{z} \end{bmatrix} \tag{11b}$$

Write the inverse of $\mathbf{J}$ in the following form

$$\mathbf{J}^{-1} = \begin{bmatrix} j_{11} & j_{12} & j_{13} \\ j_{21} & j_{22} & j_{23} \\ j_{31} & j_{32} & j_{33} \end{bmatrix} \tag{12}$$

The components $j_{mn}$ $(m,n = 1,2,3)$ can be evaluated by using Eq.(11b). Therefore, the matrix $\mathbf{B}^1$ is defined as

$$\mathbf{B}^1 - \begin{bmatrix} j_{11} \\ j_{21} \\ j_{31} \end{bmatrix} \mathbf{N}_f \tag{13}$$

and the matrix $\mathbf{B}^2$ is

$$\mathbf{B}^2 = \begin{bmatrix} j_{12} \\ j_{22} \\ j_{32} \end{bmatrix} \frac{d\mathbf{N}_f}{d\eta} + \begin{bmatrix} j_{13} \\ j_{23} \\ j_{33} \end{bmatrix} \frac{d\mathbf{N}_f}{d\varsigma} \tag{14}$$

Note that Eqs.(10-14) are only the functions of nodal coordinates of elements inside the near-far-field interface. The matrix $\mathbf{C}_e^0$ is a zero matrix for elements not adjacent to reservoir bottom inside the near-far-field interface, while for those adjacent to reservoir bottom, $\mathbf{C}_e^0$ satisfies

$$\mathbf{C}_e^0 = \frac{1}{c}\left(\frac{1-\alpha_r}{1+\alpha_r}\right) H \int_{\Gamma_b} \mathbf{N}_f^T \mathbf{N}_f d\Gamma_b \tag{15}$$

where the symbol $\Gamma_b$ denotes the reservoir bottom of the near-far-field interface, i.e. the line $y = 0$ as shown in the Fig.2. Assembling all elements' $\mathbf{E}_e^0$, $\mathbf{E}_e^1$, $\mathbf{E}_e^2$, $\mathbf{C}_e^0$ and $\mathbf{M}_e^0$ can yield the global coefficient matrices $\mathbf{E}^0$, $\mathbf{E}^1$, $\mathbf{E}^2$, $\mathbf{C}^0$ and $\mathbf{M}^0$ in Eq.(9). Details about them can be found in the literatures (Wolf & Song, 1996b; Li et al., 2008).

For a vertical near-far-field interface as shown in Fig.2, as the matrix $\mathbf{E}^1$ was a zero matrix, the dynamic stiffness matrix $\mathbf{S}^\infty(\omega)$ in Eq.(9) can be re-written readily as

$$\mathbf{S}^\infty(\omega) = \sqrt{\left(\mathbf{E}^2 + i\omega\mathbf{C}^0 - \omega^2\mathbf{M}^0\right)\mathbf{E}^{0-1}}\,\mathbf{E}^0 \tag{16}$$

where $\omega$ is an excitation frequency. The $\mathbf{S}^\infty(\omega)$ can be obtained by the Schur factorization.

## 3.2 SBFEM formulation in the time domain

The corresponding SBFEM formulation of Eq.(7) in the time domain is written as (Wolf & Song, 1996b)

$$\mathbf{V}_n(t) = \int_0^t \mathbf{M}^\infty(t-\tau)\ddot{\mathbf{\Phi}}(\tau)d\tau \tag{17}$$

in which $\mathbf{M}^\infty(t)$ is the dynamic mass matrix of the far field; $\mathbf{\Phi}(t)$ and $\mathbf{V}_n(t)$ are the corresponding variables of $\mathbf{\Phi}(\omega)$ and $\mathbf{V}_n(\omega)$ in the time domain, respectively. $\mathbf{M}^\infty(t)$ and $\mathbf{S}^\infty(\omega)/(i\omega)^2$ forms a Fourier transform pair. Upon discretization of Eq.(17) with respect to time and assuming all initial conditions equal to zero, one can get the following equation

$$\mathbf{V}_n^n = \mathbf{M}_1^\infty \dot{\mathbf{\Phi}}^n + \sum_{j=1}^{n-1}\left(\mathbf{M}_{n-j+1}^\infty - \mathbf{M}_{n-j}^\infty\right)\dot{\mathbf{\Phi}}^j \tag{18}$$

in which $\mathbf{M}_{n-j+1}^\infty = \mathbf{M}^\infty\left((n-j+1)\Delta t\right)$, $\mathbf{\Phi}^j = \mathbf{\Phi}(j\Delta t)$ and $\mathbf{V}_n^n = \mathbf{V}_n(n\Delta t)$ where $\Delta t$ denotes an increment in time step.

Applying the inverse Fourier transformation to Eq. (9) with $\mathbf{E}^1 = 0$ yields

$$\int_0^t \mathbf{m}^\infty(t-\tau)\mathbf{m}^\infty(\tau)d\tau - \frac{t^3}{6}\mathbf{e}^2 - \frac{t^2}{2}\mathbf{c}^0 - t\mathbf{m}^0 = 0 \tag{19}$$

where $t$ is time and

$$\mathbf{m}^\infty(t) = \mathbf{U}^{-1T}\mathbf{M}^\infty(t)\mathbf{U}^{-1} \tag{20}$$

$$\mathbf{e}^2 = \mathbf{U}^{-1T}\mathbf{E}^2\mathbf{U}^{-1} \tag{21}$$

$$\mathbf{m}^0 = \mathbf{U}^{-1T}\mathbf{M}^0\mathbf{U}^{-1} \tag{22}$$

$$\mathbf{c}^0 = \mathbf{U}^{-1T}\mathbf{C}^0\mathbf{U}^{-1} \tag{23}$$

in which $\mathbf{U}$ satisfies

$$\mathbf{E}^0 = \mathbf{U}^T\mathbf{U} \tag{24}$$

A procedure (Wolf & Song, 1996b) was presented to evaluate the dynamic mass matrix $\mathbf{M}^\infty(t)$ at different time $t$ governed by the convolution integral Eq.(19). In that procedure, discretization of Eq.(19) with respect to time was implemented, and an algebraic Riccati equation for evaluating $\mathbf{M}^\infty(t = \Delta t)$ at first time step and a Lyapunov equation for evaluating $\mathbf{M}^\infty(t = j\Delta t)$ at other jth time steps were formed, respectively. The $\mathbf{M}^\infty(t = j\Delta t)$ at any time was obtained by utilizing Schur factorization to solve these two types of equations. When the coefficient matrix $\mathbf{c}^0 = 0$, a simple diagonal procedure (Li, 2009) can be adopted to evaluate the $\mathbf{M}^\infty(t)$, which can avoid Schur factorization and solving Riccati equation and Lyapunov equation.

## 4. FEM-SBFEM coupling formulation of reservoir

To obtain the response of dam-reservoir system, the near-field fluid domain is discretized into an assemblage of finite elements. The corresponding finite-element governing equation of Eq.(1) for the near-field domain can be expressed as

$$\begin{bmatrix} \mathbf{m}_{11} & \mathbf{m}_{12} & \mathbf{m}_{13} \\ \mathbf{m}_{21} & \mathbf{m}_{22} & \mathbf{m}_{23} \\ \mathbf{m}_{31} & \mathbf{m}_{32} & \mathbf{m}_{33} \end{bmatrix}\begin{Bmatrix} \ddot{\Phi}_1 \\ \ddot{\Phi}_2 \\ \ddot{\Phi}_3 \end{Bmatrix} + \begin{bmatrix} \mathbf{k}_{11} & \mathbf{k}_{12} & \mathbf{k}_{13} \\ \mathbf{k}_{21} & \mathbf{k}_{22} & \mathbf{k}_{23} \\ \mathbf{k}_{31} & \mathbf{k}_{32} & \mathbf{k}_{33} \end{bmatrix}\begin{Bmatrix} \Phi_1 \\ \Phi_2 \\ \Phi_3 \end{Bmatrix} = \begin{Bmatrix} \mathbf{V}_{n1} \\ \mathbf{V}'_{n2} \\ \mathbf{V}_{n3} \end{Bmatrix} \tag{25}$$

where the global mass matrix $\mathbf{m}$, the global stiffness matrix $\mathbf{k}$ and the global vector $\mathbf{V}_n$ are treated in the standard manner as in the traditional FE procedures; the subscripts 1 and 2 refer to nodal variables at the dam-reservoir interface and the near-far-field interface, respectively, while the subscript 3 refers to other interior nodal variables in the near-field fluid. At the near-far-field interface, the near-field FEM-domain couples with the far-field SBFEM-domain. The kinematic continuity condition requires that both fields have the same normal velocity at the near-far-field interface. Hence, one has

$$-\mathbf{V}'_{n2} = \mathbf{V}_n \tag{26}$$

In the frequency domain, using Eqs.(7, 16, 25, 26) yields

$$
\begin{bmatrix} \mathbf{m}_{11} & \mathbf{m}_{12} & \mathbf{m}_{13} \\ \mathbf{m}_{21} & \mathbf{m}_{22} & \mathbf{m}_{23} \\ \mathbf{m}_{31} & \mathbf{m}_{32} & \mathbf{m}_{33} \end{bmatrix} \begin{Bmatrix} \ddot{\Phi}_1 \\ \ddot{\Phi}_2 \\ \ddot{\Phi}_3 \end{Bmatrix} +
$$

$$
\begin{bmatrix} \mathbf{k}_{11} & \mathbf{k}_{12} & \mathbf{k}_{13} \\ \mathbf{k}_{21} & \mathbf{k}_{22} + \sqrt{\left(\mathbf{E}^2 + i\omega\mathbf{C}^0 - \omega^2\mathbf{M}^0\right)\mathbf{E}^{0-1}}\,\mathbf{E}^0 & \mathbf{k}_{23} \\ \mathbf{k}_{31} & \mathbf{k}_{32} & \mathbf{k}_{33} \end{bmatrix} \begin{Bmatrix} \Phi_1 \\ \Phi_2 \\ \Phi_3 \end{Bmatrix} = \begin{Bmatrix} \mathbf{V}_{n1} \\ 0 \\ \mathbf{V}_{n3} \end{Bmatrix} \tag{27}
$$

For a harmonic response with an exciting frequency $\omega$,

$$
\Phi = \bar{\Phi}e^{i\omega t} \tag{28}
$$

Substituting Eq.(28) into Eq.(27) leads to the FEM-SBFEM coupling equation of a reservoir to solve the harmonic response of a reservoir, i.e.

$$
\left( -\omega^2 \begin{bmatrix} \mathbf{m}_{11} & \mathbf{m}_{12} & \mathbf{m}_{13} \\ \mathbf{m}_{21} & \mathbf{m}_{22} & \mathbf{m}_{23} \\ \mathbf{m}_{31} & \mathbf{m}_{32} & \mathbf{m}_{33} \end{bmatrix} + \begin{bmatrix} \mathbf{k}_{11} & \mathbf{k}_{12} & \mathbf{k}_{13} \\ \mathbf{k}_{21} & \mathbf{k}_{22} + \sqrt{\left(\mathbf{E}^2 + i\omega\mathbf{C}^0 - \omega^2\mathbf{M}^0\right)\mathbf{E}^{0-1}}\,\mathbf{E}^0 & \mathbf{k}_{23} \\ \mathbf{k}_{31} & \mathbf{k}_{32} & \mathbf{k}_{33} \end{bmatrix} \right) \begin{Bmatrix} \bar{\Phi}_1 \\ \bar{\Phi}_2 \\ \bar{\Phi}_3 \end{Bmatrix} e^{i\omega t} = \begin{Bmatrix} \mathbf{V}_{n1} \\ 0 \\ \mathbf{V}_{n3} \end{Bmatrix} \tag{29}
$$

Eq.(29) can be solved for any frequency $\omega$.

In the time domain, using Eqs.(17, 18, 25, 26) yields the FEM-SBFEM coupling equation of a reservoir to solve the transient response of a reservoir, i.e.

$$
\begin{bmatrix} \mathbf{m}_{11} & \mathbf{m}_{12} & \mathbf{m}_{13} \\ \mathbf{m}_{21} & \mathbf{m}_{22} & \mathbf{m}_{23} \\ \mathbf{m}_{31} & \mathbf{m}_{32} & \mathbf{m}_{33} \end{bmatrix} \begin{Bmatrix} \ddot{\Phi}_1^n \\ \ddot{\Phi}_2^n \\ \ddot{\Phi}_3^n \end{Bmatrix} + \begin{bmatrix} 0 & 0 & 0 \\ 0 & \mathbf{M}_1^\infty & 0 \\ 0 & 0 & 0 \end{bmatrix} \begin{Bmatrix} \dot{\Phi}_1^n \\ \dot{\Phi}_2^n \\ \dot{\Phi}_3^n \end{Bmatrix}
$$

$$
+ \begin{bmatrix} \mathbf{k}_{11} & \mathbf{k}_{12} & \mathbf{k}_{13} \\ \mathbf{k}_{21} & \mathbf{k}_{22} & \mathbf{k}_{23} \\ \mathbf{k}_{31} & \mathbf{k}_{32} & \mathbf{k}_{33} \end{bmatrix} \begin{Bmatrix} \Phi_1^n \\ \Phi_2^n \\ \Phi_3^n \end{Bmatrix} = \begin{Bmatrix} \mathbf{V}_{n1}^n \\ -\sum\limits_{j=1}^{n-1}\left(\mathbf{M}_{n-j+1}^\infty - \mathbf{M}_{n-j}^\infty\right)\dot{\Phi}_2^j \\ \mathbf{V}_{n3}^n \end{Bmatrix} \tag{30}
$$

where the superscript $n$ denotes the instant at time $t = n\Delta t$. Note that a damping matrix appears on the left hand side of Eq.(30). It can be regarded as the damping effect derived from the far-field medium and imposed on the dam-reservoir system. As the near-field domain is modeled by FEM, Eqs.(29, 30) are suitable for a reservoir with any arbitrary geometry shape.

## 5. Numerical examples

### 5.1 Harmonic response of reservoir

Two-dimensional dam-reservoir systems subjected to horizontal harmonic ground accelerations $a = \bar{a}e^{i\omega t}$ in the upstream direction were studied. For simplicity, here the dam was assumed to be rigid.

### 5.1.1 Vertical dam

For a rigid dam-reservoir system with a vertical upstream face as shown in Fig.3, the whole reservoir was flat so that the whole reservoir was modeled by the far field alone. This example's aim was only to test the correctness and efficiency of the SBFEM in Eqs.(7, 8, 16) of the far field. The whole reservoir was discretized by the SBFEM alone using 10 and 20 3-noded SBFEM elements, respectively. The hydrodynamic pressure acting on the dam-reservoir interface from a reflection coefficient $\alpha_r = 0.95$ and these two mesh densities was plotted in Fig.4. The coefficient $C_p$ was defined as $|p|/(\rho\bar{a}H)$ and $\omega_1 = c\pi/(2H)$, where $p$ denoted the amplitude of hydrodynamic pressure acting on the dam-reservoir interface.

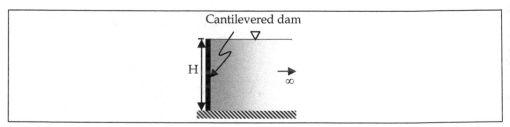

Fig. 3. Vertical dam-reservoir system

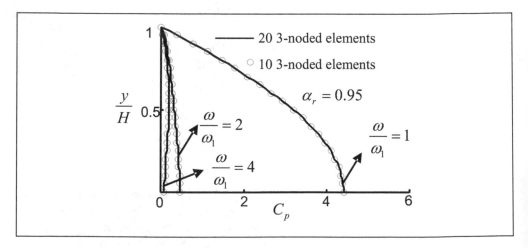

Fig. 4. Hydrodynamic pressure on vertical dam-reservoir interface from different meshes

Results from different mesh densities were the same. The hydrodynamic pressure obtained by using 10 3-noded SBFEM elements and the corresponding analytical solutions (Weber, 1994) corresponding to different $\alpha_r$ were plotted in Fig.5. The SBFEM solutions were the exact same to the analytical solutions. Furthermore, a $\omega-C_p$ figure of a point located at $y = 0.6H$ corresponding to $\alpha_r = 0.8$ was shown in Fig.6. The SBFEM solution and the analytical solution (Weber, 1994) were the same.

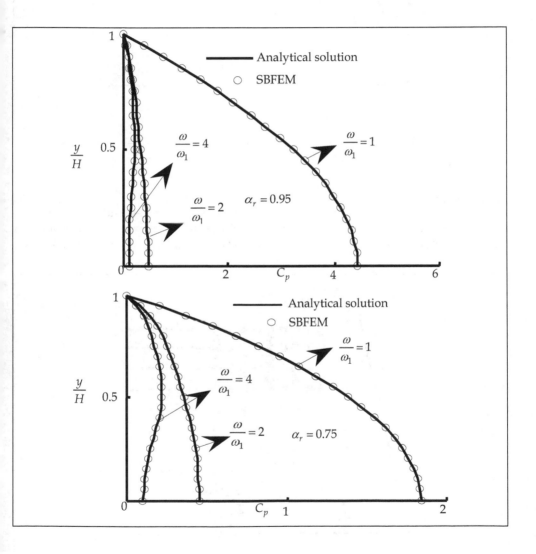

Fig. 5. Hydrodynamic pressures on vertical dam-reservoir interface caused by different $\alpha_r$

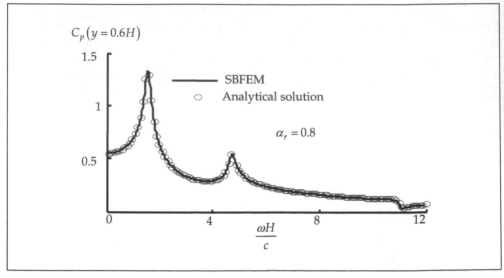

Fig. 6. $C_p(y = 0.6H)$ for different $\omega$

### 5.1.2 Gravity dam

A gravity dam shown in Fig.7 was considered to verify the correctness and efficiency of the FEM-SBFEM coupling formulation in Eq.(29). The near field was chosen as the domain with a very small distance $L = 0.001H$ away from the heel of dam and was discretized by 8-noded

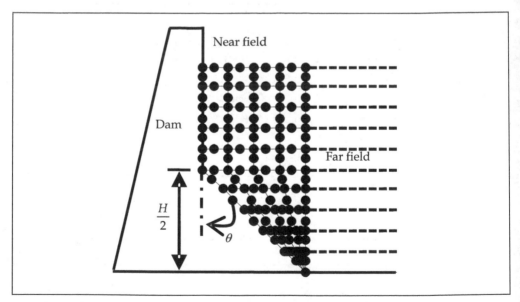

Fig. 7. Meshes of gravity dam with multi-sloping faces and $\theta = 45^0$

isoparametric acoustic fluid finite elements, while the far field was still modeled by 10 3-noded SBFEM elements. Their meshes were shown in Fig.7. Solutions from Eq.(29) and the literature (Sharan, 1992) were plotted in Fig.8. Results obtained by Eq.(29) were in excellent agreement with Sharan's results.

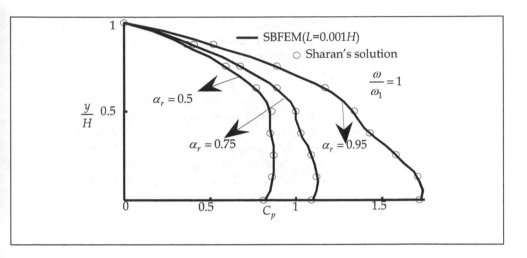

Fig. 8. Hydrodynamic pressure acting on gravity dam

## 5.2 Transient response of dam-reservoir system

Consider transient responses of dam-reservoir systems where dams were subjected to horizontal ground acceleration excitations shown in Fig.9. In the transient analysis, only the linear behavior was considered, the free surface wave effects and the reservoir bottom absorption were ignored, and the damping of dams was excluded. Dams were discretized by the FEM, while the response of the reservoir was solved by Eq.(30). The FE equation of dam and Eq.(30) was solved by Newmark's time-integration scheme with Newmark integration parameters $\alpha = 0.25$ and $\delta = 0.5$. An iteration scheme (Fan et al., 2005) was adopted to obtain the response of the dam-reservoir interaction problems.

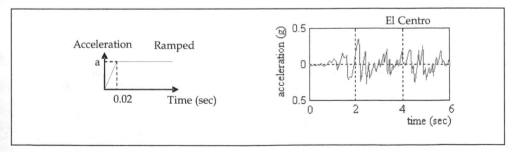

Fig. 9. Horizontal acceleration excitations

### 5.2.1 Vertical dam

As the cross section of the vertical dam-system as shown in Fig.3 was uniform, a near-field fluid domain was not necessary and the whole reservoir was modeled by a far-field domain alone. Sound speed in the reservoir is 1438.656$m/s$ and the fluid density $\rho$ is 1000$kg/m^3$. The weight per unit length of the cantilevered dam was 36000$kg/m$. The height of the cantilevered dam $H$ was 180$m$. The dam was modeled by 20 numbers of simple 2-noded beam elements with rigidity $EI$ (=9.646826×10$^{13}Nm^2$), while the whole fluid domain was modeled by 10 numbers of 3-noded SBFEM elements, whose nodes matched side by side with nodes of the dam. In this problem, the shear deformation effects were not included in the 2-noded beam elements. Time step increment was 0.005$sec$. The pressure at the heel of dam subjected to the ramped horizontal acceleration shown in Fig.9 was plotted in Fig.10 and Fig.11. Analytical solutions of deformable and rigid dams were from the literature (Tsai et al., 1990) and the literature (Weber, 1994), respectively. In Fig.11, analytical solutions (Weber, 1994), solutions from the SBFEM in the full matrix form (Wolf & Song, 1996b) and solutions from the SBFEM in the diagonal matrix form (Li, 2009) were plotted with circles, rectangles and solid line, respectively. Solutions from the SBFEM and analytical solutions were the same. In the literature (Li, 2009), it was found that diagonal SBFEM formulations need much less computational costs than those in the full matrix.

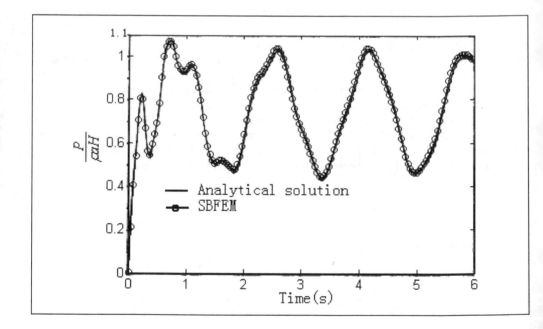

Fig. 10. Pressure at the heel of deformable dam subjected to ramped horizontal acceleration

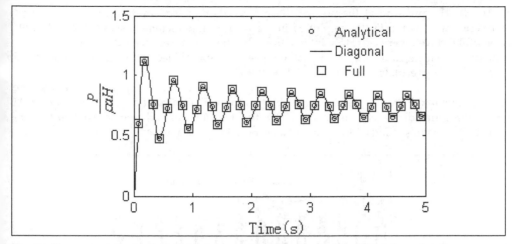

Fig. 11. Pressure at the heel of rigid dam subjected to ramped horizontal acceleration

### 5.2.2 Gravity dam

This example was analyzed to verify the accuracy and efficiency of the FEM-SBFEM coupling formulation for a dam-reservoir system having arbitrary slopes at the dam-reservoir interface. The density, Poisson's ratio and Young's modulus of the deformable dam are $2400kg/m^3$, 0.2 and $2.5×10^{10}N/m^2$, respectively. The fluid density $\rho$ is $1000kg/m^3$ and wave speed in the fluid is $1438.656m/s$. The height of the dam $H$ is $120m$. A typical gravity-dam-reservoir system and its FEM and SBFEM meshes were shown in Fig.12. The dam and the near-field fluid were discretized by FEM, while the far-field fluid was discretized by the SBFEM. 40 numbers and 20 numbers of 8-noded elements were used to model the dam and the near-field fluid domain, respectively, while 10 numbers of 3-noded SBFEM elements were employed to model the whole far-field fluid domain. Note that the size of the near-field fluid domain can be very small compared to those used in other methods. In this example, the distance between the heel of the dam and the near-far-field interface was $6m$

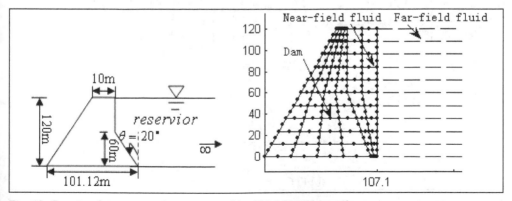

Fig. 12. Gravity dam-reservoir system and its FEM-SBFEM mesh

(=0.05*H*). The pressure at the heel of the gravity dam caused by the horizontal ground acceleration shown in Fig.9 was plotted in Fig.13. The time increment was 0.002*sec*. Results from SBFEM were very close to solutions from the sub-structures method (Tsai & Li, 1991). The displacements at the top of vertical and gravity dams subjected to a ramped horizontal acceleration were plotted in Fig.14. The displacement solutions of vertical dam from the SBFEM were the same with analytical solutions (Tsai et al., 1990). Fig.15 showed the displacement at the top of gravity dam subjected to the El Centro horizontal acceleration. At early time, the displacements obtained by the present method agreed well with sub-structure method's results (Tsai et al., 1990), especially at early time.

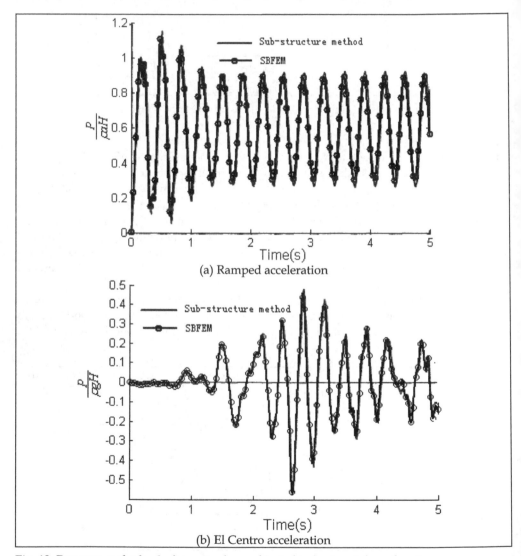

(a) Ramped acceleration

(b) El Centro acceleration

Fig. 13. Pressure at the heel of gravity dam subjected to horizontal acceleration

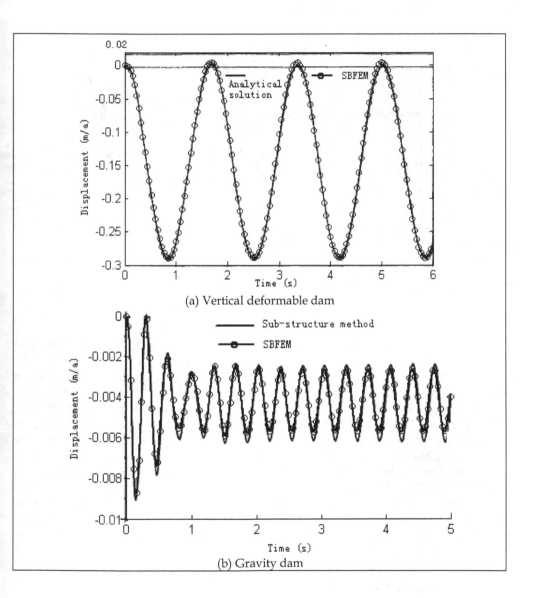

Fig. 14. Displacement at top of dam subjected to ramped horizontal acceleration

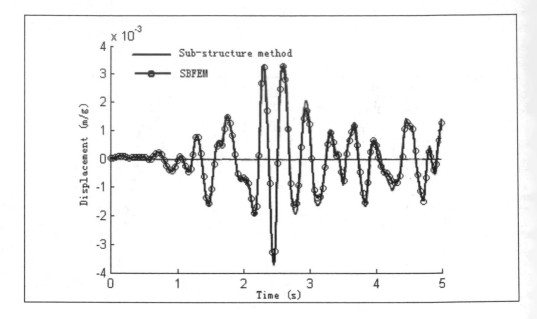

Fig. 15. Displacement at top of gravity dam subjected to El Centro horizontal acceleration

## 6. Conclusion

Aiming for dam-reservoir system problems subjected to horizontal ground motions, this chapter presented the SBFEM formulations in the frequency and time domain and its corresponding FEM-SBFEM coupling formulations to evaluate the hydrodynamic pressure of the reservoir through dividing the reservoir into a near field and far field, where the dam and the near field were modeled by FEM and the far field was discretized by the SBFEM. The SBFEM uses the dynamic stiffness matrix and the dynamic mass matrix to describe the dynamic characteristics of the far field in the frequency and time domain, respectively. The merits of the SBFEM in representing the semi-infinite reservoir were illustrated through comparisons against benchmark solutions. Numerical results showed that its accuracy and efficiency of the FEM-SBFEM formulation to obtain the harmonic and transient analysis of a dam-reservoir system. Of note, the SBFEM is a semi-analytical method. Its solution in the radial direction is analytical so that only a near field with a small volume is required. Compared to the sub-structure method, its formulations are in a simpler mathematical form and can be coupled with FEM easily and seamlessly.

## 7. Acknowledgments

This research is supported by the National Natural Science Foundation of China (No. 10902060) and China Postdoctoral Science Foundation (201003123), for which the author is grateful.

# 8. References

Bazyar, M.H. & Song, C.M. (2008). A continued-fraction-based high-order transmitting boundary for wave propagation in unbounded domains of arbitrary geometry. *International Journal for Numerical Methods in Engineering*, Vol.74, pp.209-237

Czygan, O. & Von, Estorff, O. (2002). Fluid-structure interaction by coupling BEM and nonlinear FEM. *Engineering Analysis with Boundary Elements*, Vol.26, pp.773-779

Dominguez, J ; Gallego, R. & Japon, B.R. (1997). Effects of porous sediments on seismic response of concrete gravity dams. *Journal of Engineering Mechanics – ASCE*, Vol.123, pp.302-311

Dominguez, J. & Maeso, O. (1993). Earthquake analysis of arch dams II. Dam-water-foundation interaction. *Journal of Engineering Mechanics – ASCE*, Vol.119, pp.513-530

Ekevid, T. & Wiberg, N.E. (2002). Wave propagation related to high-speed train - A scaled boundary FE-approach for unbounded domains. *Computer Methods in Applied Mechanics and Engineering*, Vol.191, pp.3947-3964

Fan, S.C. ; Li, S.M. & Yu, G.Y. (2005). Dynamic fluid-structure interaction analysis using boundary finite element method-finite element method. *Journal of Applied Mechanics - Transactions of the ASME*, Vol.72, pp.591-598

Fan, S.C. & Li, S.M. (2008). Boundary finite-element method coupling finite-element method for steady-state analyses of dam-reservoir systems. *Journal of Engineering Mechanics – ASCE*, Vol.134, pp.133-142

Gogoi, I. & Maity, D. (2006). A non-reflecting boundary condition for the finite element modeling of infinite reservoir with layered sediment. *Advances in Water Resources*, Vol.29, pp.1515-1527

Kucukarslan, S .; Coskun, S.B. & Taskin, B. (2005). Transient analysis of dam-reservoir interaction including the reservoir bottom effects. *Journal of Fluids and Structures*, Vol.20, pp.1073-1084

Li, S.M. & Fan, S.C. (2007). Parametric analysis of a submerged cylindrical shell subjected to shock waves. *China Ocean Engineering*, Vol.21, pp.125-136

Li, S.M. ; Liang, H. & Li, A.M. (2008). A semi-analytical solution for characteristics of a dam-reservoir system with absorptive reservoir bottom. *Journal of Hydrodynamics*, Vol.20, pp.727-734.

Li, S.M. (2009). Diagonalization procedure for scaled boundary finite element method in modelling semi-infinite reservoir with uniform cross section. *International Journal for Numerical Methods in Engineering*, Vol.80, pp.596-608

Maity, D. & Bhattacharyya, S.K. (1999). Time-domain analysis of infinite reservoir by finite element method using a novel far-boundary condition. *Finite Elements in Analysis and Design*, Vol.32, pp.85-96

Paronesso, A. & Wolf, J.P. (1995). Global lumped-parameter model with physical representation for unbounded medium. *Earthquake Engineering and Structural Dynamics*, Vol.24, pp.637-654

Paronesso, A. & Wolf, J.P. (1998). Recursive evaluation of interaction forces and property matrices from unit-impulse response functions of unbounded medium based on balancing approximation. *Earthquake Engineering and Structural Dynamics*, Vol.27, pp.609-618

Sharan, S.K. (1992). Efficient finite element analysis of hydrodynamic pressure on dams. *Computers and Structures*, Vol.42, No.5, pp.713-723

Sharan, S.K. (1987). Time-domain analysis of infinite fluid vibration. *International Journal for Numerical Methods in Engineering*, Vol.24, pp.945-958

Song, C.M. & Bazyar, M.H. (2007). A boundary condition in Pade series for frequency-domain solution of wave propagation in unbounded domains. *International Journal for Numerical Methods in Engineering*, Vol.69, pp.2330-2358

Song, C.M. & Wolf, J.P. (1996). Consistent infinitesimal finite-element cell method: Three-dimensional vector wave equation. *International Journal for Numerical Methods in Engineering*, Vol.39, pp.2189-2208

Song, C.M. & Wolf, J.P. (2002). Semi-analytical representation of stress singularities as occurring in cracks in anisotropic multi-materials with the scaled boundary finite-element method. *Computers and Structures*, Vol.80, pp.183-197

Song, C.M. (2004). A super-element for crack analysis in the time domain. International *Journal for Numerical Methods in Engineering*, Vol.61, pp.1332-1357

Tsai, C.S.; Lee, G.C. & Ketter, R.L. (1990). A semi-analytical method for time-domain analyses of dam-reservoir interactions. *International Journal for Numerical Methods in Engineering*, Vol.29, pp.913-933

Tsai, C.S. & Lee, G.C. (1987). Arch dam fluid interactions - by FEM-BEM and sub-structure concept. *International Journal for Numerical Methods in Engineering*, Vol.24, pp.2367-2388

Tsai, C.S. & Lee, G.C. (1991). Time-domain analyses of dam-reservoir system II. Sub-structure method. *Journal of Engineering Mechanics – ASCE*, Vol.117, pp.2007-2026

Weber, B. (1994). *Rational transmitting boundaries for time-domain analysis of dam-reservoir interaction*, Birkhauser Verlag, ISBN-10, 0817651233, Basel, Boston

Wolf, J.P. & Song, C.M. (1996a). Consistent infinitesimal finite element cell method: Three dimensional scalar wave equation. *Journal of Applied Mechanics - Transactions of the ASME*, Vol.63, pp.650-654

Wolf, J.P. & Song, C.M. (1996b). *Finite-Element Modeling of Unbounded Media*, Wiley, ISBN 978-0-471-96134-5, Chichester

Yan, J.Y.; Zhang, C.H. & Jin, F. (2004). A coupling procedure of FE and SBFE for soil-structure interaction in the time domain. *International Journal for Numerical Methods in Engineering*, Vol.59, pp.1453-1471

Yang, Z.J. & Deeks, A.J. (2007). Fully-automatic modelling of cohesive crack growth using a finite element-scaled boundary finite element coupled method. *Engineering Fracture Mechanics*, Vol.74, pp.2547-2573

# 3

# Hydrodynamic Control of Plankton Spatial and Temporal Heterogeneity in Subtropical Shallow Lakes

Luciana de Souza Cardoso[1], Carlos Ruberto Fragoso Jr.[3],
Rafael Siqueira Souza[2] and David da Motta Marques[2]
*Universidade Federal do Rio Grande do Sul (UFRGS)*
*[1]Instituto de Biociências*
*[2]Instituto de Pesquisas Hidráulicas (IPH)*
*[3]Universidade Federal de Alagoas (UFAL)*
*Centro de Tecnologia*
*Brazil*

## 1. Introduction

During the last 200 years, many lakes have suffered from eutrophication, implying an increase of both nutrient loading and organic matter (Wetzel, 1996). An aspect that has often been neglected in freshwater systems is the fact that phytoplankton is often not evenly distributed horizontally in space in shallow lakes. Although the occurrence of phytoplankton patchiness in marine systems has been known for a long time (e.g., Platt et al., 1970; Steele, 1978; Steele & Henderson, 1992), phytoplankton in shallow lakes is often assumed to be homogeneously distributed. However, there are various mechanisms that may cause horizontal heterogeneity in shallow lakes. For example, grazing by aggregated zooplankton and other organisms may cause spatial heterogeneity in phytoplankton (Scheffer & De Boer, 1995). Submerged macrophyte beds may be another mechanism, through reduction of resuspension by wave action and allopathic effects on the algal community (Van den Berg et al., 1998). For large shallow lakes, wind can be a dominant factor leading to both spatial and temporal heterogeneity of phytoplankton (Carrick et al., 1993), either indirectly by affecting the local nutrient concentrations due to resuspended particles, or directly by resuspending algae from the sediment (Scheffer, 1998). In the management of large lakes, prediction of the phytoplankton distribution can assist the manager to decide on an optimal course of action, such as biomanipulation and regulation of the use of the lake for recreation activities or potable water supply (Reynolds, 1999). However, it is difficult to measure the spatial distribution of phytoplankton. Mathematical modeling of a phytoplankton can be an important alternative methodology in improving our knowledge regarding the physical, chemical and biological processes related to phytoplankton ecology (Scheffer, 1998; Edwards & Brindley, 1999; Mukhopadhyay & Bhattacharyya, 2006).

Over the past decade there has been a concerted effort to increase the realism of ecosystem models that describe plankton production as a biological indicator of eutrophication. Most of this effort has been expended on the description of phytoplankton in temperate lakes; thus, multi-nutrient, photo-acclimation models are now not uncommon (e.g., Olsen & Willen, 1980; Edmondson & Lehman, 1981; Sas, 1989; Fasham et al., 2006; Mitra & Flynn, 2007; Mitra et al., 2007). In subtropical lakes, eutrophication has been intensively studied, but only with a focus on measuring changes in nutrient concentrations (e.g., Matveev & Matveeva, 2005; Kamenir et al., 2007). A wide variety of phytoplankton models have been developed. The simplest models are based on a steady state or on the assumption of complete mixing (Schindler, 1975; Smith, 1980; Thoman & Segna, 1980). Phytoplankton models based on more complex vertical 1-D hydrodynamic processes give a more realistic representation of the stratification and mixing processes in deep lakes (Imberger & Patterson, 1990; Hamilton et al., 1995a; Hamilton et al., 1995b). However, the vertical 1-D assumption might be too restrictive, especially in large shallow lakes that are poorly stratified and often characterized by significant differences between the pelagic and littoral zones. In these cases, a horizontal 2-D model with a complete description of the hydrodynamic and ecological processes can offer more insight into the factors determining local water quality.

Currently, computational power no longer limits the development of 2-D and 3-D models, and these models are being used more frequently. Of the wide diversity of 2-D and 3-D hydrodynamic models, most were designed to study deep-ocean circulation or coastal, estuarine and lagoon zones (Blumberg & Mellor, 1987; Casulli, 1990). However, only a few models are coupled with biological components (Rajar & Cetina, 1997; Bonnet & Wessen, 2001).

In this chapter, we present the results of comparative modeling of two subtropical shallow lakes where the wind, and derived hydrodynamics, and river flow act as the main factors controlling plankton dynamics on temporal and spatial scales. The basic hypothesis is that wind and wind derived-hydrodynamics are the main factor determining the spatial and temporal distribution of plankton communities (Cardoso et al. 2003; Cardoso & Motta Marques, 2003, 2004a, 2004b, 2004c, 2009), in association with point incoming river flows.

The spatial heterogeneity of phytoplankton in Lake Mangueira is influenced by hydrodynamic patterns, and identifying zones with a higher potential for eutrophication and phytoplankton patchiness (Fragoso Jr. et al., 2008). The spatial patterns of chlorophyll-a concentrations generated by the model were validated both with a field data set and with a cloud-free satellite image provided by a Terra Moderate Resolution Imaging Spectroradiometer (MODIS) with a spatial resolution of 1.0 km.

## 1.1 Study areas

Itapeva Lake is the first (N→S) in a system of interconnected fresh-water coastal lakes on the northern coast of the state of Rio Grande do Sul, Brazil (Fig. 1). The lake has an elongated shape (30.8 km × 7.6 km) and a surface area of ≈125 km², and is shallow, with a maximum depth of 2.5 m. The lake is oriented according to the prevailing wind direction (NE – SW quadrants), where the northern part is more constricted and consequently the water is more confined. Two rivers enter the lake: Cardoso River, in the northern part, and Três Forquilhas River in the southern part. The former is small and the flow was not important for the input; however, the contribution of the latter river was modeled and influenced the spatial pattern. Lake Mangueira (33°1'48"S 52°49'25"W) is a large freshwater ecosystem in southern Brazil (Fig. 2), covering a total area of 820 km², with a mean depth of 2.6 m and maximum depth of

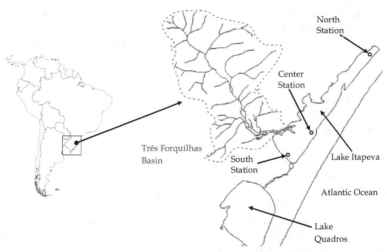

Fig. 1. Itapeva Lake in southern Brazil, with the three sampling points (North, Center and South).

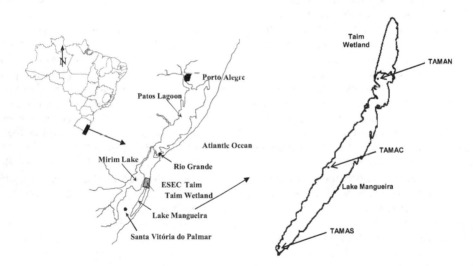

Fig. 2. Lake Mangueira in southern Brazil. The meteorological and sampling stations in the North, Center and South parts of Lake Mangueira are termed TAMAN, TAMAC and TAMAS, respectively.

6.5 m. Its trophic state ranges from oligotrophic to mesotrophic (annual mean $PO_4$ concentration 35 mg m$^{-3}$, varying from 5 to 51 mg m$^{-3}$). This lake is surrounded by a variety of habitats including dunes, pinus forests, grasslands, and two wetlands. This heterogeneous landscape harbors an exceptional biological diversity, which motivated the Brazilian federal authorities to protect part of the entire hydrological system as the Taim Ecological Station in 1991 (Garcia et al., 2006). The watershed (ca. 415 km$^2$) is primarily used for rice production, and many of the local waterbodies are used for irrigation, with a total water withdrawal of approximately 2 L s$^{-1}$ ha$^{-1}$ on 100 individual days within a 5-month period, and a high input of nutrients from the watershed during the rice-production period.

## 2. Data base

The data from Itapeva Lake were gathered over more than one year (August 1998 – August 1999), at three fixed sampling stations (North, Center and South). Lake Mangueira has been sampled for several years, although for the modeling exercise reported here we used data also collected at three fixed sampling stations (North, Center and South) from 2000 to 2001.

The sampling protocol as well as some results were published previously by Cardoso & Motta Marques (2003, 2004a, 2004b, 2004c, 2009) for Itapeva Lake, and by Crossetti et al. (2007) and Fragoso Jr. et al. (2008) for Lake Mangueira.

Environmental data (air temperature, precipitation, wind velocity and wind direction) from the meteorological station (DAVIS, Weather Wizard III, Weather Link) installed at the Center point were recorded every 30 min (beginning 24 h before each sampling event) throughout the period. Based on the prevailing wind direction in each season, the effective fetch (Lf km) was calculated (Håkanson, 1981) for each sampling point using the map of the region on a 1:250 000 scale. An estimate was also made of the height of waves produced and the bottom dynamics, from wind velocity, depth and fetch (Håkanson, 1981).

Four sections were chosen to study seiches in the lake, one section for each region (North, Center and South), and one section running in the longest and most central direction. It was considered that the seiches occur at time intervals of over 120 min; to obtain this value, the length of the lake (fetch) in the direction of the seiche, the mean wind speed, and the time needed by the wind to cover this distance were considered. This time is the minimum time for seiche occurrence, i.e., it is the time needed by the wind to cover the fetch. In addition to evaluating the existence of seiches, the period was also studied using simulated values and an empirical equation. The period calculated empirically is based on the formula for a rectangular shape (Lopardo, 2002 as cited in Cardoso & Motta Marques, 2003).

Data for turbidity, temperature, dissolved oxygen, pH, and electrical conductivity from the YSI (Yellow Springs Instruments 6000 upg3) multiprobe installed at the three sampling points were recorded every 5 minutes during each seasonal campaign. Water level, direction and velocity were recorded every 15 minutes at the same locations.

Samples were collected during five seasonal campaigns: winter/98 (August 24–25/1998), spring (December 15–20/1998), summer (March 2–7 /1999), autumn (May 21–26/1999) and winter (August 14–19/1999). The water samples for plankton analyses were collected at three depths (surface, middle and bottom) during four shifts throughout the day (06:00, 10:00, 14:00 and 18:00 h), at 24-h intervals during the three days of each seasonal sampling. The water samples for analyses of solids, nitrogen and phosphorus (APHA, 1992) were collected and integrated into the water-column data during the same periods as the plankton sampling.

### 2.1 Modeling in Itapeva Lake

Modeling in Itapeva Lake was divided into two parts: watershed and lake modeling. First, we used two different hydrological models: a) to estimate the input from the Três Forquilhas basin, and b) to estimate the output from Itapeva Lake to the river downstream.

Subsequently, we used a 2-D hydrodynamic model to evaluate the roles of the Três Forquilhas inflow and wind effects on the hydrodynamics and mixture processes of the lake. For the watershed analysis we used the IPH2 model, a rainfall, runoff-lumped model developed at the Instituto de Pesquisas Hidráulicas (IPH). Its mathematical basis is the continuity equation composed of the following algorithms: (a) losses by evapotranspiration and interception by leaves or stems of plants; (b) evaluation of infiltration and percolation by Horton; and (c) evaluation of surface and groundwater flows (Tucci, 1998). The model works by regarding a drainage basin as series of storage tanks, with rainfall entering at the top, and being split between what is returned to the atmosphere as evaporation, and what emerges from the basin as runoff (stream flow). Depending on the number of tanks and the number of parameters controlling the passage of water between them, it can be made more or less complex.

For the output analysis we used the MOLABI model (Ecoplan, 1997), a water-budget model based on the Puls method, which represents the continuity equation applied to the whole lake. The input data is rainfall over the lake, inflows from the watershed, and groundwater flux estimated by the Darcy equation. Evaporation was estimated using the Penman method.

The hydrodynamic patterns in Itapeva Lake were modeled using the model IPH-A (2D; Borche, 1996). The main inputs of the model were: water inflow, wind, rainfall and evaporation, spatial maps (including waterbody, bathymetry, bottom and surface stress coefficient). This model was applied because Itapeva Lake is a polymictic environment with no significant difference among depths (surface, middle and bottom), neither for the physical and chemical data nor for the plankton communities. The model was run from January to December 1999.

## 2.2 Modeling in Lake Mangueira

For Lake Mangueira, we applied a dynamic ecological model describing phytoplankton growth, called IPH-TRIM3D-PCLake (Fragoso Jr. et al., 2009). Although this model can represent the three-dimensional flows and the entire trophic structure dynamically, in this study case we used a simplified version of the model consisting of three modules: (a) a detailed horizontal 2-D hydrodynamic module for shallow water, which deals with wind-driven quantitative flows and water levels; (b) a nutrient module, which deals with nutrient transport mechanisms and some conversion processes; and (c) a biological module, which describes phytoplankton growth in a simple way. An overview of the modeling processes is given in Fig. 3.

The hydrodynamic model is based on the shallow-water equations derived from Navier-Stokes, which describe dynamically a horizontal two-dimensional flow:

$$\frac{\partial \eta}{\partial t} + \frac{\partial \left[ (h+\eta)u \right]}{\partial x} + \frac{\partial \left[ (h+\eta)v \right]}{\partial y} = 0 \tag{1}$$

$$\frac{\partial u}{\partial t} + u\frac{\partial u}{\partial x} + v\frac{\partial u}{\partial y} = -g\frac{\partial \eta}{\partial x} - \gamma u + \tau_x + A_h\nabla^2 u + fv \tag{2}$$

$$\frac{\partial v}{\partial t} + u\frac{\partial v}{\partial x} + v\frac{\partial v}{\partial y} = -g\frac{\partial \eta}{\partial y} - \gamma v + \tau_y + A_h\nabla^2 v - fu \tag{3}$$

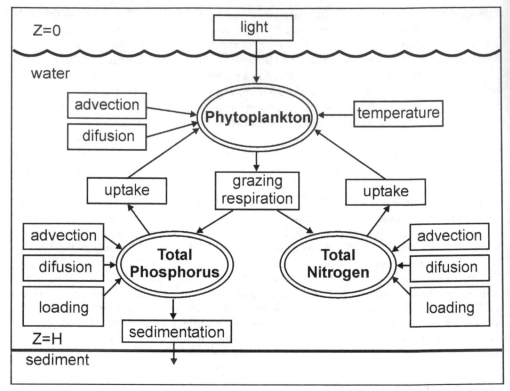

Fig. 3. Simplified representation of the interactions involving the state variables (double circle), and the processes (rectangle) used for the modeling of Lake Mangueira.

where $u(x,y,t)$ and $v(x,y,t)$ are the water velocity components in the horizontal $x$ and $y$ directions; $t$ is time; $\eta$ $(x,y,t)$ is the water surface elevation relative to the undisturbed water surface; $g$ is the gravitational acceleration; $h(x,y)$ is the water depth measured from the undisturbed water surface; $f$ is the Coriolis force; $\tau_x$ and $\tau_y$ are the wind stresses in the $x$ and $y$ directions; $\nabla = \partial/\partial x \cdot \vec{i} + \partial/\partial x \cdot \vec{j}$ is a vector operator in the $x$-$y$ plane (known as nabla operator, or del operator); $A_h$ is the coefficient of horizontal eddy viscosity; and

$\gamma = \dfrac{g\sqrt{u^2 + v^2}}{C_z}$ (Daily & Harlerman, 1966) where $C_z$ is the Chezy friction coefficient.

Usually, the wind stresses in the $x$ and $y$ directions are written as a function of wind velocity (Wu, 1982):

$$\tau_x = C_D \cdot W_x \cdot \|W\| \qquad (4)$$

$$\tau_y = C_D \cdot W_y \cdot \|W\| \qquad (5)$$

where $C_D$ is the wind friction coefficient; and $W_x$ and $W_y$ are the wind velocity components (m.s$^{-1}$) in the $x$ and $y$ directions, respectively. Wind velocity is measured at

10 m above the water surface; and $\|W\| = \sqrt{W_x^2 + W_y^2}$ is the norm of the wind velocity vector. We solved the partial differential equations numerically by applying an efficient semi-implicit finite differences method to a regular grid, which was used in order to assure stability, convergence and accuracy (Casulli, 1990; Casulli & Cheng, 1990; Casulli & Cattani, 1994).

The nutrient module includes the advection and diffusion of each substance, inlet and outlet loading, sedimentation, and resuspension through the following equation:

$$\frac{\partial(HC)}{\partial t} + \frac{\partial(uCH)}{\partial x} + \frac{\partial(vCH)}{\partial y} = \frac{\partial}{\partial x}\left(K_h \frac{\partial(HC)}{\partial x}\right) + \frac{\partial}{\partial y}\left(K_h \frac{\partial(HC)}{\partial y}\right) + \text{source or sink} \qquad (6)$$

where $C$ is the mean concentration in the water column; $H = \eta + h$ is the total depth; and $K_h$ is the horizontal scalar diffusivity assumed as 0.1 m$^2$ day$^{-1}$ (Chapra, 1997). Equation 6 was applied to model total phosphorus, total nitrogen and phytoplankton. All these equations are solved dynamically, using a simple numerical semi-implicit central finite differences scheme (Gross et al., 1999a; 1999b) (Fig. 2). Thus, the mass balances involving phytoplankton and nutrients can be written as:

$$\frac{\partial(Ha)}{\partial t} + \frac{\partial(uHa)}{\partial x} + \frac{\partial(vHa)}{\partial y} = \mu_{eff} Ha + \frac{\partial}{\partial x}\left(K_h \frac{\partial(Ha)}{\partial x}\right) + \frac{\partial}{\partial y}\left(K_h \frac{\partial(Ha)}{\partial y}\right) + \text{inlet/outlet} \qquad (7)$$

$$\frac{\partial(Hn)}{\partial t} + \frac{\partial(uHn)}{\partial x} + \frac{\partial(vHn)}{\partial y} = -a_{na}\mu_{eff} Ha + \frac{\partial}{\partial x}\left(K_h \frac{\partial(Hn)}{\partial x}\right) + \frac{\partial}{\partial y}\left(K_h \frac{\partial(Hn)}{\partial y}\right) + \text{inlet/outlet} \qquad (8)$$

$$\frac{\partial(Hp)}{\partial t} + \frac{\partial(uHp)}{\partial x} + \frac{\partial(vHp)}{\partial y} = -a_{pa}\mu_{eff} Ha - k_{phos}p + \frac{\partial}{\partial x}\left(K_h \frac{\partial(Hp)}{\partial x}\right) + \frac{\partial}{\partial y}\left(K_h \frac{\partial(Hp)}{\partial y}\right) + \qquad (9)$$
$$+ \text{inlet / outlet}$$

where $a$, $n$ and $p$ are the chlorophyll-a, total nitrogen and total phosphorus concentrations, respectively; $a_{na}$ is the N/Chla ratio equal to 8 mg N mg Chla$^{-1}$; $a_{pa}$ is the P/Chla ratio equal to 1.5 mg N mg Chla$^{-1}$, inlet/outlet represents the balance between all inlets and outlets in a control volume $\partial x\,\partial y\,\partial z$; and kphos is the settling coefficient.

The effective growth rate itself is not a simple constant, but varies in response to environmental factors such as temperature, nutrients, respiration, excretion and grazing by zooplankton:

$$\mu_{ef} = \mu_P(T, N, I)\, a - \mu_L a \qquad (10)$$

where $\mu_P(T, N, I)$ is the primary production rate as a function of temperature ($T$), nutrients ($N$), and light ($I$); $\mu_L$ is the loss rate due to respiration, excretion, and grazing by zooplankton; and $a$ is the chlorophyll-a concentration.

The hydrodynamic module was calibrated by tuning the model parameters within their observed ranges taken from the literature (Table 1). Nonetheless, the hydraulic resistance caused by the presence of emerged macrophytes in the Taim Wetland was represented by a smaller Chezy's resistance factor than was used in other lake areas (Wu et al., 1999). Calibration and validation of the hydrodynamic parameters were done using two different time-series of water level and wind produced for two locations in Lake Mangueira (North and South).

| Parameter | Description | Unit | Values/Ref. |
|---|---|---|---|
| Hydrodynamic: | | | |
| 1 $A_h$ | Horizontal eddy viscosity coefficient | $m^{1/2}$ $s^{-1}$ | 5 – 15 [1] |
| 2 $C_D$ | Wind friction coefficient | - | 2e-6 – 4e-6 [2] |
| 3 $C_Z$ | Chezy coefficient | - | 50 – 70 [3] |
| Biological: | | | |
| 1 $G_{max}$ | Maximum growth rate algae | $day^{-1}$ | 1.5 – 3.0 [4] |
| 2 $I_S$ | Optimum light intensity for the algae | cal $cm^{-2}dia^{-1}$ | 100 – 400 [5] |
| 3 $k'_e$ | Light attenuation coefficient in the water | $m^{-1}$ | 0.25 – 0.65 [5] |
| 4 $\theta_T$ | Temperature effect coefficient | - | 1.02 – 1.14 [6] |
| 5 $\theta_R$ | Respiration and excretion effect coefficient | - | 1.02 – 1.14 [5] |
| 6 $k_P$ | Half-saturation for uptake phosphorus | mg P $m^{-3}$ | 1 – 5 [7] |
| 7 $k_N$ | Half-saturation for uptake nitrogen | mg N $m^{-3}$ | 5 – 20 [7] |
| 8 $k_{re}$ | Respiration and excretion rate | $day^{-1}$ | 0.05 – 0.25 [8] |
| 9 $k_{gz}$ | Zooplankton grazing rate | $day^{-1}$ | 0.10 – 0.20 [8] |

Sources: [1] White (1974); [2] Wu (1982); [3] Chow (1959); [4] Jørgensen (1994); [5] Schladow & Hamilton (1997); [6] Eppley (1972); [7] Lucas (1997); [8] Chapra (1997)

Table 1. Hydrodynamic and biological parameters description and its values range.

For the parameters of the phytoplankton module, we used the mean values for the literature range given in Table 1. To evaluate its performance, we simulated another period of 86 days, starting 12/22/2002 at 00:00 hs (summer). Solar radiation and water temperature data were taken from the TAMAN meteorological station, situated in the northern part of Lake Mangueira. Photosynthetically active radiation (PAR) at the Taim wetland was assigned as 20% of the total radiation, in order to represent the indirect effect of the emergent macrophytes on the phytoplankton growth rate according to experimental studies of emergent vegetation stands in situ. For the lake areas, we assumed that the percentage of PAR was 50% of the total solar radiation (Janse, 2005).

The resulting phytoplankton patterns were compared with satellite images from MODIS, which provides improved chlorophyll-a measurement capabilities over previous satellite sensors. For instance, MODIS can better measure the concentration of chlorophyll-a associated with a given phytoplankton bloom. Unfortunately, there were no detailed chlorophyll-a and nutrient data available for the same period. Therefore, we compared only the median simulated values with field data from another period (2001 and 2002).

## 3. Modeling results

### 3.1 Itapeva Lake

In Itapeva Lake, wind action generated oscillations of the water level between the North and South parts of the lake. The meteorological and hydrological variables were characterized for daily and seasonal periods.

Simulations using a mathematical bidimensional horizontal hydrodynamic model succeeded in reproducing this phenomenon, and helped to calculate the synthesis of velocity and direction of the water. It was possible to confirm the complexity of the circulation in the lake and to distinguish different behaviors among the South, Center and North. Besides the similarities in morphometry between the Center and South parts, the flow from the Três Forquilhas River enters the center part of the lake and the prevailing N-NE winds move water toward the South part of the lake (Figs 4 and 5).

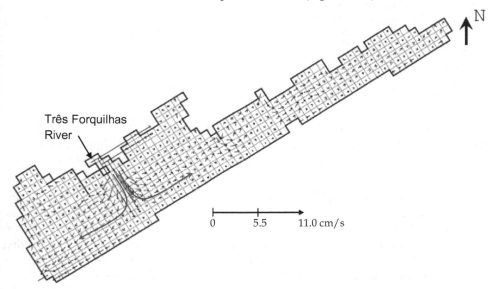

Fig. 4. Numerical simulation of the vertically averaged velocity field in Itapeva Lake during the combination of high-flow condition in the Três Forquilhas River and low wind speed. Red arrows indicate the direction of the prevailing currents.

The hydrological variables analyzed and modeled showed a quite characteristic seasonal behavior at each sampling point in Itapeva Lake, closely related to the velocity and direction of the wind. The water level responded to wind action in a very direct manner, since NE winds displaced water from north to south, along the main lake axis. Winds from the SW quadrant produced the opposite effect.

The environmental sources selected for the analysis were suspended solids and turbidity, due to their influence on many physical, chemical and biological factors. The results for hydrodynamics, such as water column and water velocity generated by the model for the sampling points, were used as the basis for the study of the environmental variations. The waves generated by wind action were the third source used to explain the variations of suspended solids and turbidity in the lake (Figs 6 and Table 2).

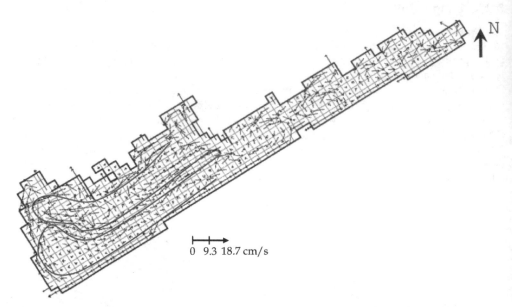

Fig. 5. Numerical simulation of the vertically averaged velocity field in Itapeva Lake during the combination of low-flow condition in the Três Forquilhas River and high wind speed from the Northeast quadrant (20/05/1999 - 21/05/1999). Red arrows indicate the direction of the prevailing currents.

The hydrodynamic variable explained 70% to 95% of the environmental variations for each seasonal campaign, using mean values for four-hour periods. Considering the entire lake and all seasonal campaigns, this explained 68% of the variation for turbidity and 49% for suspended solids. The hydrodynamic and environmental study were capable of evaluating that the changes in the water level as a function of runoff occur slowly, compared with the changes in the water level and seiches created by the effect of the wind on the lake. These variations in water levels and wind speeds have significant effects on the variability of the environmental variable tested.

| Water Quality Variable | Hydrological Variable | South, Center and North | | | | | |
|---|---|---|---|---|---|---|---|
| | | Mean | R | C1 | C2 | C3 | error |
| Turbidity | 12 | 4 hours | 0.68 | 0.42 | 0.47 | - | 12.54 |
| Suspended Solids | 12 | 4 hours | 0.49 | 0.43 | 0.27 | - | 63.63 |

Legend: Hydrological Variable – 1. water level (N), 2. water velocity (V); 3. wave height (H); 12- N-V; 13. N-H; 23. V-H; 123. N-V-H; R – correlation factor; C1, C2, C3 – N, V and H coefficients, respectively.

Table 2. Results of multiple regression between water quality variable (dependent variable) and hydrological variables (independent variable) in Itapeva Lake considering the three monitoring stations: South, Center and North.

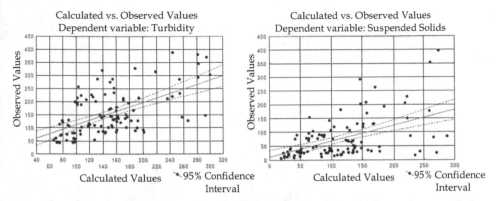

Fig. 6. Comparison between observed and calculated values for turbidity and suspended solids through multiple regression model using the three monitoring station along the Itapeva Lake.

## 3.2 Lake Mangueira

The simulated and observed values of water levels at two stations of Lake Mangueira during the calibration and validation period are shown in Fig. 7. An independent validation data set showed a good fit to the hydrodynamic module ($R^2 \geq 0.92$). The model was able to reproduce the water level well in both extremities of Lake Mangueira. Wind-induced currents can be considered the dominant factor controlling transport of substances and phytoplankton in Lake Mangueira, producing advective movement of superficial water masses in a downwind direction. For instance, a southwest wind, with magnitude approximately greater than $4 \text{ms}^{-1}$, can causes a significant transport of water mass and substances from south to north of Lake Mangueira, leading to a almost instantaneous increase of the water level in the northeastern parts and, hence the decrease of water level in southwestern areas (Fig. 7).

Our model results showed two characteristic water motions in the lake: oscillatory (seiche) and circulatory. Lake Mangueira is particularly prone to wind-caused seiches because of its shallowness, length (ca. 90 km), and width (ca. 12 km). These peculiar morphological features lead to significant seiches of up to 1 m between the south and north ends, caused by moderate-intensity winds blowing constantly along the longitudinal axis of the lake (NE-SW). Depending on factors such as fetch length and the intensity and duration of the wind, areas dominated by downwelling and upwelling can be identified (Fig. 8). For instance, if northeast winds last longer than about 6 h, the surface water moves toward the south shore, where the water piles up and sinks. Subsequently a longitudinal pressure gradient is formed and produces a strong flow in the deepest layers (below 3 m) toward the north shore, where surface waters are replaced by water that wells up from below. Such horizontal and vertical circulatory water motions may develop if wind conditions remain stable for a day or longer. The model was also used to determine the spatial distribution of chlorophyll-a and to identify locations with higher growth and phytoplankton biomass in Lake Mangueira. Fig. 9 shows the spatial distribution of the phytoplankton biomass for different times during the simulation period.

Specifically, in Lake Mangueira there is a strong gradient of phytoplankton productivity from the littoral to pelagic zones (Fig. 9). Moreover, the model outcome suggests that there is a significant transport of phytoplankton and nutrients from the littoral to the pelagic zones through hydrodynamic processes. This transport was intensified by several large sandbank formations that are formed perpendicularly to the shoreline of the lake, carrying nutrients and phytoplankton from the shallow to deeper areas.

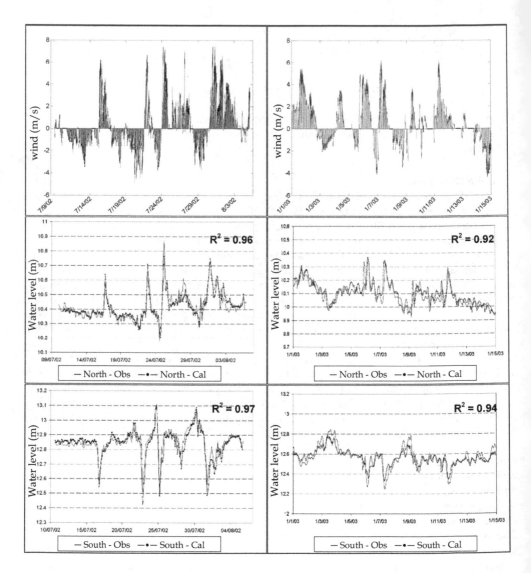

Fig. 7. Time series of wind velocity and direction on Lake Mangueira, and water levels fitted for the North and South parts of Lake Mangueira into the calibration and validation periods (solid line - observed, dotted line - calculated). Source: Fragoso Jr. et al. (2008).

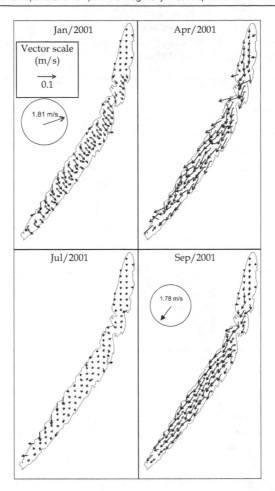

Fig. 8. Simulated instantaneous currents in the surface (black arrows) and bottom (gray arrows) layers of Lake Mangueira at four different instants. A wind sleeve (circle), in each frame, indicates the instantaneous direction and the intensity of the wind. Source: Fragoso Jr. et al. (2011).

In addition, it was possible to identify zones with the highest productivity. There is a trend of phytoplankton aggregation in the southwest and northeast areas, as the prevailing wind directions coincide with the longitudinal axis of Lake Mangueira. The clear water in the Taim wetland north of Lake Mangueira was caused by shading of emergent macrophytes, modeled as a fixed reduction of PAR.

After 1,200 hours of simulation (50 days), the daily balance between the total primary production and loss was negative. That means that daily losses such as respiration, excretion and grazing by zooplankton exceeded the primary production in the photoperiod, leading to a significant reduction of the chlorophyll-a concentration for the whole system (Fig. 9d; 9e).

We verified the modeled spatial distribution of chlorophyll-a with the distribution estimated by remote sensing (Fig. 10). The simulated patterns had a reasonably good similarity to the patterns estimated from the remote-sensing data (Fig. 10a, b). In both figures, large phytoplankton aggregations can be observed in both the southern and northern parts of Lake Mangueira, as well as in the littoral zones.

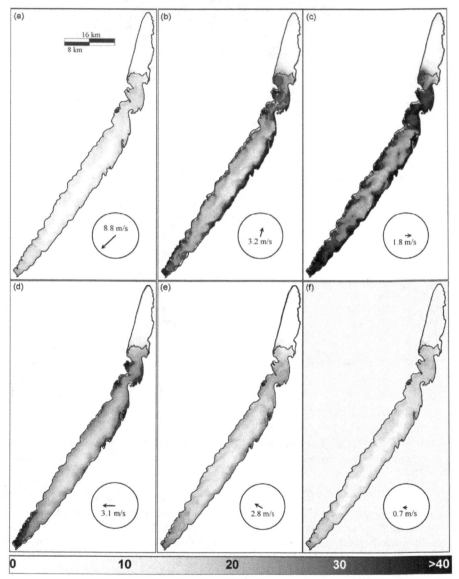

Fig. 9. Phytoplankton dry weight concentration fields in μg l⁻¹, for the whole system at different times: (a) 14 days; (b) 28 days; (c) 43 days; (d) 57 days; (e) 71 days; and (f) 86 days. The color bar indicates the phytoplankton biomass values. A wind sock in each frame indicates the direction and intensity of the wind. The border between the Taim wetland and Lake Mangueira is shown as well.

Unfortunately we did not have independent data for phytoplankton in the simulation period. Therefore we could only compare the median values of simulated and observed chlorophyll-a, total nitrogen and total phosphorus for three points in Lake Mangueira, assuming that the median values were comparable between the years. The fit of these variables was reasonable, considering that we did not calibrate the biological parameters of the phytoplankton module (see results in Fragoso Jr. et al., 2008). The lack of spatially and temporally distributed data for Lake Mangueira made it impossible to compare simulated and observed values in detail. However, the good fit in the median values of nutrients and phytoplankton indicated that the model is a promising step toward a management tool for subtropical ecosystems.

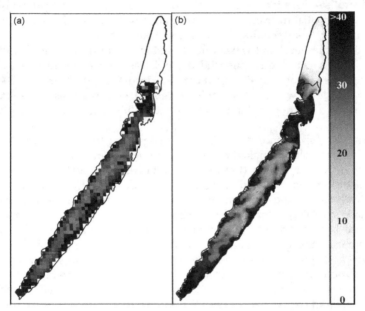

Fig. 10. Lake Mangueira: (a) MODIS-derived chlorophyll-a image with 1-km spatial resolution, taken on February 8, 2003; and (b) simulated chlorophyll-a concentration for the same date.

## 3.3 Hydrodynamic versus plankton

Hydrodynamic processes and biological changes occurred over different spatial and temporal scales in these two large and long subtropical lakes. Itapeva Lake (31 km long) is almost one-third the size of Lake Mangueira (90 km long), and therefore the hydrodynamic response is faster in Itapeva Lake. On a time scale of hours, we can see the water movement from one end of the lake to the other (e.g., from N to S during a NE wind and in the opposite direction during a SW wind). Because of this rapid response, the plankton communities showed correspondingly rapid changes in composition and abundance, especially the phytoplankton when the resources (light and nutrients) responded promptly to wind action. This interaction between wind on a daily scale (hours) and the shape of Itapeva Lake was a determining factor for the observed fluctuations in the rates of change for phytoplankton (Cardoso & Motta Marques, 2003) as well as for the spatial distribution of plankton communities (Cardoso & Motta Marques, 2004a, 2004b, 2004c). The rate of change in the phytoplankton was very high, indicating the occurrence of intense, rapid environmental changes, mainly in spring.

Marked changes in the spatial and temporal gradients of the plankton communities occurred during the seasons of the year, in response to resuspension events induced by the wind (Cardoso & Motta Marques, 2009). These responses were most intense precisely at the sites where the fetch was longest. The increase in changes occurred as the result of population replacements in the plankton communities. Resuspension renders diatoms and protists dominant in the system, and they are replaced by cyanobacteria and rotifers when the water becomes calm again (Cardoso & Motta Marques, 2003, 2004a, 2004b). Thus, diatoms and protists were the general indicator groups for lake hydrodynamic, with fast responses in their spatial distribution. Wind-induced water dynamics acted directly on the plankton community, resuspending species with a benthic habit.

In Itapeva Lake, water level and water velocity induced short-term spatial gradients, while wind action (affecting turbidity, suspended solids, and water level) was most strongly correlated with the seasonal spatial gradient (Cardoso & Motta Marques, 2009). In Lake Mangueira, water level was most strongly correlated with the seasonal spatial gradient, while wind action (affecting turbidity, suspended solids, and nutrients) induced spatial gradients.

The Canonical Correspondence Analysis (CCA) suggested that some aspects of plankton dynamics in Itapeva Lake are linked to suspended matter, which in turn is associated with the wind-driven hydrodynamics (Cardoso & Motta Marques, 2009). Short-term patterns could be statistically demonstrated using CCA to confirm the initial hypothesis. The link between hydrodynamics and the plankton community in Itapeva Lake was revealed using the appropriate spatial and temporal sampling scales. As suggested by our results, the central premise is that different hydrodynamic processes and biological responses may occur at different spatial and temporal scales. A rapid response of the plankton community to wind-driven hydrodynamics was recorded by means of the sampling scheme used here, which took into account combinations of spatial scales (horizontal) and time scale (hours).

In both lakes, the central zone of the lake takes on intermediate conditions, sometimes closer to the North part and sometimes closer to the South, depending on the duration, direction and velocity of the wind. This effect is very important for the horizontal gradients evaluated, in relation to the physical and chemical water conditions as well as to the plankton communities. In Lake Mangueira, the South zone is characterized by high water transparency whereas the North zone is more turbid, because the latter is adjacent to the wetland and is influenced by substances originated from the aquatic macrophyte decomposition. In Itapeva Lake it was not possible to distinguish such clear spatial differences. The spatial variation is directly related to wind action, because the lake is smaller and shallower than Lake Mangueira. In addition, the prevailing NE winds and the influence of the Três Forquilhas River on Itapeva Lake make the central zone often similar to the South part. The high turbidity in Itapeva Lake is an important factor affecting the composition and distribution of the plankton communities. However, in Lake Mangueira the marked spatial differences between the North and South zones were important for the composition and distribution of the plankton, and the influence of the wind was more evident in the Center zone than in the two ends of the lake.

In Lake Mangueira, wind-driven hydrodynamics creates zones with particular water dynamics (Fragoso Jr. et al., 2008). The velocity and direction of currents and water level changed quickly. Depending on factors such as fetch and wind, areas dominated by downwelling and upwelling could be identified in the deepest parts. We observed a significant horizontal spatial heterogeneity of phytoplankton associated with hydrodynamic patterns from the south to the north shore (littoral-pelagic-littoral zones) over the winter and summer periods. Our results suggest that there are significant horizontal gradients in many variables during the entire year. In general, the simulated depth-averaged chlorophyll-a concentration increased from the pelagic to the littoral zones. This indicated that a higher zooplankton biomass can exist in the littoral zones, leading, eventually, to stronger top-down control on the phytoplankton in this part of the lake.

Moreover, as expected for a wind-exposed shallow lake, Lake Mangueira did not show marked vertical gradients. The field campaigns showed that the lake is practically unstratified, emphasizing the shallowness and vertical mixing caused by the wind-driven hydrodynamics. This complete vertical mixing, as expected, was noted for both the pelagic and littoral zones. However, we are still of the opinion that incorporation of horizontal spatially explicit processes associated with the hydrodynamics is essential to understand the dynamics of a large shallow lake. The occurrence of hydrodynamic phenomena such as the seiches between the extreme ends, in a very long and narrow lake, is important, since seiches function as a conveyor belt, accounting for the vertical mixing and transportation of materials between the two ends of the lake and between the wetlands in the North and South areas in Lake Mangueira. Seiches was very important to explain much of the spatial changes in Itapeva Lake.

## 4. Conclusion

Recognition of the importance of spatial and temporal scales is a relatively recent issue in ecological research on aquatic food webs (Bertolo et al., 1999; Woodward & Hildrew, 2002; Bell et al., 2003; Mehner et al., 2005). Among other things, the observational or analytical resolution necessary for identifying spatial and temporal heterogeneity in the distributions of populations is an important issue (Dungan et al., 2002). Most ecological systems exhibit heterogeneity and patchiness over a broad range of scales, and this patchiness is fundamental to population dynamics, community organization and stability. Therefore, ecological investigations require an explicit determination of spatial scales (Levin, 1992; Hölker & Breckling, 2002), and it is essential to incorporate spatial heterogeneity into ecological models to improve understanding of ecological processes and patterns (Hastings, 1990; Jørgensen et al., 2008).

Water movement in aquatic systems is a key factor which drives resources distribution, resuspend and carries particles, reshape the physical habitat and makes available previously unavailable resources. As such processes, and communities change along and patterns are created in time and space. Ecological models incorporating hydrodynamics and trophic structure are poised to serve as thinking pads allowing discovering and understanding patterns in different time and space scales of aquatic ecosystems. In lake ecosystem simulations, the horizontal spatial heterogeneity of the phytoplankton and the hydrodynamic processes are often neglected. Our simulations showed that it is important to consider this spatial heterogeneity in large lakes, as the water quality, community structures and hydrodynamics are expected to differ significantly between the littoral and the pelagic zones, and between differently shaped lakes. Especially for prediction of the water quality (including the variability due to wind) in the littoral zones of a large lake, the incorporation of spatially explicit processes that are governed by hydrodynamics is essential. Such information may be also important for lake users and for lake managers.

## 5. Acknowledgment

We are grateful to the Brazilian agencies FAPERGS (Fundação de Amparo à Pesquisa do Estado do Rio Grande do Sul) and CNPq (Conselho Nacional de Pesquisa) / PELD (Programa de Ecologia de Longa Duração) for grants in support of the authors.

## 6. References

APHA (American Public Health Association). (1992). *Standard Methods for Examination of Water and Wastewater* (18. ed.), APHA, ISBN-10 0875532071, ISBN-13 978-0875532073, Washington.

Bell, T.; Neill, W.E. & Schluter, D. (2003). The effect of temporal scale on the outcome of trophic cascade experiments. *Oecologia*, Vol.134, pp. 578-586, ISSN Print 0029-8549, ISSN Online 1432-1939.

Bertolo, A.; Lacroix, G. & Lescher-Moutoue, F. (1999). Scaling food chains in aquatic mesocosms: do the effects of depth override the effects of planktivory? *Oecologia*, Vol.121, pp. 55-65, ISSN Print 0029-8549, ISSN Online 1432-1939.

Bonnet, M.P. & Wessen, K. (2001). ELMO, a 3-D water quality model for nutrients and chlorophyll: first application on a lacustrine ecosystem. *Ecological Modelling*, Vol.141, pp. 19-33, ISSN 0304-3800.

Borche, A. (1996). *IPH-A Aplicativo para modelação de estuários e lagoas. Manual de utilização do sistema*. UFRGS, Porto Alegre, Brazil.

Blumberg, A. F. & Mellor, G. L. (1987). A description of a three-dimensional coastal ocean circulation model, In: *Three-Dimensional Coastal Ocean Models*, N. S. Heaps (Ed.), 1-16, American Geophysical Union (AGU), ISBN 0875902537, Washington, DC.

Cardoso, L. de S. & Motta Marques, D.M.L. (2003). Rate of change of the phytoplankton community in Itapeva Lake (North Coast of Rio Grande do Sul, Brazil), based on the wind driven hydrodynamic regime. *Hydrobiologia*, Vol.497, pp. 1–12, ISSN Print 0018-8158, ISSN Online 1573-5117.

Cardoso, L. de S. & Motta Marques, D.M.L. (2004a). Seasonal composition of the phytoplankton community in Itapeva Lake (North Coast of Rio Grande do Sul - Brazil) in function of hydrodynamic aspects. *Acta Limnologica Brasiliensia*, Vol.16, pp. 401–416, ISSN 0102-6712.

Cardoso, L. de S. & Motta Marques, D.M.L. (2004b). Structure of the zooplankton community in a subtropical shallow lake (Itapeva Lake - South of Brazil) and its relationship to hydrodynamic aspects. *Hydrobiologia*, Vol.518, pp. 123-134, ISSN Print 0018-8158, ISSN Online 1573-5117.

Cardoso, L. de S. & Motta Marques, D.M.L. (2004c). The influence of hydrodynamics on the spatial and temporal variation of phytoplankton pigments in a large, subtropical coastal lake (Brazil). *Arquivos de Biologia e Tecnologia*, Vol.47, pp. 587–600, ISSN 1516-8913.

Cardoso, L. de S. & Motta Marques, D.M.L. (2009). Hydrodynamics-driven plankton community in a shallow lake. *Aquatic Ecology*, Vol.43, pp. 73–84, ISSN Print 1386-2588, ISSN Online 1573-5125.

Cardoso, L. de S.; Silveira, A.L.L. & Motta Marques, D.M.L. (2003). A ação do vento como gestor da hidrodinâmica na lagoa Itapeva (litoral norte do Rio Grande do Sul-Brasil). *Revista Brasileira de Recursos Hídricos*, Vol.8, pp. 5-15, ISSN 1807-1929.

Carrick, H.J.; Aldridge, F.J. & Schelske, C.L. (1993). Wind Influences Phytoplankton Biomass and Composition in a Shallow, Productive Lake. *Limnology and Oceanography*, Vol.38, pp. 1179-1192, ISSN 0024-3590.

Casulli, V. (1990). Semi-Implicit Finite-Difference Methods for the 2-Dimensional Shallow-Water Equations. *Journal of Computational Physics*, Vol.86, pp. 56-74, ISSN 0021-9991.

Casulli, V. & Cattani, E. (1994). Stability, Accuracy and Efficiency of a Semiimplicit Method for 3-Dimensional Shallow-Water Flow. *Computers & Mathematics with Applications*, Vol. 7, pp. 99-112, ISSN 0898-1221.

Casulli, V. & Cheng, R.T. (1990). Stability Analysis of Eulerian-Lagrangian Methods for the One-Dimensional Shallow-Water Equations. *Applied Mathematical Modelling*, Vol.14, pp. 122-131, ISSN 0307-904X.

Chapra, S.C. (1997). *Surface water-quality modeling*. McGraw-Hill series in water resources and environmental engineering, ISBN 9780070113640, Boston.

Chow, V.T. (1959). *Open Channel Hydraulics*, McGraw-Hill, ISBN 0073397873, New York.

Crossetti, L.; Cardoso, L. de S.; Callegaro, V.L.M.; Silva, S.A.; Werner, V.; Rosa, Z. & Motta Marques, D.M.L. (2007). Influence of the hydrological changes on the phytoplankton structure and dynamics in a subtropical wetland-lake system. *Acta Limnologica Brasiliensia*, Vol.19, pp. 315–329, ISSN 0102-6712.

Dungan, J.L.; Perry, J.N.; Dale, M.R.T.; Legendre, P.; Citron-Pousty, S.; Fortin, M.J.; Jakomulska, A.; Miriti, M. & Rosenberg, M.S. (2002). A balanced view of scale in spatial statistical analysis. *Ecography*, Vol.25, pp. 626-640, ISSN 0906-7590.

Ecoplan Engenharia Ltda (1997). *Avaliação da disponibilidade hídrica superficial e subterrânea do litoral norte do Rio Grande do Sul, englobando todos os corpos hídricos que drenam o Rio Tramandaí*. Porto Alegre, Brazil.

Edmondson, W.T. & Lehman, J.T. (1981). The Effect of Changes in the Nutrient Income on the Condition of Lake Washington. *Limnology and Oceanography*, Vol.26, pp. 1-29, ISSN 0024-3590.

Edwards, A.M. & Brindley, J. (1999). Zooplankton mortality and the dynamical behaviour of plankton population models. *Bulletin of Mathematical Biology*, Vol.61, pp. 303-339, ISSN Print 0092-8240, ISSN Online 1522-9602.

Eppley, R.W. (1972). Temperature and Phytoplankton Growth in the Sea. *Fishery Bulletin* (Wash DC), Vol.70, pp. 1063-1085, ISSN 0090-0656.

Fasham, M.J.R.; Flynn, K.J.; Pondaven, P.; Anderson, T.R. & Boyd, P.W. (2006). Development of a robust marine ecosystem model to predict the role of iron in biogeochemical cycles: A comparison of results for iron-replete and iron-limited areas, and the SOIREE iron-enrichment experiment. *Deep-Sea Research Part I-Oceanographic Research Papers*, Vol. 53, pp.333-366, ISSN 0967-0637.

Fragoso Jr, C.R.; Motta Marques, D.M.L. , Collischonn, W.; Tucci, C.E.M. & Nes, E.H.V. (2008). Modelling spatial heterogeneity of phytoplankton in Lake Mangueira, a large shallow lake in South Brazil. *Ecological Modelling*, Vol.219, pp. 125-137, ISSN 0304-3800.

Fragoso Jr., C.R.; van Nes, E.H.; Janse, J.H. & Motta Marques, D.M.L. (2009). IPH-TRIM3D-PCLake: A three-dimensional complex dynamic model for subtropical aquatic ecosystems. *Environmental Modelling & Software*, Vol.24, pp. 1347–1348, ISSN 1364-8152.

Fragoso Jr, C.R.; Motta Marques, D.M.L. ; Ferreira, T.F. ; Janse, J.H. ; van Nes, E.H. (2011). Potential effects of climate change and eutrophication on a large subtropical shallow lake. Environmental Modelling & Software, Vol. 26, pp. 1337-1348, ISSN 1364-8152.

Garcia, A.M.; Hoeinghaus, D.J.; Vieira, J.P.; Winemiller, K.O.; Marques, D. & Bemvenuti, M.A. (2006). Preliminary examination of food web structure of Nicola Lake (Taim Hydrological System, south Brazil) using dual C and N stable isotope analyses. *Neotropical Ichthyology*, Vol.4, pp. 279-284, ISSN 1679-6225.

Gross, E.S.; Koseff, J.R. & Monismith, S.G. (1999a). Evaluation of advective schemes for estuarine salinity simulations. *Journal of Hydraulic Engineering-Asce*, Vol.125, pp. 32-46, ISSN Print 0733-9429, ISSN Online 1943-7900.

Gross, E.S.; Koseff, J.R. & Monismith, S.G. (1999b). Three-dimensional salinity simulations of south San Francisco Bay. *Journal of Hydraulic Engineering-Asce*, Vol.125, pp. 1199-1209, ISSN Print 0733-9429, ISSN Online 1943-7900.

Håkanson, L. (1981). *A Manual of Lake Morphometry*. Springer-Verlag, ISBN-10 3540104801, ISBN-13 978-3540104803, Berlin.

Hamilton, D.; Schladow, S. & Zic, I. (1995a). *Modelling artificial destratification of prospect and nepean reservoirs: final report*, Centre for Water Research, University of Western Australia.

Hamilton, D.P.; Hocking, G.C. & Patterson, J.C. (1995b). Criteria for selection of spatial dimensionality in the application of one and two dimensional water quality models. *Proceedings of International Congress on Modelling and Simulation*, ISBN 0725908955, The University of Newcastle, Australia, june/2005.

Hastings, A. (1990). Spatial heterogeneity and ecological models. *Ecology*, Vol.71, pp. 426-428, ISSN 0012-9658.

Hölker, F. & Breckling, B. (2002). Scales, hierarchies and emergent properties in ecological models: conceptual explanations. In: *Scales, hierarchies and emergent properties in ecological models*, F. Hölker (Ed.), 7-27, Peter Lang, ISBN 3631389248, Frankfurt.

Imberger, J. & Patterson, J.C. (1990). Physical Limnology. *Advances in Applied Mechanics*, Vol.27, pp. 303-475, ISSN 00652156.

Janse, J.H. (2005). *Model studies on the eutrophication of shallow lakes and ditches*, Wageningen University, Wageningen.

Jørgensen, S.E. (1994). *Fundamentals of Ecological Modelling Developments in Environmental Modelling* (3rd ed), Elsevier, ISBN 0080440150, Amsterdam.

Jørgensen, S.E.; Fath, B.D.; Grant, W.E.; Legovic, T. & Nielsen, S.N. (2008). New initiative for thematic issues: An invitation. *Ecological Modelling*, Vol.215, pp. 273-275, ISSN 0304-3800.

Kamenir, Y.; Dubinsky, Z.; Alster, A. & Zohary, T. (2007). Stable patterns in size structure of a phytoplankton species of Lake Kinneret. *Hydrobiologia*, Vol.578, pp. 79-86, ISSN Print 0018-8158, ISSN Online 1573-5117.

Levin, S.A. (1992). The problem of pattern and scale in ecology. *Ecology*, Vol.73, pp. 1943-1967, ISSN 0012-9658.

Lucas, L.V. (1997). *A numerical investigation of Coupled Hydrodynamics and phytoplankton dynamics in shallow estuaries*, Univ. of Stanford, Stanford.

Matveev, V.F. & Matveeva, L.K. (2005). Seasonal succession and long-term stability of a pelagic community in a productive reservoir. *Marine and Freshwater Research*, Vol.56, pp. 1137-1149, ISSN 1323-1650.

Mehner, T.; Holker, F. & Kasprzak, P. (2005). Spatial and temporal heterogeneity of trophic variables in a deep lake as reflected by repeated singular samplings. *Oikos*, Vol.108, pp. 401-409, ISSN Print 0030-1299, ISSN Online 1600-0706.

Mitra, A. & Flynn, K.J. (2007). Importance of interactions between food quality, quantity, and gut transit time on consumer feeding, growth, and trophic dynamics. *American Naturalist*, Vol.169, pp. 632-646, ISSN Print 00030147, ISSN Online 15375323.

Mitra, A.; Flynn, K.J. & Fasham, M.J.R. (2007). Accounting for grazing dynamics in nitrogen-phytoplankton-zooplankton models. *Limnology and Oceanography*, Vol.52, pp. 649-661, ISSN 0024-3590.

Mukhopadhyay, B. & Bhattacharyya, R. (2006). Modelling phytoplankton allelopathy in a nutrient-plankton model with spatial heterogeneity. *Ecological Modelling*, Vol.198, pp. 163-173, ISSN 0304-3800.

Olsen, P. & Willen, E. (1980). Phytoplankton Response to Sewage Reduction in Vattern, a Large Oligotrophic Lake in Central Sweden. *Archiv fur Hydrobiologie*, Vol.89, pp. 171-188, ISSN 0003-9136.

Platt, T.; Dickie, L.M. & Trites, R.W. (1970). Spatial Heterogeneity of Phytoplankton in a near-Shore Environment. *Journal of the Fisheries Research Board of Canada*, Vol.27, pp. 1453-1465, ISSN 0015-296X.

Rajar, R. & Cetina, M. (1997). Hydrodynamic and water quality modelling: An experience. *Ecological Modelling*, Vol.101, pp. 195-207, ISSN 0304-3800.

Reynolds, C.S. (1999). Modelling phytoplankton dynamics and its application to lake management. *Hydrobiologia*, Vol.396, pp. 123-131, ISSN Print 0018-8158, ISSN Online 1573-5117.

Sas, H. (1989). *Lake restoration by reduction of nutrient loading: expectations, experiences, extrapolations*, Academia Verlag Richarz, ISBN-10 388345379X, ISBN-13 978-3883453798, St. Augustin.

Scheffer, M. (1998). *Ecology of Shallow Lakes. Population and Community Biology*, Chapman and Hall, ISBN 0412749203, London.

Scheffer, M. & De Boer, R.J. (1995). Implications of spatial heterogeneity for the paradox of enrichment. *Ecology*, Vol.76, pp. 2270-2277, ISSN 0012-9658.

Schindler, D.W. (1975). Modelling the eutrophication process. *Journal of the Fisheries Research Board of Canada*, Vol.32, pp. 1673-1674, ISSN 0015-296X.

Schladow, S.G. & Hamilton, D.P. (1997). Prediction of water quality in lakes and reservoirs .2. Model calibration, sensitivity analysis and application. *Ecological Modelling*, Vol.96, pp. 111-123, ISSN 0304-3800.

Smith, R.A. (1980). The Theoretical Basis for Estimating Phytoplankton Production and Specific Growth-Rate from Chlorophyll, Light and Temperature Data. *Ecological Modelling*, Vol.10, pp. 243-264, ISSN 0304-3800.

Steele, J.H. (1978). *Spatial Pattern in Plankton Communities*, Plenum Press, ISBN 030640057X, New York.

Steele, J.H. & Henderson, E.W. (1992). A simple model for plankton patchiness. *Journal of Plankton Research*, Vol.14, pp. 1397-1403, ISSN 0142-7873.

Thoman, R.V. & Segna, J.S. (1980). Dynamic phytoplankton-phosphorus model of Lake Ontario: ten-year verification and simulations. In: *Phosphorus management strategies for lakes*, C. Loehr; C.S. Martin & W. Rast (Eds.), 153-190, Amr Arbor Science Publishers, ISBN 0250403323, Michigan.

Tucci, C.E.M. (1998). *Modelos Hidrológicos*. Coleção ABRH de Recursos Hídricos, UFRGS, ISBN8570258232, Porto Alegre, Brazil.

Van den Berg, M.S.; Coops, H.; Meijer, M.L.; Scheffer, M. & Simons, J. (1998). Clear water associated with a dense Char a vegetation in the shallow and turbid Lake Veluwemeer, the Netherlands. In: *Structuring Role of Submerged Macrophytes in Lakes*, E. Jeppesen; M. Søndergaard & K. Kristoffersen (Eds.), 339-352, Springer-Verlag, ISBN 0387982841, New York.

White, F.M. (1974). *Viscous Fluid Flow*, McGraw-Hill, ISBN 0070697124, New York.

Woodward, G. & Hildrew, A.G. (2002). Food web structure in riverine landscapes. *Freshwater Biology*, Vol. 47, pp. 777-798, ISSN Print 0046-5070. ISSN Online 1365-2427.

Wu, F.C.; Shen, H.W. & Chou, Y.J. (1999). Variation of roughness coefficients for unsubmerged and submerged vegetation. *Journal of Hydraulic Engineering-Asce*, Vol.125, pp. 934-942, ISSN Print 0733-9429, ISSN Online 1943-7900.

Wu, J. (1982). Wind-Stress Coefficients over Sea-Surface from Breeze to Hurricane. *Journal of Geophysical Research-Oceans and Atmospheres*, Vol.87, pp. 9704-9706, ISSN 0196-2256.

# 4

# A Study Case of Hydrodynamics and Water Quality Modelling: Coatzacoalcos River, Mexico

Franklin Torres-Bejarano, Hermilo Ramirez and Clemente Rodríguez
*Mexican Petroleum Institute*
*Mexico*

## 1. Introduction

The common basis of the modeling activities is the numerical solution of the momentum and mass conservation equations in a fluid. For hydrodynamic modeling, the Navier-Stokes equations are usually simplified according to the specific water body properties, obtaining, for example, the shallow water equations, so called because the horizontal scale is much larger than the vertical. Therefore, in cases where the river has a relation width-depth of 20 or more and for many common applications, variations in the vertical velocity are much less important than the transverse and longitudinal direction (Gordon et al., 2004). In this sense, the equations can be averaged to obtain the vertical approach in two dimensions in the horizontal plane, which adequately describes the flow field for most of the rivers with these characteristics.

At the same time, the contaminant transport models have evolved from simple analytical equations based on idealized reactors to sophisticated numerical codes to study complex multidimensional systems. Since the introduction of the classic Streeter-Phelps model in the 1920 to evaluate the Biochemical Oxygen Demand and dissolved oxygen in a steady state current, contaminant transport and water quality models have been developed to characterize and assist the analysis of a large number of water quality problems.

This chapter presents the numerical solution of the two-dimensional Saint-Venant and Advection-Diffusion-Reaction equations to calculate the free surface flow and contaminant transport, respectively. The solution of both equations is based on a second order Eulerian-Lagrangian method. The advective terms are solved using the Lagrangian scheme, while the Eulerian scheme is used for diffusive terms. The specific application to the Coatzacoalcos River, Mexico is discussed, having as a main building block the water quality assessment supported on mathematical modelling of hydrodynamics and contaminants transport. The solution method here proposed for the two-dimensional equations, yields appropriate results representing the river hydrodynamics and contaminant behaviour and distribution when comparing whit field measurements.

In this work is presented the structure of a numerical model giving an overview of the program scope, the conceptual design and the structure for each hydrodynamic, pollutants transport and water quality modules that includes ANAITE/2D model (Torres-Bejarano and Ramirez, 2007). The numerical solution scheme is detailed explained for both Saint-Venant and the Advection-Diffusion-Reaction equation. To validate the model, some comparisons were made between model results and different field measurements.

## 2. The numerical model

The developed model is a scientific numerical hydrodynamic and water quality model written in FORTRAN; the model has been named ANAITE/2D. The current version of this model solves the Saint-Venant equations for hydrodynamics representation and the Advection-Diffusion-Reaction (A-D-R) equation using a two-dimensional approach to simulate the pollutants fate.

### 2.1 The hydrodynamic module

In order to obtain a better representation of the hydrodynamics reproduced by the ANAITE model (Torres-Bejarano & Ramirez, 2007), this work presents the solution to change from one-dimensional steady state approach to unsteady two-dimensional flow approximation, solving the two-dimensional Saint-Venant equations (eqs. 1, 2 and 3); these equations describe two-dimensional unsteady flow vertically averaged, representing the principles of conservation of mass and momentum and are obtained from the Reynolds averaged Navier-Stokes equations under certain simplifications (Chaudhry, 1993). These equations have a wide applicability in the study of free surface flow. For example, the flow in open channels with steep slopes (Salaheldin et al., 2000), flows over rough infiltrating surfaces (Wang et al., 2002), propagation of flood waves Rivers (Ying et al., 2003), dam break flow (Mambretti et al., 2008), among others.

$$\frac{\partial h}{\partial t} + \frac{\partial (hu)}{\partial x} + \frac{\partial (hv)}{\partial y} = 0 \tag{1}$$

$$\frac{\partial u}{\partial t} = -u\frac{\partial u}{\partial x} - v\frac{\partial u}{\partial y} - g\frac{\partial h}{\partial x} + v_t\left(\frac{\partial^2 u}{\partial x^2} + \frac{\partial^2 u}{\partial y^2}\right) + g\left(S_{ox} - S_{fx}\right) \tag{2}$$

$$\frac{\partial v}{\partial t} = -u\frac{\partial v}{\partial x} - v\frac{\partial v}{\partial y} - g\frac{\partial h}{\partial y} + v_t\left(\frac{\partial^2 v}{\partial x^2} + \frac{\partial^2 v}{\partial y^2}\right) + g\left(S_{oy} - S_{fy}\right) \tag{3}$$

where:
$Sf$ = friction slope, (·)
$h$  = water depth, (m)
$u$  = longitudinal velocity, x direction (m/s)
$v$  = transversal velocity, y direction (m/s)
$vt$ = turbulent viscosity, (m²/s)
$g$  = acceleration due to gravity, (m/s²)
In this equations system, it was assumed that the effect of the Coriolis force and tensions due to wind at the free surface are negligible given the nature of the problems that focus this work, although their inclusion in the numerical scheme can be done without difficulty.

### 2.2 The water quality module

The water quality model, adapted to the main stream of Coatzacoalcos river, simulates the behaviour and concentration distributions for different water quality parameters. The water quality module solves the following parameters, grouped according to the chemical properties:

- **Physics**: Temperature, Salinity, Suspended Solids, Electric Conductivity.
- **Biochemical**: Dissolved Oxygen (DO), Biochemical Oxygen Demand (BOD), Fecal Coliforms (FC).
- **Eutrophication**: Ammonia Nitrogen (NH₃), Nitrates (NO₃), Organic Nitrogen (N_org.), Inorganic phosphorous (phosphate, PO₄), organic Phosphorous (P_org.).
- **Metals**: Cadmium, Chromium, Nickel, Lead, Vanadium, Zinc.
- **HAPs**: Acenaphthene, Phenanthrene, Fluoranthene, Benzo(a)anthracene, Naphthalene.

The transport and transformation of the different environmental parameters was carried out by applying the two-dimensional approach of A-D-R (eq. 4):

$$\frac{\partial C}{\partial t} + u\frac{\partial C}{\partial x} + v\frac{\partial C}{\partial y} = \frac{\partial}{\partial x}\left(Ex\frac{\partial C}{\partial x}\right) + \frac{\partial}{\partial y}\left(Ey\frac{\partial C}{\partial y}\right) \pm \Gamma_C \tag{4}$$

where:
$C$ = Concentration of any water quality parameter, (mg/L)
$Ex$ = Coefficient of longitudinal dispersion, (m²/s)
$Ey$ = Coefficient of transversal dispersion (m²/s)
$\Gamma c$ = Reaction mechanism (specified for each parameter) (m⁻¹)

The reaction mechanism, $\Gamma c$, is used to represent the water quality parameters, and it is solved individually for each of them.

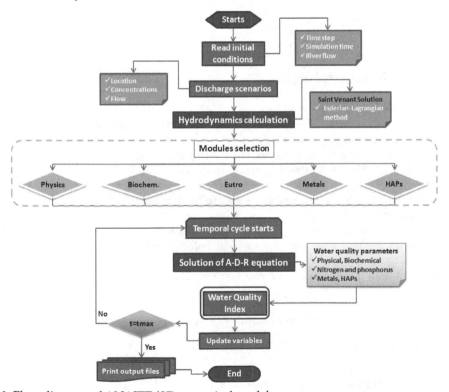

Fig. 1. Flow diagram of ANAITE/2D numerical model

## 2.3 Numerical solutions

The Saint-Venant and A-D-R equations are numerically solved using a second order Eulerian-Lagrangean method; a detailed explanation of the solution is presented in this work. The method separates the equations into its two main components: advection and diffusion, which are solved using a combination of Lagrangian and Eulerian techniques, respectively. In this way, the entire equations are solved. Fig. 1 shows the flow diagram for the model general solution.

### 2.3.1 Numerical grid

A numerical grid Staggered Cell type is used (Fig. 2). In this grid the scalars are evaluated in the center of the cell and vector magnitudes on the edges.

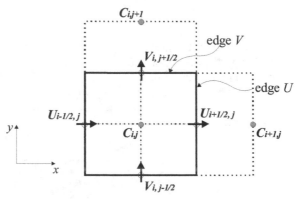

Fig. 2. Notation for Staggered cell grid

### 2.3.2 Advection (Lagrangian method)

The advection solution uses a Lagrangian method whose interpolation/extrapolation principle is base on the characteristics method. In the characteristics method is assigned to each node at $t^{n+1}$ a particle that does not change its value as it moves along a characteristic line defined by the flow. It locates its position in the previous time $t^n$ by the interpolation of adjacent values of a characteristic value, in this case the Courant number, which is assigned to node $t^{n+1}$. For simplicity, the method is exemplified in one dimension, but is similar for two or three dimensions (Fig. 3).

Assuming that the value of the variable at point P ($\varphi_p{}^n$), can be calculated by interpolating between the values $\varphi^n{}_{i-1}$ y $\varphi_i{}^n$ from the adjacent points $x_0$ and $x_1$ respectively (Rodriguez et al., 2005).

If a particle at the point $P$ travels at a constant velocity $U$ will move a distance $x + U \Delta t$ at time $t + \Delta t$, so it is:

$$\varphi(x,t) = \varphi(x + u\Delta t, t + \Delta t) = \varphi(x + p\Delta x, t + \Delta t) \tag{5}$$

if we apply the modified Gregory-Newton interpolation:

$$f(P) \equiv f(x + p\Delta x) = f_1 - p[f_1 - f_0] + \frac{p(p-1)}{2!}[f_2 - 2f_1 + f_0] - \Lambda \tag{6}$$

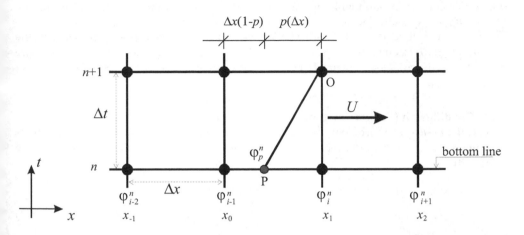

Fig. 3. Notation used, shown in one-dimensional mesh

Where $f(P)$, $f_1$, $f_0$ y $f_2$ are the values at points $P$, $x_1$, $x_0$, and $x_2$ respectively, $p$ is a weight coefficient which positions the point $P$ with respect to $\varphi_i^n$ and $\varphi_{i-1}^n$. Since the polynomial is a second-degree interpolation, the three terms of equation (6) are used. Substituting the known values for the three points:

$$\varphi_p^n = \left(1 - p^2\right)\varphi_i^n + \left(0.5p^2 + 0.5p\right)\varphi_{i-1}^n + \left(0.5p^2 - 0.5p\right)\varphi_{i+1}^n$$

As the value at point P is required in a two dimensional grid, the solution is expressed as shown in equation (7).

$$
\begin{aligned}
\varphi_{i-a,j-b} = &\left(1-p^2\right)\left[\left(1-q^2\right)\varphi_{i,j}^n + \left(0.5q^2 + 0.5q\right)\varphi_{i,j-1}^n + \left(0.5q^2 - 0.5q\right)\varphi_{i,j+1}^n\right] \\
&+ \left(0.5p^2 + 0.5p\right)\left[\left(1-q^2\right)\varphi_{i-1,j}^n + \left(0.5q^2 + 0.5q\right)\varphi_{i-1,j-1}^n + \left(0.5q^2 - 0.5q\right)\varphi_{i-1,j+1}^n\right] \\
&+ \left(0.5p^2 - 0.5p\right)\left[\left(1-q^2\right)\varphi_{i+1,j}^n + \left(0.5q^2 + 0.5q\right)\varphi_{i+1,j-1}^n + \left(0.5q^2 - 0.5q\right)\varphi_{i+1,j+1}^n\right]
\end{aligned}
\tag{7}
$$

where $p$ and $q$, are the Courant numbers in $x$ and $y$ directions, respectively. The calculation of the Courant numbers for $u$ and $v$ is as follows:

$$C_u(p) = \frac{\overset{*}{u}_{i,j}\Delta t}{\Delta x_i} \; ; \; C_v(q) = \frac{\overset{*}{v}_{i,j}\Delta t}{\Delta y_j} \tag{8}$$

where:

$$\overset{*}{u}_{i,j} = \left(1-\alpha\right)u_{i,j} + \frac{\alpha}{4}\left(u_{i+1,j+1} + u_{i-1,j+1} + u_{i+1,j-1} + u_{i-1,j-1}\right) \tag{9}$$

$$\overset{*}{v}_{i,j} = \left(1-\alpha\right)v_{i,j} + \frac{\alpha}{4}\left(v_{i+1,j+1} + v_{i-1,j+1} + v_{i+1,j-1} + v_{i-1,j-1}\right) \tag{10}$$

α is coefficient of relaxation with typical values between 0 – 1. In this work 0.075 was used. $u^*_{i,j}$ and $v^*_{i,j}$ are spatial velocities in $x$ and $y$ direction respectively.

The solution method is applied in a similar way to the advective terms presents in the continuity (Eq. 1), momentum (Eqs. 2 and 3) and A-D-R (Eq. 4) equations; $\varphi = h, u, v$ y $C$, are the advective variables, obtained by the lagrangian method in its second order approximation.

### 2.3.3 The diffusion (Eulerian method)

*Turbulent diffusion.* The turbulent viscosity coefficient present in the Saint-Venant equations is evaluated with a cero order model o mixing length model in two dimensions (vertically averaged):

$$v_t = l_m^2 \sqrt{2\left(\frac{\partial u}{\partial x}\right)^2 + 2\left(\frac{\partial v}{\partial y}\right)^2 + \left(\frac{\partial u}{\partial y} + \frac{\partial v}{\partial x}\right)^2 + \left(2.34\frac{u_f}{\kappa h}\right)^2} \quad \therefore \quad l_m \approx 0.267\kappa h \tag{11}$$

where:

$l_m$ = mixing length

$u_f$ = friction velocity, $u_f = \sqrt{ghS}$

$\kappa$ = von Kármán constant

Thus, the diffusion terms in $x$ and $y$ are solved respectively by the following formulas:

$$Dift\_u = v_t \left\{ \left[ \left(\frac{u_{i+1,j} - u_{i,j}}{\Delta x_i}\right) - \left(\frac{u_{i,j} - u_{i-1,j}}{\Delta x_{i-1}}\right) \right]^n + \left[ \left(\frac{u_{i,j+1} - u_{i,j}}{\Delta y_j}\right) - \left(\frac{u_{i,j} - u_{i,j-1}}{\Delta y_{j-1}}\right) \right]^n \right\} \tag{12}$$

$$Dift\_v = v_t \left\{ \left[ \left(\frac{v_{i+1,j} - v_{i,j}}{\Delta x_i}\right) - \left(\frac{v_{i,j} - v_{i-1,j}}{\Delta x_{i-1}}\right) \right]^n + \left[ \left(\frac{v_{i,j+1} - v_{i,j}}{\Delta y_j}\right) - \left(\frac{v_{i,j} - v_{i,j-1}}{\Delta y_{j-1}}\right) \right]^n \right\} \tag{13}$$

Detailed analysis of turbulence, their interpretation and mathematical treatment can be found at Rodi, (1980), Rodríguez et al., (2005).

*Longitudinal and transverse dispersion in rivers.* The A-D-R equation evaluates the dispersion process by $Ex$ and $Ey$ coefficients.

In this work the longitudinal dispersion coefficient proposed by Seo and Cheong (1998) has been implemented (Eq. 14):

$$\frac{Ex}{hu_f} = 5.915\left(\frac{W}{h}\right)^{0.620}\left(\frac{u}{u_f}\right)^{1.428} \tag{14}$$

The expression for estimating the transverse dispersion coefficient in a river is given by (adapted from Martin and McCutcheon, 1999):

$$Ey = 0.023hU^* \tag{15}$$

The 0.023 value was specifically obtained for the studied river. The dispersion terms in equation (4) are numerically solved as follow:

$$Disp\_C = \frac{\Delta t}{\Delta x_i}\left[\frac{Ex_{i+1,j}}{\Delta x_i}\left(C_{i+1,j}-C_{i,j}\right)-\frac{Ex_{i,j}}{\Delta x_{i-1}}\left(C_{i,j}-C_{i-1,j}\right)\right]^n$$
$$+\frac{\Delta t}{\Delta y_j}\left[\frac{Ey_{i,j+1}}{\Delta y_j}\left(C_{i,j+1}-C_{i,j}\right)-\frac{Ey_{i,j}}{\Delta y_{j-1}}\left(C_{i,j}-C_{i,j-1}\right)\right]^n$$

(16)

### 2.3.4 The pressure terms

This is the term that takes into account the external forces in the Saint-Venant equations, in this case the gravitational forces. It is solved with a centered difference of depth values in the calculation grid (Eq. 17).

$$Pres\_u = -g\left[\frac{h_{i+1,j}-h_{i-1,j}}{2\Delta x}\right]^n \; ; \; Pres\_v = -g\left[\frac{h_{i,j+1}-h_{i,j-1}}{2\Delta y}\right]^n$$

(17)

### 2.3.5 The continuity equation

Expanding the derivative and rearranging terms in equation (1), the continuity equation is as follows:

$$h_{i,j}^{n+1} = Advec\_h - \Delta t\left[h_{i,j}^*\left(\frac{u_{i+1,j}-u_{i-1,j}}{2\Delta x_i}\right)+h_{i,j}^*\left(\frac{v_{i,j+1}-v_{i,j-1}}{2\Delta y_j}\right)\right]^n$$

(18)

where:

$$h_{i,j}^* = \frac{1}{4}\left(h_{i+1,j+1}+h_{i-1,j+1}+h_{i+1,j-1}+h_{i-1,j-1}\right)$$

(19)

### 2.3.6 General solution for velocities

Grouping the obtained terms, the following general equations for velocities are reached:

$$\left[u_{i,j}\right]^{n+1} = Advec\_u + \Delta t\left[Dift\_u \; - \; Pres\_u\right]^n - \Delta tg\left[S_{ox}+S_{fx_{i,j}}\right]^n$$

(20)

$$\left[v_{i,j}\right]^{n+1} = Advec\_v + \Delta t\left[Dift\_v \; - \; Pres\_v\right]^n - \Delta tg\left[S_{oy}-S_{fy_{i,j}}\right]^n$$

(21)

The first element of the last term is the free surface slope, which multiplied by the gravity represents the action of gravitational forces. This term can be expressed as:

$$S_{ox} = -\partial z_0 / \partial x \; ; \; S_{oy} = -\partial z_0 / \partial y$$

(22)

The second element of the last term represents the bottom stress, which causes a nonlinear effect of flow delay and is calculated by the Manning formula:

$$S_{fx} = \frac{n^2 u \sqrt{u^2 + v^2}}{h^{4/3}} \; ; \; S_{fy} = \frac{n^2 v \sqrt{u^2 + v^2}}{h^{4/3}} \tag{23}$$

where:
$S_f$ = friction slope, $(\cdot)$
$n$  = Manning roughness coefficient

### 2.3.7 General solution for A-D-R equation

Finally, grouping term for A-D-R equation the solution is obtained with:

$$\left[C_{i,j}\right]^{n+1} = Advec\_c + \Delta t \left[Disp\_C\right]^n \pm \Delta t \left[\Gamma_C\right]^n \tag{24}$$

As mentioned in section 2.2, $\Gamma c$ is solved individually for each parameter.

### 2.3.8 Stability requirements

Because a finite difference scheme is used, should be considered the linear stability criteria. The selection of the time step and space must satisfy the condition of Courant-Friedrichs-Lewy (CFL) for a stable solution. The CFL condition for two-dimensional Saint-Venant equations can be written as (Bhallamudi and Chaudhry, 1992):

$$C_n = \frac{\left(R + \sqrt{gh}\right) \cdot \Delta t}{\Delta x \cdot \Delta y} \sqrt{\Delta x^2 + \Delta y^2} < 1 \tag{25}$$

where: $R$ = magnitude of the resultant velocity, (m/s)

## 3. Study case: Coatzacoalcos River

The last stretch of the Coatzacoalcos River located in the Minatitlan-Coatzacoalcos Industrial Park (MCIP), with about 40 km length, is part of an area of vast natural diversity, where the high population concentration creates important environmental changes, due to pressures arising mainly from consumption and industrial activities. Currently, insufficient information exists for MCIP regarding hydrodynamics and water quality at the river stretch. This is one of the most polluted rivers in Mexico and is consequently a critical area in terms of industrial pollution.

Coatzacoalcos is a commercial and industrial port that offers the opportunity to operate a transportation corridor for international traffic, the site is the development basis for industrial, agricultural, forestry and commercial in this region; by the volume of cargo is considered the third largest port in the Gulf of Mexico (Fig. 4). The area of Coatzacoalcos river mouth has had a rapid urban and industrial growth in the last three decades. In this area, the largest and most concentrated industrial chemical complex, petrochemical and derivatives has been developed in Latin-America.

The importance of this industrial park, formed mainly by the Morelos petrochemical complex, Cangrejera, Cosoloacaque and Pajaritos, is such that 98% of the petrochemicals used throughout the country are produced there. Fig. 5 shows the industrial and petrochemical facilities located in this area.

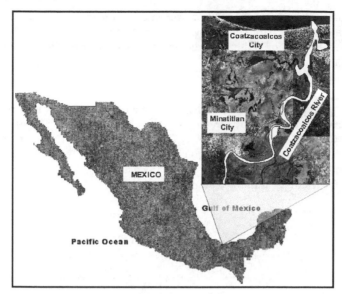

Fig. 4. Location of Coatzacoalcos River

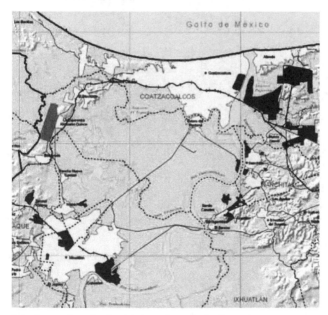

Fig. 5. Minatitlan-Coatzacoalcos industrial park and study zone

## 4. Results and discussion

We have developed a numerical model that solves the two-dimensional Saint-Venant equations and the Advection-Diffusion-Reaction equation to study the pollutants transport.

The model results show agreement with measurements of velocity direction and magnitude, as well with water quality parameters. Therefore, it is considered that the developed model can be implemented and applied to different situations for this study area and others rivers with similar characteristics.

### 4.1 Model validation
For validation purpose, a sampling and measurement campaign was carried out in the Coatzacoalcos river stretch from upstream of Minatitlan city (17° 57' 00" N - 94° 33' 00" W) to its mouth in Gulf of Mexico (18° 09' 32" N – 94° 24' 41.33" W). The main objective was to obtain velocities, bathymetry and water quality at 10 points of the Coatzacoalcos River (Fig. 6). The information obtained through direct measurements and chemical analysis is primarily used for testing and numerical model validation.

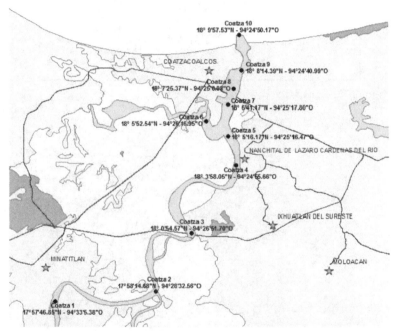

Fig. 6. Measurements and sampling sites

Table 1 shows the water velocity magnitude, direction and location measured on the ten stations. These values were compared with the model results. Fig. 7 and Fig. 8 show a comparison between measured and calculated water velocities.

As shown in Fig. 7 and 8, the hydrodynamics numerical results correspond fairly well with field measurements. The model results show agreement in direction and magnitude of the measured velocity, which demonstrate that the model results are consistent and reliable to the real river behaviour. Therefore, it is considered that the developed model can be implemented and applied to different situations for the studied area.

Likewise, the water quality modules were validated by comparison with field measurements, observing that the model results are consistent with these measurements and they are in the same order of magnitude. Fig. 9 shows some of the obtained results (each point represents a measurement station and the solid line the model result).

| Station | Latitude | Longitude | East Vel. (cm/s) | North Vel. (cm/s) | Resultant (m/s) | Direction (°) |
|---------|----------|-----------|------------------|-------------------|-----------------|---------------|
| 1 | 17.96468 | -94.5529350 | -24.06 | -42.34 | 0.49 | 60.39 |
| 2 | 17.97063 | -94.4749317 | 4.75 | 23.65 | 0.24 | 78.65 |
| 3 | 18.01482 | -94.4479867 | -43.56 | -8.69 | 0.44 | 11.28 |
| 4 | 18.06698 | -94.4156000 | -11.32 | 58.24 | 0.59 | -79.00 |
| 5 | 18.08841 | -94.4215717 | -10.74 | 39.62 | 0.41 | -74.83 |
| 6 | 18.1023 | -94.4374683 | 9.38 | -5.57 | 0.11 | -30.69 |
| 7 | 18.1117 | -94.4217833 | 36.78 | -58.26 | 0.69 | -57.74 |
| 8 | 18.12396 | -94.4148883 | 34.06 | -1.98 | 0.34 | -3.33 |
| 9 | 18.13607 | -94.4112700 | 6.62 | -16.16 | 0.17 | -67.72 |
| 10 | 18.16384 | -94.4154150 | -7.14 | 1.02 | 0.07 | -8.15 |

Table 1. Measurements of water velocity

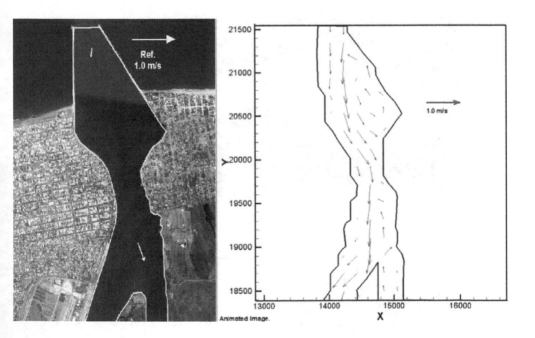

Fig. 7. Comparison between measured and calculated velocities for the river mouth

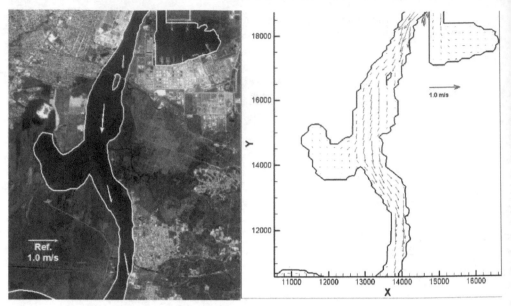

Fig. 8. Comparison between measured and calculated velocities for the river middle part

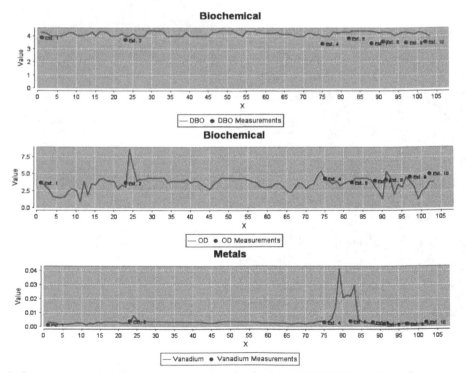

Fig. 9. Concentration profiles for measured and calculated DO, BOD and Vanadium

## 4.2 Numerical modeling

The initial step in the methodology implemented was the numerical grid generation, using specialized software. Initial and boundary conditions (tide, the level of free water surface, and hydrodynamic condition) were imposed; the model was set up with information gathered in the measurement campaigns, as well as water balances to determine the river dynamics.

The mesh or numerical grid was created using the program ARGUS ONE (http://www.argusint.com), Fig. 10 shows the calculation grid system for Coatzacoalcos river stretch, which has a length of about 25 km, spacing of $\Delta x = \Delta y = 100$ m. The grid has 163 element in the X direction and 211 points in Y direction, giving a total of 34393 elements.

Fig. 10. Grid configuration

Two simulation scenarios were performed representing dry season and rain season. The input data required are shown in Table 2: Manning roughness coefficient, hydrological flow, cross-sectional area, flow velocity and direction.

| Parameter | Dry season | Rain season |
|---|---|---|
| Flow rate (m³/s) | 405 | 1104.9 |
| Cross section (m²) | 812.4 | 2216.7 |
| Flow velocity (m/s) | 0.5 | 0.5 |
| Flow direction (°) | 58.78° | 60.39 |
| Manning coefficient | 0.025 | 0.025 |

Table 2. Initial data for simulations

## 4.3 Results of hydrodynamics simulations

The simulation represents 30 days corresponding to dry and rain season. The numerical integration time or time step was, $\Delta t = 2.0$ s. Fig. 11 shows the obtained result for resultant velocity.

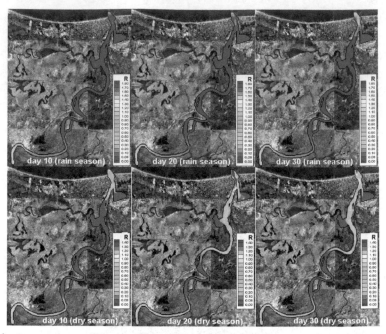

Fig. 11. Velocity contours of Coatzacoalcos River (rain and dry season) over time

## 4.4 Modeling of pollutant transport

This section presents some of the simulated parameters. Initially we describe the simulation scenario solved to provide a better idea and understand clearly what the simulations are representing.

### 4.4.1 Simulation scenario

To assess the water quality in Coatzacoalcos River, a representative discharge scenario was initially defined, highlighting the main activities and characteristics over this area (Fig. 12). The idea was study the river environmental behaviour, influenced by the oil activity developed on this area, six discharges where located representing the industrial activity and the influence of urban areas where such facilities industry are seated.

The discharge conditions for every simulated water quality parameter are presented in Table 3.

| Discharge | Temp. (°C) | DO (mg/L) | BOD (mg/L) | Faecal Col. (NMP/100 ml) |
|---|---|---|---|---|
| Minatitlan | 26.70 | 3.70 | 290.71 | 21,5000 |
| Refinery | 25.70 | 3.70 | 190.86 | 21,3000 |
| Uxpanapa R. | 24.70 | 3.70 | 190.70 | 21,7000 |
| Nanchital | 27.70 | 3.70 | 290.43 | 31,5000 |
| Pajaritos P.C. | 26.70 | 3.70 | 290.83 | 22,4100 |
| Coatzacoalcos | 25.70 | 3.70 | 110.60 | 21,4300 |

Table 3. Discharges conditions

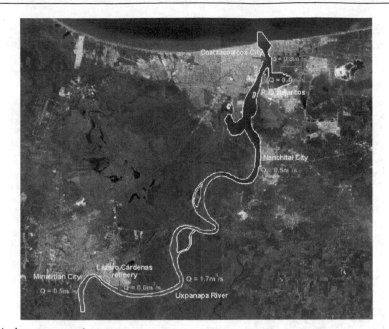

Fig. 12. Discharge scenario

The following figures show the model results for these water quality parameters.

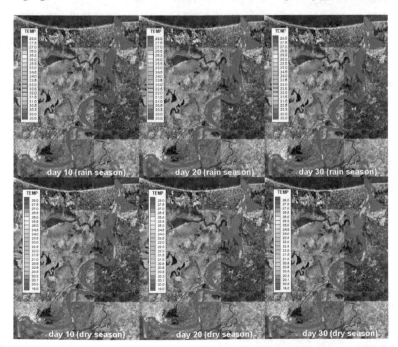

Fig. 13. Temperature simulations over time

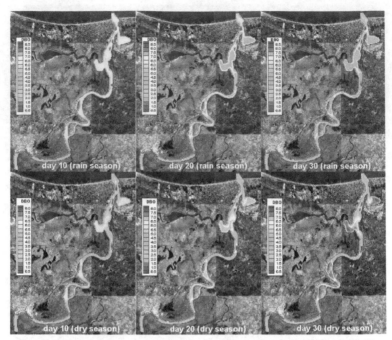

Fig. 14. Biochemical Oxygen Demand simulations over time

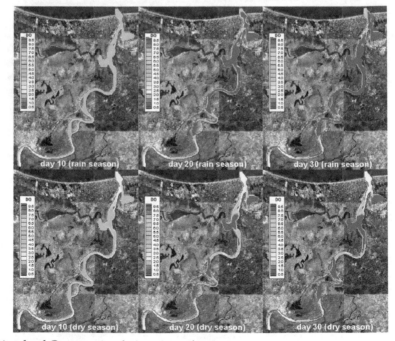

Fig. 15. Dissolved Oxygen simulations over time

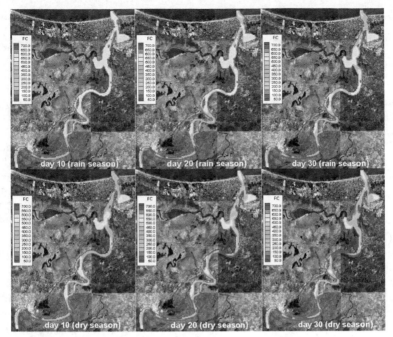

Fig. 16. Faecal Coliforms simulations over time

## 5. Conclusions

The solution obtained for the two-dimensional Saint-Venant and A-D-R equations using an Eulerian-Lagrangian method has great versatility, obtaining consistent and satisfactory results for different types of flow and open channel conditions. The considered scheme provides numerical stability that avoids numerical oscillations of the obtained solutions and also allows significant larger time steps ($\Delta t$). The combination with the Eulerian solution for diffusive terms is always guaranteed satisfying the C-F-L condition.

About the hydrodynamics study of Coatzacoalcos river, it was determined that the river behaviour is influenced by several factors, being the most important the hydrological aspect, which varies depending on the time of the year. Because of this, it was observed that dry season presents an important tide penetration towards the mainland of the river, while for rain season when the river flow increase, the penetration is less significant and the water mainly flows downstream to the mouth in the Gulf of Mexico.

On the other hand, the pollutants transport is dominated strongly by the hydrodynamics, and the difference for the two simulated seasons was observed. This simulation shows higher concentrations and also a more significant dispersion in dry season, because the tide penetration occurs intermittently upstream and downstream in the area near to the river mouth. While for rain season there is no significant contaminant dispersion, with a local effect of the simulated discharges.

Thus, a solution algorithm has been proposed to the study open channel hydrodynamics, which together with the A-D-R equation solution allows the study of transport, transformation and reaction of pollutants, being the basis of the water quality model proposed.

## 6. References

Bhallamudi, M.S. y Chaudhry, M. H. (1992). Two dimensional modelling of supercritical and subcritical flow in channel transitions, *Journal Hydraulic Engineering*, ASCE, Vol.30, No.1, pp. 77-91.

Chaudhry, M. H. (1993). Open Channel Flow. Prentice Hall, New Jersey.

Gordon N., McMahon T., Finlayson B., Gippel C. & Nathan R. (2004). *Stream Hydrology: An Introduction for Ecologists*. (2nd Ed.), John Wiley & Sons, ISBN: 0-470-84357-8. USA.

Mambretti S., Larcan E., Wrachien D. (2008). 1D modelling of dam-break surges with floating debris. *Biosystem Engineering*, vol. 100 (2008) 297 – 308. ISSN: 15375110.

Martin JL, McCutcheon STC. (1999). *Hydrodynamics and Transport for Water Quality Modeling*. Lewis, Boca Raton, FL.

Rodi, W. (1980). Turbulence models and their application in hydraulics: a state of the art review, *Book publication of international association for hydraulic research*, Delft, Netherlands.

Rodriguez, C., Serre E., Rey, C. and Ramirez, H. (2005). A numerical model for shallow-water flows: dynamics of the eddy shedding. *WSEAS Transactions on Environment and Development*. Vol. 1, pp. 280-287, ISSN: 1790-5079

Salaheldin T., Imran J. & Chaudhry., M. (2000). Modeling of Open-Channel Flows with Steep Gradients. *Ingeniería del Agua*, vol. 7 ( 4), , pp. 391-408. ISSN: 1134-2196.

Seo II W. & Cheong TS. (1998). Predicting longitudinal dispersion coefficient in natural streams. *Journal of Hydraulic Engineering*. 124(1):25 – 32.

Torres-Bejarano F. & Ramirez H., (2007). The ANAITE model for studying the hydrodynamics and water quality of natural rivers with soft slope. *International Journal of Environmental Contamination* 23 (3) (2007), 115-127. ISSN: 01884999.

Wang G., Chen S., Boll J., Stockle C., McCool D. (2002). Modelling overland flow based on Saint-Venant equations for a discretized hillslope system. *Hydrological Processes*, vol. 16 (12) (2002), pp. 2409 – 2421.

Ying X., Wang S. & Khan A. (2003). Numerical Simulation of Flood Inundation Due to Dam and Levee Breach, *Proceedings of ASCE World Water and Environmental Resources Congress 2003*, Philadelphia, USA, June 2003.

# 5

# Challenges and Solutions for Hydrodynamic and Water Quality in Rivers in the Amazon Basin

Alan Cavalcanti da Cunha[1], Daímio Chaves Brito[1],
Antonio C. Brasil Junior[2], Luis Aramis dos Reis Pinheiro[2], Helenilza
Ferreira Albuquerque Cunha[1], Eldo Santos[1] and Alex V. Krusche[3]
[1]Federal University of Amapá - Environmental Science Department and Graduated
Program in Ecological Sciences of Tropical Biodiversity
[2]Universidade de Brasilia. Laboratory of Energy and Environment
[3]Environmental Analysis and Geoprocessing Laboratory CENA
Brazil

## 1. Introduction

This research is part of a multidisciplinary research initiative in marine microbiology whose goal is to investigate microbial ecology and marine biogeochemistry in the Amazon River plume. Aspects related to Amazon River fluvial sources impacts on the global carbon cycle of the tropical Atlantic Ocean are investigated within the ROCA project (River-Ocean Continuum of the Amazon). This project is intended to provide an updated and integrated overview of the physical, chemical and biological properties of the continuous Amazon River system, starting at *Óbidos*, located 800 km from the mouth of the river, and interacting to the discharge influence region at the Atlantic Ocean (Amazon River plume). This geographic focal region includes the coast of the State of *Amapá* and the north of *Marajó* archipelago in Northeast Brazilian Amazon.

The ROCA project is focused on the connection between the terrestrial Amazon River and the ocean plume. This plume extends for hundreds of kilometres from the river delta towards the open sea. This connection is vital for the understanding of the regional and global impacts of natural and anthropogenic changes, as well as possible responses to climate change (Richey et al. 1986; Richey et al. 1990; Brito, 2010). Different phenomena of interest are typically linked to the quantity and quality of river water (flows of carbon and nutrient dynamics) and the dynamics of sediments. All of them are strongly influenced by substances transport characteristics and water bodies physical properties and physical properties in the water bodies, constrained by spatial distribution of water flow (influenced by bottom topography and coastline of river mouth archipelago) and the unsteady interaction with tides and ocean currents. These very complex phenomena at the Amazon mouth are still not fully understood.

Based on this framework, river and ocean plume hydrodinamics are fundamental components in the complex interactions between physical and biotic aspects of river-ocean interaction. They drive biogeochemical processes (carbon and nutrient flows), variations in water quality (physical-chemical and microbiological). They drive biogeochemical processes (river bottom and suspended sediments) (Richey et al., 1990; Van Maren & Hoekstra, 2004, Shen et al. 2010; Hu & Geng, 2011). The understanding of the Amazon River mouth flows is an important and opened question to be investigated in the context of the river-ocean integrated system.

In Brazil, the National Water Agency (ANA) monitors water flows at numerous locations throughout the Amazon basin (Abdo et al. 1996; Guennec & Strasser, 2009). However, the last monitoring station located on the Amazon River and nearest to the ocean is *Obidos* (1°54'7.36"S, 55°31'10.43"W). There are no systematically recorded data available in downriver locations towards the mouth. The Amapá State coast is, *geographically*, an ideal site for such future systematic experimental flow measurements, since about 80% of the net discharge of the Amazon River flows in the North Channel located in front of the city of Macapá (0° 1'51.41"N, 51° 2'56.88"W) (ANA, 2008). The fact that this flow is not continuous and varies with ocean tides, creating an area of inflow-outflow transition makes this region a challenging subject for water research.

This research focus on two main issues: a) to establish an overview of physical aspects over transect $T_2$ in the North Channel of the Amazon River, where measurements were performed for quantification of liquid discharge and additional sampling procedures for assessing water quality and quantify concentration of $CO_2$ in the air and water; b) to evaluate typical local effects of river flow interacting with the shore and small rivers, based on turbulent fluid flow modeling and simulation.

## 2. Main driving forces of the Amazon river mouth discharge

Tidal propagation in estuaries is mainly affected by friction and freshwater discharge, together with changes in channel depth and morphology, which implies damping, tidal wave asymmetry and variations in mean water level. Tidal asymmetry can be important as a mechanism for sediment accumulation while mean water level changes can greatly affect navigation depths. These tidal distortions are expressed by shallow water harmonics, overtides and compound tides (Gallo, 2004). The Amazon estuary presents semidiurnal overtides, where the most important astronomic components are the M2 (lunar component) and S2 (solar component), consequently, the most common overtide is the M4 (M2 + M2) and the main compound tide is the Msf (relative to fluvial flow). Amplitude characteristics of the mouth of the Amazon River is represented by tidal components M2 and S2, of 1,5m and 0,3m, respectively, corresponding to North Station Bar, Amapá State (Galo, 2004; Rosman, 2007).

Form factor (F) expressing the importance of scale on components of the diurnal and semi-diurnal tides, the Amazon estuary can be classified as a typical semi-diurnal tide (0 < F < 0.25). However, this classification does not considers the effects of river discharge. River discharge certainly contribut to friction and to balance the effect of convergence in the lower estuary and also to what happens between the platform edge of the ocean station and the the mouth of the Amazon River.

There is evidence of nonlinearity in tidal propagation, which can be observed by the gradual redistribution of power between M2 and its first harmonic M4. Considering tides as the sum of discrete sinusoids, the asymmetry can be interpreted through the generation of harmonics in the upper estuary (Galo, 2004; Rosman, 2007). In the case of a semi-diurnal tide, with its main components M2 and M4 first harmonic, the phase of high frequency harmonic wave on the original controls the shape of the curve and therefore the asymmetry.

Three major effects characterize the amount and behaviour of flow it the mouth of the Amazon River: (a) relative discharge contributions from sub-basins of the main channel; b) tidal cycles and; (c) regional climate dynamics.

According to Gallo (2004) the Amazon River brings to the Atlantic Ocean the largest flow of freshwater in the world. Based on *Óbidos* records, there is an average flow of approximately $1.7 \times 10^5$ m$^3$/s, with a maximum of approximately $2.7 \times 10^5$ m$^3$/s and a minimum of $0.6 \times 10^5$ m$^3$/s. According to ANA (2008), the flow reaches a net value of approximately 249,000 m$^3$/s, with a maximum daily difference of 629,880 m$^3$/s (ebb) and a minimum of -307,693 m$^3$/s (flood). The most important contributions come from the *Tapajós* River with an average flow of approximately $1.1 \times 10^4$ m$^3$/s, the *Xingu* River with an average of approximately $0.9 \times 10^4$ m$^3$/s and *Tocantins* River, at the southern end of the platform, with an approximate average flow of $1.1 \times 10^4$ m$^3$/s.

Penetration of a tidal estuary is result of interaction between river flow and oscillating motion generated by the tide at the mouth river, where long tidal waves are damped and progressively distorted by the forces generated by friction on river bed, turbulent flow characteristics of river and channel geometry. Gallo (2004) describes that propagation of the tide in estuaries is affected mainly by friction with river bed and river flow, as well as changes in channel geometry, generating damping asymmetry in the wave and modulation of mean levels. Such distortions can be represented as components of shallow water, over-tides and harmonic components. The Amazon River estuary can be classified as macrotidal, typically semi-diurnal, whose most important astronomical components are M$_2$ (principal lunar semidiurnal) and S$_2$ (Principal solar semidiurnal) and therefore the main harmonics generated are high frequency, M$_4$ (lunar month) and the harmonic compound, Msf (interaction between lunar and solar waves) (Bastos, 2010; Rosman, 2007).

In the Amazon the most important climatic variables are convective activity (formation of clouds) and precipitation. The precipitation regime of the Amazon displays pronounced annual peaks during the austral summer (December, January and February - DJF) and autumn (March, April and May - MAM), with annual minima occurring during the austral winter months (June, July and August - JJA) and spring (September, October and November - SON). The rainy season in Amapá occurs during the periods of DJF and MAM (Souza, 2009; Souza & Cunha, 2010).

The variability of rainfall during the rainy season is directly dependent on the large-scale climatic mechanisms that take place both in the Pacific and the Atlantic Oceans (Souza, 2009). In the Pacific Ocean, the dominant mechanism is the well-known climatic phenomenon El Niño / Southern Oscillation (ENSO), which has two extreme phases: El Niño and La Niña. The conditions of El Niño (La Niña) are associated with warming (cooling) anomalies in ocean waters of the tropical Pacific, lasting for at least five months between the summer and autumn. In the Atlantic Ocean, the main climatic mechanism is called the Standard Dipole or gradient anomalies of Sea Surface Temperature (SST) in the intertropical Atlantic (Souza & Cunha, 2010).

This climate is characterized by a simultaneous expression of SST anomalies spatially configured with opposite signs on the North and South Basins of the tropical Atlantic. This inverse thermal pattern generates a thermal gradient (inter-hemispheric and meridian) in the tropics, with two opposite phases: the positive and negative dipole. The positive phase of the dipole is characterized by the simultaneous presence of positive / negative SST anomalies, setting the north / south basins of the tropical Atlantic Ocean. The dipole negative phase of the configuration is essentially opposed. Several observational studies showed that the phase of the dipole directly interferes with north-south migration of the Intertropical Convergence Zone (ITCZ). The ITCZ is the main inducer of the rain weather system in the eastern Amazon, especially in the states of Amapá and Pará, at its southernmost position defines climatologically the quality of the rainy season in these states (Souza & Cunha, 2010). The behavior of the climate is important because it significantly influences the hydrological cycle and, therefore, the hydrodynamic and mixing processes in the water.

According to Van Maren & Hoesktra (2004) the mechanisms of intra-tidal mixing depend strongly on seasonally varying discharge (climate) and therefore hydrodynamics. In this case, during the dry season, there is a breakdown of stratification during the tidal flood that occurs in combination with the movements of tides and advective processes. Intra-tidal mixing is probably greater in semi-diurnal than in diurnal tides, because the semi-diurnal flow velocity presents a non-linear relationship with the mixture generated in the river bed and the mean velocity.

A second, Hu & Geng (2011), studying water quality in the Pearl River Delta (PRD) in China, found that coupling models of physical transport and sediments could be used to study the mass balance of water bodies. Thus, most of the flows of water and sediment occur in wet season, with approximately 74% of rainfall, 94% water flow and 87% of suspended sediment flow. Moreover, although water flow and sediment transport are governed primarily by river flow, tidal cycle is also an important factor, especially in the regulation of seasonal structures of deposits in river networks (deposition during the wet season and erosion in the dry season). As well as net discharge there are several types of physical forces involved in these processes, including: monsoon winds, tides, coastal currents and movements associated with gravitational density gradients. Together these forces seem to jointly influence the control of water flow and sediment transport of that estuary.

A third example, according to Guennec & Strasser (2009), hydrodynamic modeling along a stretch of the Manacapuru-Óbidos river in the upper Amazon a stretch of the *Manacapuru-Óbidos* river in the upper Amazon revealed that the ratio of liquid flow that passes through the floodway changes from 100% during the low water period to 76% (on average) during the high water period. Expressed in volume, this means that about 88% of the total volume available during a hydrological cycle moves through the floodway of the river, and only 12% moves through the mid portion. The volume that reaches the fringe of the flood plain is approximately 4% and appears to be temporarily stored.

Based on the climatic characteristics of the State of Amapá, one of the main challenges for both hydrological and hydrodynamic studies is to integrate meteorological information from the Amazon Basin and include these forces when evaluating the responses of aquatic ecosystems in the Lower Amazon River estuary (Brito, 2010; Bastos, 2010; Cunha et al., 2006; Rickey et al., (1986), Rosman (2007), Gallo (2004), ANA (2008) and Nickiema et al., (2007).

## 3. River flow measurements in Amazon North Channel

In the Amazon River (North Channel) two up to date measurements of net discharges were made. The measuring process, consists of: 1) performing a series of flow measures over a minimum period of 12.30 h, using ADCP with an average of 12 experimental measurements;

2) interpolate the temporal evolution of flow and velocity from these measurements;  3) integrate the values with the tidal cycle to obtain the average flow rate (or velocity); 4) analyze the maximum and minimum flow, and the relationship between flow/velocity and level, as described by ANA (2008), Cunha et al. (2006) and Silva & Kosuth (2001).

Fig. 1 shows the location of Transect T2 (blue line) of the North Channel and Matapi River, both studied by Brito (2010) and Cunha (2008) nearly to city of *Macapá*, respectively .The requirement for local knowledge of the river bathymetry is demonstrated by the geometric complexity of the channels and variations in the average depths of the channel. Cunha (2008) observed depths ranging from 3 m (minimum) to approximately 77 m (maximum) in the section indicated.

Brito (2010) has studied the water quality sampled water quality and participated in the quantification of the measurements of liquid discharge in the North Channel. The width of the North Channel is approximately 12.0 kilometres (30/11/2010). The width of the South Channel was approximately 13.0 kilometres (12/02/2010).

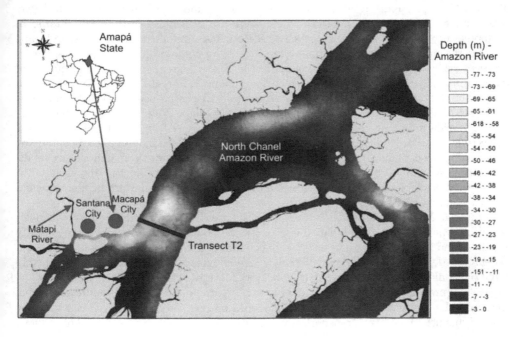

Fig. 1. Features in river sections close to Transect T2 located in the North Channel of the Amazon River – Amapá State (S0 03 32.2  W51 03 47.7)

### 3.1 Methodological approach for discharge measurement in large rivers

Muste & Merwade (2010) describe recent advances in the instrumentation used for investigations of river hydrodynamics and morphology including acoustic methods and remote sensing. These methods are revolutionizing the understanding, description and modelling of flows in natural rivers.

Stone & Hotchkiss (2007) report that accurate field measurements of shallow river flows are needed for many applications including biological research and the development of numerical models. Unfortunately, data quantifying the velocity of river current are difficult to obtain due to the limitations of traditional measurement techniques. These authors comment that the mixing processes and transport of sediment are among the most important impacts on aquatic habitat. The velocity of large rivers is typically measured in either stationary or moving boats with reels or ADCP (Acoustic Doppler Current Profiler).

ADCPs are designed to measure the velocity of the current in a section of a watercourse, producing a velocity profile of the section based on the principle of sound waves from the Doppler effect. This effect is a result of the change in the frequency of the echo (wave) which varies with the motion of the emitting source or reflector. Using this technique, it is possible to measure more accurately net discharge i) in sites and ii) on occasions where the task of measuring flow is more difficult with traditional techniques. At the same time results from ADCP are comparable with results from techniques using traditional methods and can be used to evaluate the qualitative discharge of suspended sediments. In both cases, the technique can be applied in specific monitoring programs (Abdo et al., 1996).

The ADCP has some technical advantages over more traditional techniques (e.g. quantitative net discharge) in places where there are difficulties in applying traditional methods, such as large rivers, during the wet season, discontinuous river sections, and some authors recommend that its use should become  more common in estuaries (Guennec & Strasser, 2009). The main advantages of using ADCP are: a greater quantity and quality of data, improved accuracy (5%); measurements are obtained in real time, with a high rate of reproductibility. The technique for measuring liquid discharge using ADCP technique is also faster is also faster than conventional methods and can be used in large and small water bodies. Furthermore, it requires less effort, does not need alignment, allows for the correction of detours in discrete river sections, and estimating the motion of sediment on the river bed. It also demonstrates a good correlation with the more conventional methods, permitting to obtain of section profiles, river width, flow velocity, the qualitative distribution of suspended sediments, measurement time, boat speed, water temperature and salinity (Guennec & Strasser, 2009).

According to Muste & Merwade (2010) recent advances in instrumentation for the analysis of river flows include the combination of acoustic methods with remote sensing to quantify variables and hydrodynamic and morphological parameters in natural bodies of water, and notably the degree of importance of these new technologies is more evident when applied to large rivers under tidal influence (Abdo et al. 1996; Martoni & Lessa, 1999).

These instruments can be quick and efficient in providing detailed multidimensional measures that contribute to the investigation of complex processes in rivers, especially hydrodynamics, sediment transport, availability of habitats and ecology of aquatic ecosystems.

Muste & Merwade (2010) describe how to quantify the hydrodynamic characteristics and morphology of complex channels. In addition to the ability to extract information that is available through conventional methods in the laboratory, the ADCP and MBES (Multibeam Echosounder) can provide additional information that is critical to the understanding and development of modelling processes in rivers, for example providing a 3D view of river hydrodynamics that was previously unavailable to hydrological studies of large rivers.

A major challenge for studies involving large rivers in the Amazon is the operation of flow meters. For example, the United States Geological Survey (USGS) operates more than 7,000 net discharge monitoring stations in the U.S. These stations provide a near real-time data flow from all stations on the Internet, generating data of water column depth, or the flow stage discharge-curves at each location.

However, a significant challenge for large rivers is the fact that the channel bed, sediment and/or sand banks move location over time. Thus, the discharge-curves need to be updated through regular measurements of the depth, breadth and speed of the river at the monitoring stations. Such difficulties have led to the replacement of conventional techniques with measurements of velocity and net discharge. But at the same time, a vast storage capacity for the data obtained is required, especially when the ADCP data are combined with topography. For example, the data storage requirements have increased from the order of hundreds of thousands to the order of millions of bathymetric and hydrodynamic information points (Muste & Merwade, 2010).

This obstacle requires massive investments in instruments with extraordinary data processing abilities in order to store, group, process and quickly distribute data in a myriad of different formats to fulfil information needs of users. On the other hand, the numerical models developed to accommodate the 3D information of hydrodynamics and bathymetry is only available with the use of intensive techniques like ADCP or MBEs.

Dinerhart & Burau (2005) used the ADCP in the Sacramento River (CA/USA) in diurnal tidal rivers for mapping velocity vectors and indicators of suspended sediment. They observed that in surface waters, the ADCP is particularly useful for quickly measuring the current discharge of large rivers with non-permanent flows, presenting several advantages such as visualisations of time based flows and sediment dynamics in tidal rivers.

Another important parameter in the biogeochemical cycle of aquatic ecosystem is the longitudinal dispersion. The longitudinal dispersion coefficient (D) is an important parameter needed to describe solute transport along river currents (Shen et al., 2010). This parameter is usually estimated with tracers. For economic and logistical reasons, the use of the latter is prohibitive in large rivers.

The same authors argue that these shortcomings can be overcome with the use of ADCP simultaneously with tracers in the stretch of river, by examining the conditions under which both methods produce similar results. Thus, ADCP appears to be an excellent alternative / addition to the traditional tracer based method, provided that care is taken to avoid spurious data in the computation of weighted average distances used in the representation of the average conditions of the river stretch in question.

Stevaux et al. (2009) studied the structure and dynamic of the flow in two large Brazilian rivers (the Ivaí and the Paraná) using eco-bathymetry and ADCP together with samples of suspended sediments. This occurred in two phases of the hydrological cycle (winter and summer). The methodology proved to be valid and easily transferable to other river systems of similar dimensions. For example, at the confluences of river estuaries with complex hydraulic interactions resulting from the integration of two or more different flows, constituting a "competition and interaction" environment. This is because continuous changes occur in flow velocity, discharge and structure, in addition to the changes in the physical and chemical properties of water quality and channel morphology. These dynamic systems are very important in river ecology, reflecting many features and limiting conditions of the environment.

According to Stevaux et al. (2009), from a hydrological perspective, the confluences can be considered as likely sites of turbulence and convergent or divergent movements, forming upward, downward or lateral vortices. These effects generate chaotic motion, generating secondary currents of differing velocities and directions, including some that feedback to the flow main current.

For these authors these dynamics induce the main movement of the sediment formed in the river bed and consequently the main source of variation and alteration in the shape of the channel bed. In this case, highlights the main factors controlling mixing in channel confluences: morphological, such as the confluence angle and asymmetry of the channel bed; and hydraulics, such as momentum and contrast in the flow density. These results also confirmed that the identification and understanding of flow mixing at river confluences is very important in studies of pollution, nutrients, dispersal of dissolved oxygen, and other ecological variables (Rosman, 2007; Bastos, 2010; Cunha, 2008; Pinheiro, 2008). The rate of movement of the flow can be used to determine flow predominance in the confluence.

## 3.2 ADCP Measurements in North Channel

According to ANA (2008) measurements occurred in the section of the channel, located between *Amapá* Coast and island of *Marajó*, North Channel of the Amazon River, with the total time to perform the measurement of the 12 hours and 40 minutes.

Due to the effect of tidal flow and the flow directions of the channel of the Amazon River at its mouth, to determine the actual flow of the river the flow is continuously measurement during a tidal cycle. Or more precisely, it is necessary to perform measurement during a wave variation of flow generated due to influence of the tide.

Considering the peculiarities of the flow measurement under the influence of the tide and large transect of the North Channel of the Amazon River at its mouth (about 11.9 km), the traditional methods of measuring flow in large rivers do not apply to flow measurement in this situation.

The measurements of flow at the mouth or the Amazon were only possible through the development of equipment for flow measurement by Doppler effect (ADCP) due to the drastic reduction of time required for measurement and reduction of risk associated. Calculation of the total duration of measurement: the total length is determinate by measurement the time difference between initial and final wave flow due to the tidal cycle, expressed in seconds.

Since the wave due to the tidal cycle can be represented by a periodic wave, the start and end times are obtained by the abscissas corresponding to the intersection points of the flow versus time curve with a straight parallel to the axis "x" vs Q (discharge), whose first derivative has the same sign, i.e. at points where the curve is increasing (positive derivative) or descending (negative derivative). The calculation of the actual flow in the section is done by determining the area under curve Q x t, which corresponds to the total volume of water that passed through the measurement section during the period of the wave flow, divided by the total time duration. Determination of the flow of the North Channel were performed on different days, the first is the simple sum of three part of effective flow rates measured. The second involves the propagation of waves flow measurements performed by the same reference time, the sum of curves "overlapping" and integrating.

The expected outcome of this type of information is the realisation of regular "in situ" hydrodynamic measurements, to understand the relationship between river hydrodynamics and biogeochemical factors along this key stretch of the Amazon River (Transect T2) near Macapá-AP. The idea is to integrate them to control upstream hydrodynamics and biogeochemistry, as well as to understand how the ecosystem may responds to anthropogenic climate change (Fig. 2).

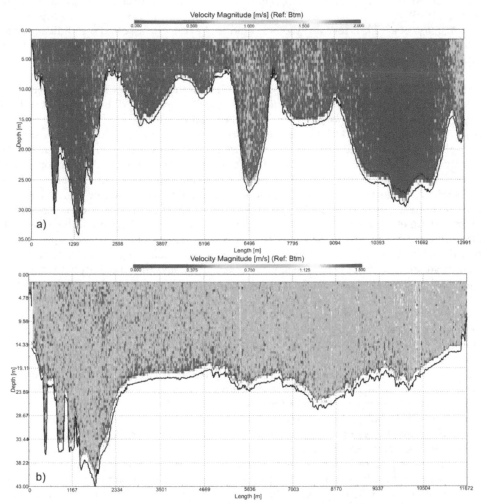

Fig. 2. Net discharge measured with ADCP (600Hz). a) North Channel profile, with
$Q = 1.3 \times 10^5$ m$^3$/s (12/05/2010), b) South Channel, with $Q = 1.2 \times 10^5$ m$^3$/s (12/07/2010).

## 4. Carbon and nutrient biogeochemistry at the Mouth of the Amazon River

Brito (2010) prepared a review based on some of the main studies from the ROCA project
that indicated that the tropical North Atlantic Ocean can be considered a source of
approximately 30 Tg C yr-1 into the atmosphere.

But Subramaniam et al (2008) found a carbon sink of similar magnitude with biologically
measures of approximately 28 Tg C yr-1 from the atmosphere to the ocean, resulting from
nitrogen and phosphorus in the river, in addition to the fixation of $N_2$ in the plume of the
Amazon River. Thus, the Amazon plume reverses the normal oceanic conditions, causing
carbon capture and sequestration of $CO_2$, defined as the net remover of carbon from the

atmosphere to the ocean (Dilling, 2003; Battin et al, 2008; Ducklow et al., 2008; Legendre and Le Freve, 1995).

Subramaniam et al (2008) revealed the importance of symbiotic associations of diazotrophic diatoms (DDAs) in nitrogen fixation in the Amazon plume and showed that the chemical outputs associated with these organisms represent a regionally significant carbon sink. DDAs or other agents of $N_2$ fixation have also been found in other tropical river systems, such as the Nile (Kemp et al., 1999), Congo (AN, 1971), the South of China Sea (Voss et al., 2006) and the Bay of Bengal (Unger et al., 2005), and it is speculated that these have global significance, as a previously neglected biological carbon pump.

These results suggest that techniques used to study inland waterways of the Amazon may be applied to other systems e.g. the Amazon plume. However, knowledge about the magnitude, spatial extent and final destination of this plume is limited. The importance of connections with the processes that occur upstream are also very poorly known. Independent measurements of net community production, diazotrophic production and flow of particles near the surface of the plume agree with the export of carbon (Subramaniam et al., 2008), but the ultimate fate of carbon and nitrogen and the sensibility of the plume front to global climate change are currently unknown.

The microbial community is a driving force behind the processing of material along the Amazon continuum, from land to sea. Cole (2007) suggested that the biosphere should be considered as a metabolically active network of sites that are interconnected by a fluvial network.

Despite indications that the organic carbon derived from soil is resistant to degradation on land, remaining stored for decades or centuries (Battin et al., 2008), once released into aquatic ecosystems, there is evidence that this carbon is dissolved rapidly in the rivers in a matter of days or weeks (Cole & Caraco, 2001).

High levels of $CO_2$ and low $O_2$ concentrations are often found in muddy rivers (Brito, 2010) which suggests that organic carbon derived from the soil represents a substantial carbon source for the heterotrophic network of the river ecosystem (Richey, 1990).

Current knowledge about the diversity and dynamics of bacterioplankton comes almost exclusively from studies of lakes (Crump et al., 1999). Various small rivers have also been sampled (Cottell et al., 2005; Crump & Hobbie, 2005), but from 25 of the world's major rivers, only four had their genetic sequences recorded in the bacterioplankton "Genbank" (National Center for Biotechnology and Information, U.S. National Library of Medicine): the Columbia River, USA (Crump et al., 1999), the Changjiang River, China (Sekigushi et al, 2002), the Mackenzie River, Canada (Galand et al., 2008) and the Paraná river, Brazil (Lemke, 2009).

The lack of information regarding bacterioplankton in large rivers limits understanding of global biogeochemical cycles and the ability to detect community responses to biotic and anthropogenic climate impacts in these critical ecosystems (Crump et al., 1999).

In less turbid areas of the continuum, the process of photosynthesis can reduce or even reverse the $CO_2$ emission rate (Dilling, 2003). Likewise, when the river meets the sea, the loss of suspended sediment increases the penetration of light sufficiently to stimulate marine primary production (Smith & Demaster, 1995). Once light has been removed as a limiting factor nutrients released by river "metabolism" allow phytoplankton blooms, whose community structure is probably dependent on concentrations and ratios of (limiting) nutrients such as nitrogen, phosphorus and iron (Dilling, 2003; Subramaniam et al., 2008).

One of the main unifying conceptual frameworks in biological oceanography is the idea that the structure of the phytoplankton community profoundly affects the export and sequestration of organic material. That is, the biological carbon pump and chemical nutrient cycles (Michaels & Silver, 1988; Wassman, 1988; Peinert et al., 1989; Legendre & Le Fevre, 1995, Ducklow et al., 2001).
Cyanobacteria are recognised as a particularly important group of organisms in the carbon cycle, occurring in a wide variety of ecosystems, especially in aquatic environments. Cyanobacteria can survive in extreme conditions and are found in habitats with wide ranges of temperature, salinity and nitrogen availability (Falconer, 2005). The abundance of cyanobacteria varies seasonally, as a consequence of changes in water temperature and solar radiation as well as weather conditions and nutrient supply (Falconer, 2005). It is known that their distribution it is not homogeneous on the surface or in water column. The distribution of cyanobacteria assemblages may vary depending on gradients such as depth, salinity, temperature, space and seasonality. However, as well as a lack of studies in this area, it is difficult to analyze the dynamics and diversity of planktonic groups such as cyanobacteria, especially when the analytic scale is at the species level, where diversity is high and the river area to be sampled is enormous.

## 4.1 Methods and preliminary biogeochemistry results of transect T2

In the North and the South Channel in Amapá, sampling of water quality was conducted with i) quarterly and ii) in the Channel North monthly frequency (Brito, 2010). Quarterly collections are used to obtain vertically and horizontally integrated samples for the calculation of dissolved and particulate loads in the water column with the use of 15 to 20 metre boats. The monthly samples are used for seasonal interpolation and are obtained from the surface and 60% of water depth. The measures routine collected are used to derive parameters relating to the ion and nutrient system, carbonates, organic material, water discharge and suspended sediment.

The depth and surface samples are obtained by immersion pumps, where water is then pumped into a graduated cylinder of 2 litters, which is flooded for at least three times prior to sampling and the overflow is maintained during the procedures sampling.

Transect T2 (Table 1) defines the main flow of the river to the north of the island of Marajo. As sea water never passes the dividing line at the mouth of the river in front of the city of Macapá (Nikieme et al., 2007), this is considered a final a purely fluvial compounent of the river.

| Point | Local | Coordinates | |
|---|---|---|---|
| Left Bank - North Channel | Amazon | S0 03 32.2 | W51 03 47.7 |
| Middle River - North Channel | Amazon | S0 04 35.9 | W51 01 46.7 |
| Right Bank - North Channel | Amazon | S0 05 01.9 | W51 00 21.9 |
| Left Bank - South Channel | Amazon | S0 09 51.8 | W50 37 48.9 |
| Middle River - South Channel | Amazon | S0 10 43.0 | W50 36 59.4 |
| Right Bank - South Channel | Amazon | S0 11 59.8 | W50 35 59.7 |

Table 1. Geographical coordinates of sampling sites

The graduated cylinder 2 liters are removed aliquots with syringes of 60 mL for routine analysis of chemical parameters Na $^+$, K $^+$, Ca$^{2+}$, Mg$^2$ $^+$, Al, Si, Cl $^-$, SO$_4$$^{2-}$, NO$_2$$^-$, NO$_3$$^-$, NH$_4$$^+$, PO$_4$$^{3-}$, total nitrogen and dissolved organic carbon (DOC). Chemical parameters for the samples are filtered using cellulose acetate filters with pore size 0.45 micrometres in bottles of high density polyethylene capacity of 60 mL containing 6 mg of thymol as a preservative. For DOC, samples are filtered in triplicate using glass fiber filters preheated (500 °C for 5 hours) GFF type with pore size 0.7 mm, in glass vials of 25 mL, also preheated. Samples are preserved with 25 mL of HCl 50%.

In situ measurements (meter used) are: electrical conductivity (Amber Science 2052); pH and temperature (Orion 290A plus), and dissolved oxygen (DO) with the YSI 55 meter.

To fill the collection tube, using the same technique of overflow, ten bottles of glass BOD (Biochemical Oxygen Demand) of 60 mL for measurements of respiration rates are used. Of these, five bottles are preserved with 0.5 mL of manganous sulfate and 0.5 mL of sodium azide and five bottles are incubated in coolers containing water from the river, staying in the dark for 24 hours.

To measure the concentration of suspended sediment in the water, are filled with twenty gallons of polyethylene-liter and transported to the laboratory for further processing to separate the coarse particles (up to 63 μm) of fine particles (between 0 and 63 μm, 45 μm) samples.

The sampling parameters of fecal coliform and Escherichia coli, coliform, sterile bags are made directly from the collector tube connected to the pump.

During the sampling process, a phytoplankton net (63μm mesh) is submerged in the side of the boat to collect coarse sediments. At the end of sampling, the content network is rinsed polycarbonate flask of 250 ml wide mouth and preserved with 250 mL of HCl 50%. To measure the isotopic composition of carbon and nitrogen.

To measure the concentration of dissolved CO$_2$, water is pumped through a closed plexiglass tube in which a small part is filled with air, this air is pumped out of the pipe to a device called a CO$_2$ analyzer IRGA (InfraRed Gas Analyzer ) Licor LI820 model, after passing through the analyzer back to the air pipe to reach the equilibrium (Cole et al, 1994. With modifications). Expected to balance the flow of air, so it made measurements of CO$_2$ dissolved in water for 5 minutes.

The flux measurements of CO$_2$ in surface water are made by the same method, but instead of the plexiglass tube to balance the flow of CO$_2$ it uses a static camera that is floating on the surface of the river connected to tubes to circulate air to the IRGA for 5 minutes.

In addition, samples are collected to characterization of phytoplankton for identification and determination of density.

Table 2 shows the preliminary results of some parameters obtained in the first gathering held in the channel north, following Brito (2010):

With preliminary data observed, we cannot make qualitative and quantitative analysis representative. But we can highlight some features of the river, such as the harmony of the data of surface and depth, the good condition of dissolved oxygen in water, acid pH which is a characteristic of the Amazonian rivers, the presence of considerable amounts of iron in water, low water hardness, the presence of nutrients in the water and the high level of carbon dioxide dissolved in water (mass transfer through air-water interface).

| Station | Depth (m) | Time | Conductivity (µS/cm) | $O_2$ CENA (mg/L) | $O_2$% CENA | $O_2$ UNIFAP (mg/L) | $O_2$% UNIFAP | Temperature (°C) | pH | $CO_2$ Air (ppm) | $CO_2$ water (ppm) | DIC (mg/L) | Alkalinity (mg/L) | Respiration (µM/hour) |
|---|---|---|---|---|---|---|---|---|---|---|---|---|---|---|
| Left Bank Surface | 0.50 | 18:55 | 45.70 | 5.80 | 76.90 | 6.05 | 80.20 | | 6.62 | 477.91 | 2059.30 | 9.02 | 8.98 | 1.223 |
| Left Bank Depth | 15.00 | 18:20 | 46.50 | 5.87 | 77.60 | 6.08 | 78.10 | 30.00 | 6.78 | - | - | 9.06 | 9.17 | |
| Half Surface | 0.50 | 16:10 | 46.60 | 7.13 | 94.30 | 6.09 | 81.40 | 30.80 | 6.95 | - | - | 9.13 | 9.38 | 1.882 |
| Half Depth | 20.00 | 16:30 | 52.10 | 5.06 | 92.10 | | | 31.30 | 6.79 | - | - | 9.25 | 9.35 | |
| Right Bank Surface | 0.50 | 12:00 | 46.70 | 6.10 | 81.40 | 6.00 | 79.70 | 30.50 | 6.91 | 415.88 | 1628.31 | 9.29 | 9.21 | 0.430 |
| Right Margin Depth | 18.00 | 11:40 | 46.70 | 5.87 | 78.20 | 5.75 | 76.30 | 30.30 | 6.84 | - | - | 9.26 | 9.50 | 0.192 |

| Station | Al (mg/L) | Fe (mg/L) | Ca (mg/L) | K (mg/L) | Mg (mg/L) | Na (mg/L) | $Cl^-$ (mg/L) | $SO_4^{2-}$ (mg/L) | Si (mg/L) | $PO_4^{3-}$ (mg/L) | $NH_4^+$ (mg/L) | $NO_3^-$ (mg/L) | $NO_2^-$ (mg/L) | TN (mg/L) | DOC (mg/L) |
|---|---|---|---|---|---|---|---|---|---|---|---|---|---|---|---|
| Left Bank Surface | 0.00 | 0.22 | 3.40 | 0.59 | 0.57 | 0.91 | 3.71 | 3.24 | 3.29 | 0.02 | 0.00 | 0.61 | 0.04 | 0.27 | 4.11 |
| Left Bank Depth | 0.00 | 0.21 | 3.69 | 0.61 | 0.60 | 0.95 | 2.65 | 3.51 | 3.31 | 0.05 | 0.00 | 0.45 | 0.04 | 0.28 | 4.42 |
| Half Surface | 0.00 | 0.22 | 3.66 | 0.64 | 0.59 | 0.90 | 2.11 | 3.73 | 3.31 | 0.04 | 0.02 | 0.61 | 0.03 | 0.26 | 3.69 |
| Half Depth | 0.00 | 0.20 | 3.73 | 0.62 | 0.60 | 0.91 | 2.23 | 4.05 | 3.32 | 0.03 | 0.02 | 0.61 | 0.03 | 0.25 | 4.42 |
| Right Bank Surface | 0.00 | 0.25 | 3.74 | 0.62 | 0.60 | 0.90 | 2.27 | 3.88 | 3.36 | 0.03 | 0.00 | 0.39 | 0.04 | 0.26 | 3.69 |
| Right Margin Depth | 0.00 | 0.11 | 3.73 | 0.62 | 0.60 | 0.91 | 2.34 | 4.08 | 3.27 | 0.03 | 0.00 | 0.65 | 0.04 | 0.25 | 3.95 |

Table 2. Preliminary results and physical-chemical in Transect T2, in the North Channel (two water sampling).

## 5. Numerical modeling

The propagation of the tidal flow in estuaries is a complex free surface problem. It is unsteady oscillatory and therefore may have reversal flow that is not uniform. The equations governing the flow (conservation of mass and movement) are not linear due to friction, the spatial variations velocity and changes in the dimension of the estuary (Gallo, 2004, Cunha, 2008). Thus, to resolve the hydrodynamic in general and the propagation of the tidal estuary, taking into account this complexity, it is necessary to use numerical models to represents the average flow.

According to Versteeg & Malalasekera (1995) a turbulence model can be considered a computational procedure applied to close the system of equations used to represent the average flow, and calculations applicable to a variety of generic problems regarding flow dynamics. The authors state that one of the models most useful in solving the set of equations to be solved for the transport of Reynolds stresses is the $k$-$\varepsilon$ model. The standard $k$-$\varepsilon$ model presents two equations, one for $k$ and one for $\varepsilon$ based on our understanding of the processes that cause relevant changes in these variables. The turbulent kinetic energy $k$ is defined as the variance of velocity fluctuations and $\varepsilon$ is the dissipation of turbulent kinetic energy (the rate at which turbulent kinetic energy is dissipated in the flow). $k$-$\varepsilon$ turbulence model and the SST (Shear Stress Tensor - a more complex variant of the $k$-$\varepsilon$ model) were used respectively for closing the Reynolds equations (Cunha, 2008; Pinheiro & Cunha, 2008). Cunha (2008) and Pinheiro & Cunha (2008) conducted two case studies with numerical models using the software CFX/ANSYS 1) coastal area on the coast of the city of Macapá, and 2) the Matapi River, near the Industrial District of the city of Santana (0° 0'32.53"S, 51°12'7.43"W) (Fig. 3a and 3b). In both cases, the objective was to study the dispersion behaviour of pollutant plumes into surface waters in the estuary of the Lower Amazon and their behaviour during a semi-diurnal tidal cycle.

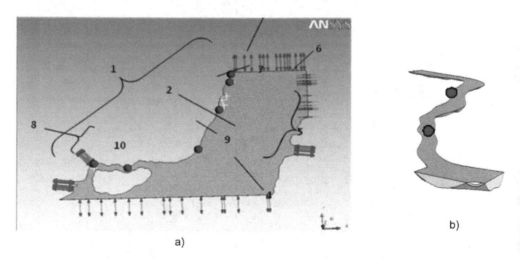

a)                                                                b)

Fig. 3. Pre-processing within a CFX (Computational Fluid Dynamics) study of the dispersion of pollutants in the estuarine zone next to Macapá: a) Macapá and Santana Coast; b) Matapi River (point 8 in Fig. 3a), near the Island and Industrial District of Santana-AP.

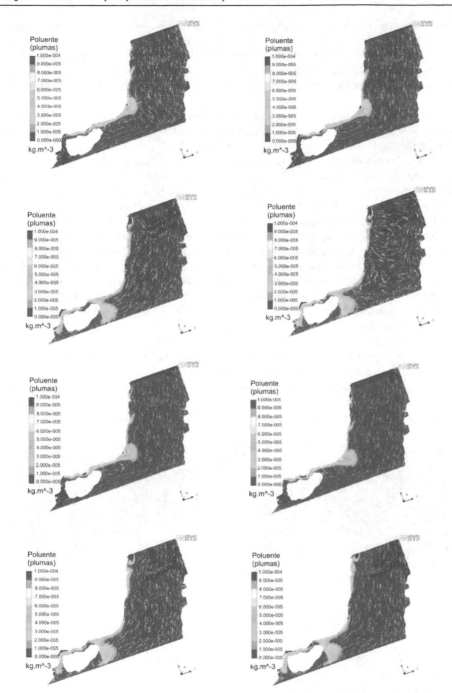

Fig. 4. Velocity and dispersion of pollutants in natural runoff - in the coast of the cities of Macapá and Santana. Representation of a semi-diurnal tidal cycle.

Fig. 3a illustrates the pre-processing step for simulation of pollutant dispersion and demonstrates the complex geometrical configuration required to represent the turbulent flow. There are six continuous pollution sources in the cities of Macapá and Santana, from the mouth of the Matapi River (southern area) to the north of the city, the upper area of the figure. The natural flux of flows passes from the bottom (left) to top (far right, in the north and northeast). The continuous point sources of pollutants are represented by red circles along the coast, which represent the main release points of untreated pollutants into the waters in Macapá and Santana cities (Pinheiro & Cunha, 2008). The same representation occur in Rio Matapi indicated by Fig 3b (Cunha, 2008).

Fig. 4 shows the results of the simulations of pollutant dispersal plumes (light blue and reddish margins) during a tidal cycle. These maps show that the plumes tend to stay close to the shore. From left to right (top row) is the initial phase of a simulated low tide (approximately 7 hours). Again, from left to right (bottom row) begins the high tide phase (approximately 5.5 hours). During the tidal cycle it was possible to simulate the complex interactions between hydrodynamics and a coupled scalar (hypothetical pollutant), with an emphasis on the dynamic plumes between mainland Santana and the island of Santana.

Case study 2 (Fig. 5), shows the phases of the dispersion of pollutant plumes (hypothetical tracer) in the Matapi River during a tidal cycle. The flow pattern (streamlines) changes significantly over a period of the semi-diurnal tide. Simulating the dispersal of pollutants indicates a remarkable complexity in the flow, depending on the geometry of the river channel and the timing of the reversal of the tidal cycle.

In Fig. 5, from left to right depicts changes in pollutant plumes during low tide (approximately 7 hours), during a complete semi-diurnal tidal cycle, where the natural flux of the tide flows from top to bottom. The reverse shows the rising tide.

In Fig. 6, from left to right, there are three different flow fields indicated: a) velocity vectors, b) streamlines (paths of constant speeds), c) dispersion pattern of the scalar from two (hypothetical) continuous point sources of pollutants.

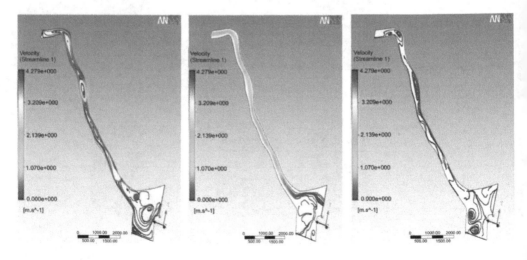

Fig. 5. Lines of transient currents in the Rio Matapi: a) low tide at t = 1h, b) end of the ebb at t = 5.5 h, c) reversal of the tide at t = 6h.

In both case studies, despite the sophistication of the numerical analysis, technical advances such as calibration and validation of models are still necessary. The complexity of the process involving modelling steps, proceedings to investigate the aquatic biogeochemistry and hydrometry of large rivers have yet to beovercome in the estuarine region of Amapá.

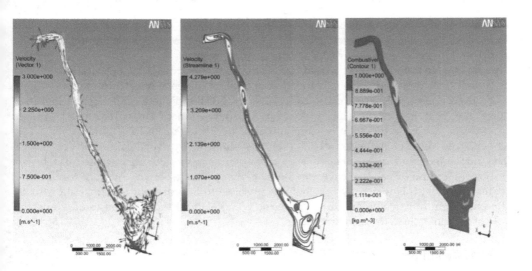

Fig. 6. Low and rising tides. t = 360 min (6.0 h). a) Velocity; b) Stream line; b) Plume Concentration.

## 6. Conclusions

The main conclusions of this research are:

In the estuarine region of the Lower Amazon River, in the state of Amapá, the measurement of net discharges of large tidal rivers is only feasible with the use of devices such as ADCP to integrate hydrodynamic processes and water quality variables (biogeochemical cycle and interaction between the plume of the Amazon River and Atlantic Ocean).

Relevant hydrodynamic parameters such as velocity profiles, stress and identification of background turbulent flow velocity components need to be determined with the aid of modern equipment whose operation must be efficient and economic for hydrometric quantification in complex estuarine environments.

Bathymetric analyses, at the scales of interest, have been a difficult hurdle to overcome because of the intricate system of channels in the Amazon River.

The logistics required for experimental studies in large rivers is a major obstacle that has inhibited research interest in this poorly studied area.

A major challenge to be overcome in systematic studies of water quality parameters is the generation of local physical parameters, such as rating curves, rates of sedimentation and resuspension of sediments, etc, which are a fundamental input for complex numerical models of water quality.

The modelling of water quality in the Amazon estuary is complex due to the absence and / or inadequacy of data describing different physical characteristics. The drainage system imposes enormous difficulties in this area. An example is the absence of long-term time series to obtain necessary parameters and coefficients to build numerical models.

Hydrodynamic simulations of flow and dispersion of pollutant plumes released into the environment are difficult to implement, and require calibration and verification with local data that are not always available.

Existing techniques in numerical modelling can become strong allies in informing public policy and the management of regional water resources.

Among the parameters of interest from numerical models, the generation of 3D computational meshes is potentially the most important source of novel information.

The development of local expertise constitutes one of the biggest challenges in the area, since the best and most efficient option for development of experimental studies in hydrodynamics and computer simulation, is the formation of local human resources. The main advantages are lower operating costs for complex experimental campaigns.

The implementation of a database accessible to the interested user would also be an important technological challenge for the systematic studies of the hydrodynamics and water quality at this region. Thus, would be possible to improve our understanding about the ecosystem functioning and to evaluate the complexity of the Amazon estuary and the role of carbon cycle in these environments.

## 7. Acknowledgements

This research is part of the ROCA project (River-Ocean Continuum of the Amazon), funded by the Gordon and Betty Moore Foundation. The main collaborating institutions are the Center for Nuclear Energy in Agriculture - CENA / USP, the National Institute of Amazonian Research - INPA, the Museu Paraense Emilío Goeldi - MPEG, Federal University of Pará - UFPA and the Federal University of Amapá - UNIFAP (Postgraduate Program in Tropical Biodiversity / PPGBIO and the Laboratories of Simulation and Modeling, Chemistry and Environmental Sanitation / Undergraduate Program in Environmental Sciences). This study was supported by the National Research Council (CNPq/MCT – Productivity Fellowship (Process N. 305657/2009-7), and by the Institute of Scientific and Technologic Research of the Amapá State (NHMET & CPAC/IEPA) and by the Tropical Biodiversity Postgraduate Program (PPGBio) of the Federal University of Amapá (UNIFAP), Brazil; SUDAM Project - "Integrated Network Management for Monitoring and Environmental Dynamics Hydroclimatic State of Amapá"; REMAM II Project/FINEP/CNPq – Extremes Hydrometeorology and Climate Phenomena; and Project CENBAM-FINEP/CNPq/FAPEAM/UNIFAP - INCT "Center for Integrated Studies of Amazonian Biodiversity".

## 8. References

Abdo, J. M. M; Benevides, V. F. S; Carvalho, D. C. & Oliveira, E. *III Curso Internacional sobre Técnicas de Medição de Descarga Líquida em Grandes Rios*. Relatórios de Trabalhos de Campo. Ministério de Minas e Energia/Secretaria de Energia. Departamento Nacional de Águas e Energia Elétrica. Coordenação de Recursos Hídricos. Manaus/Manacapuru. Agosto, 1996.

An, C. N. Atlantic Ocean Phytoplankton South of the Gulf of Guinea on Profiles Along 11 and 14S. *Oceanology 6* , 896-901, 1971.

ANA (Agência Nacional de Águas). Resultados das Medições de Vazão Realizadas com ADCP na Foz do Amazonas em Junho de 2008. Nota Técnica Conjunta nº 1/2008/NHI/SAR Documento n° 15.847/2008. 09 de julho de 2008.

Bastos, A. M. Modelagem de escoamento ambiental como subsídio à gestão de ecossistemas aquáticos no Baixo Igarapé da Fortaleza. Dissertação (Mestrado em Biodiversidade Tropical), Universidade Federal do Amapá, Macapá, 133f, 2010.

Battin, T. J., Kaplan, L. A., Findlay, S., Hopkinson, C. S., Martí, E., & Packman, A. I. Biophysical controls on organic carbon fluxes in fluvial networks. *Nat Geosci 1* , 95-100, 2008.

Brito, D. C. Biogeoquímica e ciclos de carbono e nutrientes na desembocadura do Rio Amazonas– AP. Projeto de Qualificação Doutorado em Biodiversidade Tropical. PPGBIO - Universidade Federal do Amapá – UNIFAP. Macapá, 40 P, 2010.

Cole, J. J., & Caraco, N. F. Carbon in catchments: connecting terrestrial carbon losses with aquatic metabolism. *Mar. Freshwater Res. 52* , 101-110, 2001.

Cole, J. Plumbing the global carbon cycle: Integrating inland waters into the terrestrial carbon budget. *Ecosystems, 10(1)* , 172-185, 2007.

Cottrell, M. T., Waidner, L. A., Yu, L. Y., e Kirchman, D. L. Bacterial diversity of metagenomic and PCR libraries from the Delaware River. *Environ. Microbiol. 7(12)* , 1883-1895, 2005.

Crump, B. C., & Hobbie, J. E. Synchrony and seasonality of bacterioplankton communities in two temperate rivers. *Limnol. Oceanogr. 50(6)* , 1718-1729, 2005.

Crump, B. C., Armbrust, E. V., & Barross, J. A. Phylogenetic analysis of particle-attached and free-living bacterial communities in the Columbia River, its estuary, and the adjacent coastal ocean. *Appl. Environ. Microbiol. 65(7)* , 3192-3204, 1999.

Cunha, A. C. Moviment of pollutants in the environment: experimental and numerical modeling of the surface flow on Matapi River Mouth. *Amapa/Brazil. Thesis/Pos-doc. AWU, CA/EUA* , 111, 2008.

Cunha, A. C., Brasil, Jr., Coelho, J. G., Floury, C. & Sousa, M. "Estudo numérico do escoamento superficial na foz do Rio Matapi – costa interna estuarina do Amapá". In: Proceedings of the 11th Brazilian Congress of Thermal Sciences and Engineering - ENCIT, paper CIT-0977. Braz. Soc. of Mechanical Sciences and Engineering - ABCM, Curitiba, Brazil, Dec. 5-8, 2006.

Dilling, L. The role of carbon cycle observations and knowledge in carbon management. *Annual Review of Environment and Resources 28* , 521-558, 2003.

Dinerhart, R. L. & Burau, J.R. Repeteated surveys by acoustic Doppler current profiler for flow and sediment dynamics in a tidal river. Journal of Hydrology. 1-21, 2005.

Ducklow, H. W., Steinberg, D. K., & Buesseler, K. O. Upper ocean carbon export and the biological pump. *Oceanography 14* , 50-58, 2001.

Falconer, I. R. Cyanobacterial Toxins of Drinking Water Supplies: Cylindrospermopsins and Microcystins. Florida: CRC Press, Boca Raton, 2005.

Galand, P. E., Lovejoy, C., Pouliot, J., Garneau, M. E., & Vicent, W. F. Microbial community diversity and heterotrophic production in a coastal Arctic ecosystem: A stamukhi lake and its source waters. *Limnol. Oceanogr. 53(2)*, 813-823, 2008.

Gallo, M. N. A influência da vazão fluvial sobre a propagação da maré no estuário do rio amazonas. Dissertação de Mestrado em Engenharia de Ciências Oceânicas. COPPE/UFRJ, Rio de Janeiro – RJ. 99 p, 2004.

Guennec, B E. & Strasser, M. A. O Transporte de sedimento em suspensão no rio Amazonas – o papel do leito médio e maior na avaliação da capacidade de transporte dos sedimentos finos em regime não-permanente. XVII Simpósio Brasileiro de Recursos Hídricos, 2009.

Hu, J.; Li, S. & Geng, B. Modeling the mass flux budgets of water and suspended sediments for the river network and estuary in the Pearl River Delta, China. Journal of Marine Systems. May, 2011.

Kemp, A. E., Pearce, R. B., Koisume, I., Pike, J., & Rance, S. J. The role of mat-forming diatoms in the formation of Mediterranean sapropels. *Nature 398*, 57-61, 1999.

Legendre, L. & Le Freve, J. Microbial food webs and the export of biogenic carbon in the oceans. *Aquatic Microbial Ecology 9*, 69-77, 1995.

Lemke, M. J., Description of Freshwater Bacterial Assemblages from the Upper Parana River Floodpulse System, Brazil. *Microb. Ecol. 57(1)*, 94-103, 2009.

Martoni, A. M. & Lessa, R. C. Modelagem hidrodinâmica do canal do rio Paraná, trecho Porto São José – Porto 18. Parte II: Calibragem do Modelo. Acta Scientiarum 21(4): 961-970, 1999.

Michaels, A. F., & Silver, M. W. Primary production, sinking fluxes and the microbial food web. *Deep Sea Research Part I 35*, 473-490, 1988.

Muste, M, Kim D., and Merwade V., Modern Digital Instruments and Techniques for Hydrodynamic and Morphologic Characterization of Streams, Chapter 7 in Gravel Bed Rivers 7: Developments in Earth Surface Processes, edited by Ashmore P., Bergeron N., Biron P., Buffin-Bélanger T., Church M., Rennie C. Roy A.M., Wiley, New York, NY, 2010

Nikiema, O., Devenon, J. L., & Baklouti, M. Numerical modeling of the Amazon River plume. *Continental Shelf Research 27*, 873-899, 2007.

Peinert, R., Bodungen, B. V., & Smetacek, V. S. Food web structure and loss rate. In: W. H. Berger, V. S. Smetacek, e G. Wefer, *Productivity of the Ocean: Present and Past* (pp. 35-48). New York: John Wiley and Sons Limited, 1989.

Pinheiro, L. A. R. & Cunha, A. C. "Desenvolvimento de Modelos Numéricos Aplicados à Dispersão de Poluentes na Água sob Influência de Marés Próximas de Macapá e Santana-AP". Relatório de Iniciação Científica, SETEC, Macapá – AP, 2008.

Pinheiro. L. A. R. "Simulação Computacional Aplicada à Dispersão de Poluentes e Analises de Riscos à Captação de Água na Orla de Macapá-AP". Pesquisa & Iniciação Científica Amapá, Macapá, v.1, n. 4, p. 58-61, 2º semestre, 2008.

Richey, J. E., Hedges, J. I., Devol, A. H., & Quay, P. D. Biogeochemistry of carbon in the Amazon River. *Limnology and Oceanography 35*, 352-371, 1990.

Richey, J. E., Meade, R. H., Salati, E., Devol, A. H., Nordin, C. F., & Santos, U. Water discharge and suspended sediment concentrations in the Amazon River: 1982-1984. *Water Resour. Res. 22* , 756-764, 1986.

Rosman, P. C. Referência Técnica do Sistema Base de Hidrodinâmica Ambiental. Versão 11/12/2007. 211 p. COPPE/UFRJ. Rio de Janeiro. 2007.

Sekiguchi, H., Watanabe, M., Nakahara, T., Xu, B. H., & Uchiyama, H. Succession of bacterial community structure along the Changjiang River determined by denaturing gradient gel electrophoresis and clone library analysis. *Appl. Environ. Microbiol*, 181-188, 2002.

Shen, C; Niu, J. Anderson, E. J; & Phanikumar, M. S. Estimating longitudinal dispersion in rivers using Acoustic Doppler Current Profilers. Advances in Water Resources. 33. 615-623, 2010.

Silva, M. S.; Kosuth, P. Comportamento das vazões do rio Matapi em 27.10.2000. Congresso da Associação Brasileira de Estudos do Quaternário, 8. Imbé-RS. Resumos, ABEQUA, p. 594-596. 2001.

Smith Jr, W. O. & DeMaster, D. J. Phytoplankton biomass and productivity in the Amazon River plume: correlation with seasonal river discharge. *Continental Shelf Research 6* , 227-244, 1995.

Souza, E. B. & Cunha, A. C. Climatologia de precipitação no Amapá e mecanismos climáticos de grande escala. In: A. C. Cunha, E. B. Souza, e H. F. Cunha, *Tempo, clima e recursos hídricos: Resultados do projeto REMETAP no Estado do Amapá* (pp. 177-195). Macapá-AP: IEPA, 2010.

Souza, E. B. Precipitação sazonal sobre a Amazônia Oriental no período chuvoso: observações e simulações regionais com o RegCM3. *Revista Brasileira de Meteorologia* , v. 24, n. 2, 111 – 124, 2009.

Stevaux, J. C; Franco, A. A; Etchebehere, M. L. C & Fujita, R. H. Flow structure and dynamics in large tropical river confluence: example of the Ivaí and Paraná rivers, southern Brazil. Geociências. São Paulo. V.28, n.1, p. 5-13, 2009.

Stone, M. C. & Hotchkiss, R. H. Evaluating velocity measurement techniques in shallow streams. *Journal of Hydraulic Research*. Vol. 45, No. 6, pp. 752-762, 2007.

Subramaniam, A., Yager, P. L., Carpenter, E. J., Mahaffey, C., Biörkman, K. & Cooley, S. Amazon River enhances diazotrophy and carbon sequestration in the tropical North Atlantic Ocean. *Science. 105(30)* , pp. 10460-10465, 2008.

Unger, D., Ittekkot, V., Schafer, P. & Tiemann, J. Biogeochemistry of particulate organic matter from the Bay of Bengal as discernible from hydrolysable neutral carbohydrates and amino acids. *Marine Chemistry 96* , 155, 2005.

Van Maren, D. S. & Hoekstra, P. Seasonal variation of hydrodynamics and sediment dynamics in a shallow subtropical estuary: the Ba Lat River, Vietnam. Estuarine, Coastal and Shelf Science. 60, 529e540, 2004.

Versteeg, H. K. & Malalasekera, W. An introduction to computational fluid dynamics: the finite volume method. Prentice Hall. 257p. 1995.

Voss, M., Bombar, D., Loick, N. & Dippner, J. Riverine influence on nitrogen fixation in the upwelling region off Vietnam, South China Sea. *Geophysical Research Letters, 33* , L07604, doi:10.1029/2005GL025569, 2006.

Wasman, P. Retention versus export food chains: processes controlling sinking loss from marine pelagic systems. *Hydrobiologia 363* , 29-57, 1988.

# Part 2

# Tidal and Wave Dynamics: Seas and Oceans

# 6

# Freshwater Dispersion Plume in the Sea: Dynamic Description and Case Study

Renata Archetti and Maurizio Mancini
*DICAM University of Bologna, Bologna*
*Italy*

## 1. Introduction

An interesting mesoscale feature of continental and shelf sea is the plumes produced by the continuous discharge of fresh water from a coastal buoyancy source (rivers, estuarine or channel).

The general spreading of freshwater plume depends on a large number of factors: tide, out flowing discharge, wind, local bathymetry, Coriolis acceleration, inlet width and depth.

The discharge of freshwater from coastal sources drives an important coastal dynamic, with significant gradients of salinity. These phenomena are highly dynamic and have several effects on the coastal zone, such as reducing salinity, changing continuously the vertical profiles and distribution of parameters, such as dissolved matters, pollutants and nutrients (Jouanneau & Latouche, 1982; Fichez et al., 1992; Grimes & Kingford, 1996; Duran et al., 2002; Froidefond et al., 1998; Broche et al., 1998; Mestres et al., 2003; Mestres et al., 2007).

As a result of these effects, several classification schemes based on simple plume properties have been proposed in an attempt to predict the overall shape and scale of plumes. Kourafalou et al. (1996) classified plumes as supercritical and subcritical, according to the ratio between the outflow and the shear velocity. Yankovsky and Chapman (1997) derived two length scales based on outflow properties (velocity, depth and density anomaly) and used them to discriminate between bottom-advected, intermediate and surface-advected plumes, depending on the vertical and horizontal density gradients near the plume front; in spite of the absence of external forcing mechanisms in their theory, they correctly predicted the plume type for several numerical and real cases. Garvine (1987) classified plumes as supercritical or subcritical using the ratio of horizontal discharge velocity to internal wave phase speed and he later proposed (Garvine, 1995) a classification system based on bulk properties of the buoyant discharge.

Referenced plume studies present numerical modelling of the case, in situ observations (Sherwin et al., 1997; Warrick & Stevens, 2011; Ogston et al., 2000), satellite observations (Di Giacomo et al., 2004; Nezlin and Di Giacomo, 2005; Molleri et al., 2010) or aerial photographs (Figueiredo da Silva et al., 2002; Burrage et al., 2008). In several cases two techniques are coupled (O'Donnell, 1990; Stumpf et al., 1993; Froidefond et al., 1998; Siegel et al., 1999).

In this work a freshwater dispersion by a canal harbour into open sea is described in depth with the aim of a 3D numerical model and with the validation of in situ measurements carried out with innovative instruments. The measurements appear in literature for the first time. The investigated area relates to the coastal zone near Cesenatico (Adriatic Sea, Italy). The aim of this chapter is to describe the dynamic of freshwater dispersion and to show the results of the simulation of flushing, mixing and dispersion of discharged freshwater from the harbour channel mouth under different forcing conditions.

The chapter will be organized in the following sections:

- Introduction with focus on research and works dealing with the modelling of freshwater dispersion plumes in the sea and their comparison with existing data;
- Description of the numerical model;
- Physical features of the case study;
- Plume modelling on Cesenatico (Italy) discharging area;
- Validation of model results with in situ measurement campaigns;
- Conclusions.

## 2. The numerical model

The numerical model is based on motion and continuity equations (Liu & Leendertse, 1978) tested by the authors. The model utilizes a Liu and Leendertse's scheme describing vertical water motion by calculation of the turbulence field. The model equations for conservation of momentum and continuity, written in Cartesian coordinates, in incompressible fluid conditions and under the effects of Earth's rotation are:

$$\frac{\partial u}{\partial t} + \frac{\partial(uu)}{\partial x} + \frac{\partial(uv)}{\partial y} + \frac{\partial(uw)}{\partial z} - f \cdot v + \frac{1}{\rho}\frac{\partial p}{\partial x} - \frac{1}{\rho}\left(\frac{\partial \sigma_x}{\partial x} + \frac{\partial \tau_{xy}}{\partial y} + \frac{\partial \tau_{xz}}{\partial z}\right) = 0 \qquad (1)$$

$$\sigma_x = A_x \frac{\partial u}{\partial x}; \qquad \tau_{xy} = A_x \frac{\partial u}{\partial y} \qquad (2)$$

$$\frac{\partial v}{\partial t} + \frac{\partial(vu)}{\partial x} + \frac{\partial(vv)}{\partial y} + \frac{\partial(vw)}{\partial z} + f \cdot u + \frac{1}{\rho}\frac{\partial p}{\partial y} - \frac{1}{\rho}\left(\frac{\partial \tau_{yx}}{\partial x} + \frac{\partial \sigma_y}{\partial y} + \frac{\partial \tau_{yz}}{\partial z}\right) = 0 \qquad (3)$$

$$\sigma_y = A_y \frac{\partial v}{\partial y}; \qquad \tau_{yx} = A_y \frac{\partial v}{\partial y} \qquad (4)$$

$$\frac{\partial w}{\partial t} + \frac{\partial(wu)}{\partial x} + \frac{\partial(wv)}{\partial y} + \frac{\partial(ww)}{\partial z} + \frac{1}{\rho}\frac{\partial p}{\partial z} - \frac{1}{\rho}\left(\frac{\partial \tau_{zx}}{\partial x} + \frac{\partial \tau_{zy}}{\partial y} + \frac{\partial \sigma_z}{\partial z}\right) + g = 0 \qquad (5)$$

$$\frac{\partial u}{\partial x} + \frac{\partial v}{dy} + \frac{\partial w}{dz} = 0 \qquad (6)$$

where $t$ denotes time, $x$, $y$ and $z$ are Cartesian coordinates (positive towards Est, South, up) $u$, $v$, $w$ denote velocity components in the direction of $x$, $y$, $z$, $f$ is the Coriolis parameter (assumed to be a constant), $g$ is the acceleration due to gravity, $\rho$ is the density of water, $A$ is the horizontal eddy viscosity $\sigma_i$, $\tau_{ij}$ with i, j = x, y, z components of Reynolds tensor proportional to vertical gradient of velocity. In accordance with other authors it was assumed as adequate the use of two-dimensional depth-integrated equations for conservation of mass and momentum for a typically well-mixed water column due to wind and tidal stirring. So the vertical momentum equation has been substituted by baroclinic pressure equation (8) where sea water density has been formulated according with the international thermodynamic equation of sea water based on the empirical state function of UNESCO81 which links density to Salinity, Temperature and Pressure:

$$\frac{\partial S}{\partial t} + \frac{\partial(Su)}{\partial x} + \frac{\partial(Sv)}{\partial y} + \frac{\partial(Sw)}{\partial z} - \frac{\partial}{\partial x}\left[D_x \frac{\partial S}{\partial x}\right] - \frac{\partial}{\partial y}\left[D_y \frac{\partial S}{\partial y}\right] - \frac{\partial}{\partial z}\left[k \frac{\partial S}{\partial z}\right] = 0 \tag{7}$$

$$\frac{\partial p}{\partial z} + \rho(S,T)g = 0 \tag{8}$$

$$\frac{\partial T}{\partial t} + \frac{\partial(Tu)}{\partial x} + \frac{\partial(Tv)}{\partial y} + \frac{\partial(Tw)}{\partial z} - \frac{\partial}{\partial x}\left[D_x \frac{\partial T}{\partial x}\right] - \frac{\partial}{\partial y}\left[D_y \frac{\partial T}{\partial y}\right] - \frac{\partial}{\partial z}\left[k^1 \frac{\partial T}{\partial z}\right] = 0 \tag{9}$$

S and T are salinity and temperature, respectively. $D_x$, $D_Y$ are horizontal eddy diffusivities for S and T; k and k' are vertical diffusion coefficients for mass and wheat. For vertical balance an E coefficient of vertical exchange is introduced for momentum which relates vertical Reynolds forces to the vertical gradient of horizontal components of velocities and expressed by Kolmogorov e Prandtl as:

$$E = \rho L \sqrt{e} \exp^{(-mRi)} \tag{10}$$

$$Ri = -\frac{g}{\rho e}\frac{\partial p}{\partial z}L^2 \tag{11}$$

$$L = \chi z \sqrt{1 - \frac{z}{d}} \tag{12}$$

$$\tau_{xz} = E_x \frac{\partial u}{\partial z} \tag{13}$$

$$\tau_{yz} = E_y \frac{\partial v}{\partial z} \tag{14}$$

where m is a numeric parameter and $Ri$ is the Richardson number in terms of vertical gradient of density ($\rho$) and of eddy kinetic energy (e) and L defined, according with Von Kàrman, by $\chi$ (numerical coefficient) and d distance of bottom (z=0) from surface (z=d). Vertical exchange coefficients for mass (k) and heat (k') are defined similarly to E coefficients using adequate parameters substitutive of $\rho$ and m. The adopted vertical scheme introduces eddy kinetic energy as a state function which requires its own dynamic equation for balance and conservation:

$$\frac{\partial e}{\partial t} + \frac{\partial(eu)}{\partial x} + \frac{\partial(ev)}{\partial y} + \frac{\partial(ew)}{\partial z} - \frac{\partial}{\partial x}\left[D_x \frac{\partial e}{\partial x}\right] - \frac{\partial}{\partial y}\left[D_y \frac{\partial e}{\partial y}\right] - \frac{\partial}{\partial z}\left[E_e \frac{\partial e}{\partial z}\right] + S - D_e = 0 \tag{15}$$

with horizontal and vertical exchange eddy diffusivities $D_x$ $D_y$ and $E_e$ were defined in a similar way to exchange mass coefficients. So energy sourcing into the grid is detected in the function of strain tensions induced by vertical velocity and the term of energy dissipated by shear stress at the bottom is calculated from energy in flux direction in lower level S and Chézy shear coefficient C. In the surface layer a wind effect generating turbulence by waves is considered in the higher part of the water column. Supposing constant motion for the sea, $Et$ is the total energy generated for surface units in the function of wind velocity

$$S = L\sqrt{e}\left(\frac{\partial \overline{u}}{\partial z}\right)^2 \tag{16}$$

$$D_e = a_2 \frac{e^{3/2}}{L} \tag{17}$$

$$S = g\frac{U^3}{C^2} \tag{18}$$

$$E_t = 5.610^{-9} u_w^4 \tag{19}$$

where $u$ is the mean velocity, $D_e$ the dissipation coefficient, $a_2$ a constant parameter taking account of energy transfer from high to small turbulence conditions and $uw$ the wind velocity.

The model equations are discretized and solved into a finite-difference formulation.

## 3. Physical features of the case study

The study site of the present survey is the pulmonary system of Cesenatico canal harbour (Northern Adriatic Sea, Italy) and the near coastal zone (Mancini, 2009). An aerial view of the canal harbour in Cesenatico is shown in Fig. 1A. During summer and in dry weather conditions the main part of discharged freshwater comes from waste water treatment plants (WWTP) and from drainage pipe systems. Treated and untreated wastewaters reach the canal with high hourly variations and the sea outlet provides regulated discharge into sea according to unsteady tidal flow. Generally the harbour canal, having a very low ground slope, guarantees a good thermoaline turbulent mixing, for the entire water column, if the basin structure presents a high pulmonary surface/wastewater loading ratio. Small estuaries with higher ground slope, receiving sea water by tidal oscillations within a few hundred metres from the coast-line and high flow rate coming from WWTP, present flow primarily directed to the sea where freshwater continuously flows on the surface. Cesenatico canal harbour basin reveals a condition where the pulmonary area and freshwater input from inland (300.000 EI, Equivalent Inhabitants) produce in the canal mouth vertical profiles of velocity, turbulent thermoaline mixing and depth of the overflowing layer daily and hourly varying in function of tidal type and phase (Bragadin et al., 2009). The investigated area is characterized by a low constant slope of the bottom and starts from the canal harbour mouth towards to a 2000 m radius. The near coastal zone is characterized by submerged breakwaters in a northerly direction and by emerged breakwaters in a southerly direction. The discharging plume area and its vertical profiles of thermoaline and quality parameters appear strongly conditioned by external waves and currents also in dry weather conditions. Recent monitoring investigations (Mancini, 2009) reveal vertical parameter profiles at mouth varying from an initial uniform vertical profile, also in correspondence with low tidal outgoing ranges. During these conditions, reinforced afternoon wind generates significant sea waves, which oppose the surface freshwater flow. On the contrary, during nightly major outgoing tidal phases, strong stratification is maintained in the mouth zone, as in the dispersion plume area facing the breakwaters. Morphologic, hydraulic and water quality measurements have been executed into the transition estuary of the harbour canal and near the mouth of adjacent breakwaters (a view of the breakwater is shown in Fig. 2B).

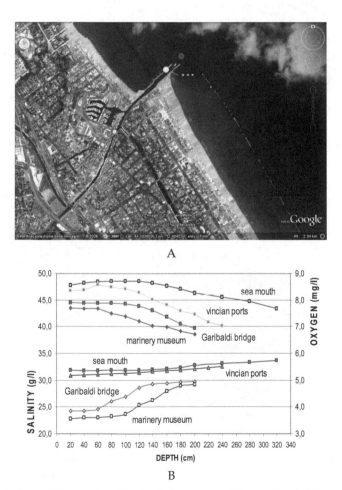

A

B

Fig. 1. A) Aerial view of Cesenatico. The bullets represent the position of the sample in Fig. 1B: from the city centre to the sea, respectively: Mariner Museum, Garibaldi Bridge, Vincian Ports and sea mouth. B) Salinity and oxygen profiles inside the harbour canal in the position plotted in nearby Fig. 1A.

Other research activities, carried out on freshwater-sea water balance for the volumes of channels of the Cesenatico Port Canal system, (Mancini, 2008) indicated that wastewater, discharged in the internal zone, has a hydraulic retention time before sea dispersion, ranging between 1 and 3 days. As a consequence, there is a freshwater storage in the most internal parts of the channel during incoming tidal phases; on the other side during the outgoing tidal phases, there is the outflow of the main part of the internal storage freshwater. Fig. 2B shows monitoring data in correspondence with an internal channel section at 4 Km from the mouth. The selected section balance between saline waters introduced by tidal oscillations combined with wastewater flux incoming from inland maintains the salinity range within a 0-3 g/l. The figure also plots oxygen and redox, showing similar behaviour.

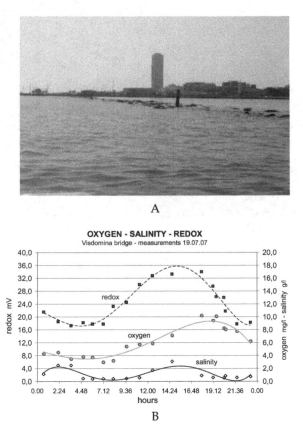

Fig. 2. A) Cesenatico northern coastal area characterized by submerged breakwaters. B) Daily behaviour of physical-chemical parameters measured in the internal section of the channel limiting transition volumes.

Before the outfall these outgoing volumes flow through the historical tract of the harbour which presents depths varying between 3 to 5 m. In this tract a typical overflowing volume upon the static higher density deep layers (Fig. 1B) can be observed. The Marinery Museum and Garibaldi Bridge are located in the canal, approx. 500 m from the mouth, the Vincian Ports are located 50 m from the sea mouth. In the last tract close to the mouth a partial mixing with sea water is permitted, that increases salinity, oxygen and pH, according to the external sea conditions.

## 4. Plume modelling on Cesenatico (Italy) discharging area

The hydrodynamic model described above has been used to simulate the evolution of the plume originating from the freshwater discharge from the harbour canal. The bathymetry and the geometry of the breakwaters and structures were modelled by a small mesh dimension. The regular mesh, shown in Fig. 3, was made by 144x170 grid points, with cell dimensions 12 m x 12 m. On the same figure it is possible to recognize the shoreline, the harbour canal (points P2, P3 and P4) and coastal defence structures, parallel to the shoreline. The points on the same Fig. 3 are the profile points where measurements, later described, have been carried out.

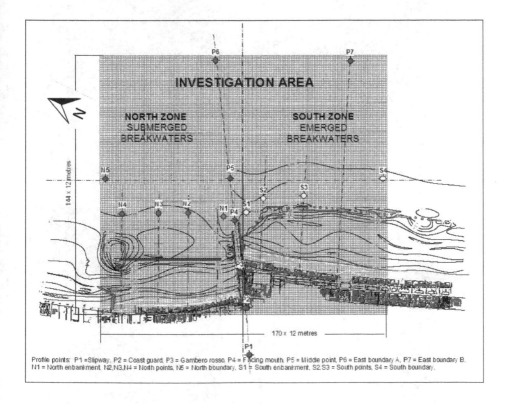

Fig. 3. Investigation area, calculation mesh utilized in model simulations and location of the fixed investigated points.

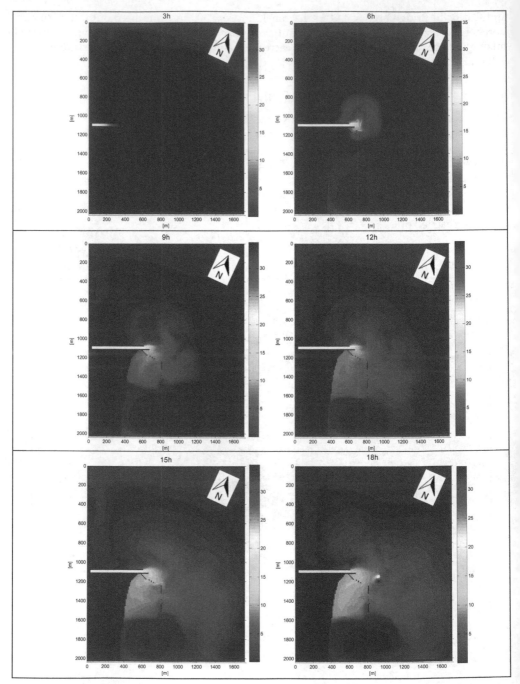

Fig. 4A. Example of simulation results of the freshwater plume dispersion in different tidal phases plotted in Fig. 4B.

A typical summer condition is shown in Fig. 4, the hydrodynamic and dispersion is forced by the freshwater outflow and by the tidal excursion at the offshore boundary. Unfortunately field data are not available for this condition, but only for different scenarios commented in the later sections.

Fig. 4A presents the results of a simulation carried out in the absence of coastal surface current and wind velocity lower than 1 knot. Simulation conditions are representative of the cycle of freshwater outfall in which tide, according with internal basin storage volumes, provides outgoing velocity from the channel mouth starting from 10.00 a.m. and ending 18 hours later at 4.00 a.m. The physical feature of the presented simulation is characterized by a first low decreasing tidal phase and low outgoing velocity typical of the last summer periods. The tidal excursion at several tidal phases is shown in Fig. 4B.

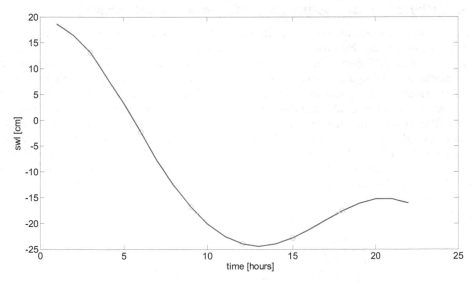

Fig. 4B. Sea water level at the offshore boundary during simulation with results in Fig. 3.

Here, in the early afternoon, variations in salinity and phytoplankton biomass are limited and restricted to the near mouth area and the surface thermoaline profile could be conditioned by wind coastal waves. Evening and nightly scenarios show static conditions for coastal sea with very low current and undefined direction, while the most part of freshwater accumulated in the internal basin is outfalled from the mouth according to the maximum tidal decreasing phase. Thermoaline stratification is guaranteed, such as in internal harbour section as in the receiver coastal sea. The simulation period shown in Fig. 4 (12h-15h-18h) covers the main decreasing tidal phase, when most freshwater, coming from WWTP and confined into the internal channel according with tidal phase, is completely discharged through the harbour canal. Evident stratification conditions are represented in coastal sea away from the breakwaters, such as in the north and south zones. The maximum decrease in sea salinity concentrations is evaluated in 7-8 g/l within the south breakwater confined shore area near the south embankment. In this zone, water volumes flowing through restricted breakwater mouths permit higher incoming surface velocity and low depth permits near the beach vertical mixing and a more homogeneous areal distribution.

The results also reveal different effects on plume areal dispersion and on thermoaline profiles between zones confined by continuous breakwaters (north shore) and by discontinuous breakwater (south shore). Comparing salinity vertical distribution in internal and external points of the north continue breakwaters, under a surface layer (50-60 cm) almost corresponding to breakwater submergence (Lamberti et al., 2005), differences in salinity and oxygen profiles become significant. Freshwater dispersion appears obstructed in the internal north confined area because continuous breakwaters produce a "wall effect" for incoming plume with mass exchange reduced for deep layers. Here, in the absence of north directed sea currents, flows are allowed only from north-south boundary mouths with vertical mixing limited to the surface layer.

## 5. Validation of model results with in situ measurement campaigns

In 2009 several field campaigns took place in order to observe the hydrodynamics at the outfall, to measure the velocities of the flow and the water quality parameters in order to validate the model. The measurements were performed with the support of a Bellingardo 550 motorboat utilizing a Geo-nav 6sun GPS system, a Navman 4431 ultrasonic transducer and an YSI556 multi-parameter probe. Morphologic, hydraulic and water quality measurements were executed into the transition estuary of the harbour canal and near the mouth. The dispersion area and profile distribution of freshwater outgoing from the harbour mouth and discharged in the coastal area was investigated and monitored. Experiments were carried out on June 2009 and September 2009. The surface currents were observed with the aim of drifters properly designed to follow the surface pollution and oil (Archetti, 2009). The drifters (Fig. 5) were equipped with a GPS to acquire the geographical position every 5 minutes and an IRIDIUM satellite system was used to send data to a server. Simultaneously, tide, waves, wind and rainfall conditions were collected.

Fig. 5. Lagrangian drifter in the sea during the experiment.

### 5.1 Experiment I: June 18, 2009.
The first experiment was carried out on June 18, 2009. The wave conditions were measured by the wave buoy located 5 nautical miles off the shore of Cesenatico (details on the wave position and data are available at http://www.arpa.emr.it/sim/?mare/boa). The significant

wave height $H_S$ was lower than 0.3 m for the whole day. The measured sea water level and wave conditions on the day of the experiment are plotted in Fig. 6A. The weather conditions were very mild, without wind and with ascending tide, so we had the opportunity to monitor a condition driven only by the tidal excursion. Figure 6B shows the swl during the experiment and the contemporary velocity and direction of the drifters launched 1 km offshore from the Cesenatico harbour canal.

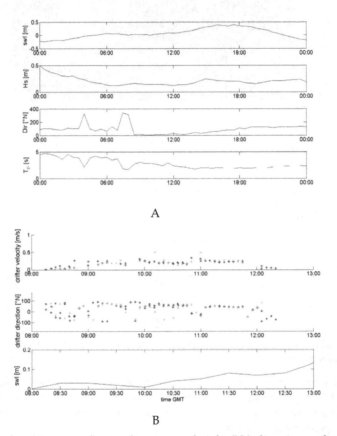

Fig. 6. A) Measured swl (top panel), significant wave height ($H_S$), direction and period ($T_P$). B), drifters' velocity (top panel), direction (central panel) and contemporary swl (bottom panel).

Clusters of three drifters were launched simultaneously at the offshore boundary. The launch position of the drifters is the offshore location in Fig. 7A. The first cluster was launched at about 9:00 a.m. just offshore from the harbour breakwaters, at a distance of 1.2 km from the beach, the second cluster was launched one hour later offshore from the northern beach and the last cluster was launched at 11:00 am offshore from the southern beach. The velocity and direction of the drifters during the experiment is plotted in Fig. 6B. The mean drifter velocity during the experiments was 0.18 m/s, with a direction perpendicular to the beach.

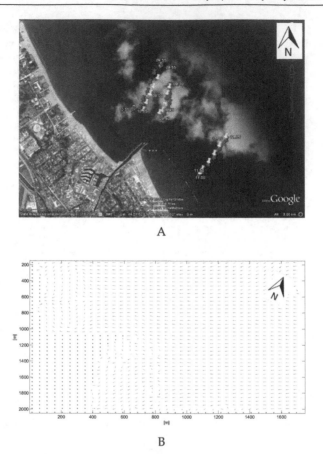

Fig. 7. A) Satellite view of the study area and pattern of two drifters launched on June 18, 2009. B). Field for experiment I of surface currents.

The observed condition was simulated by the model; the hydrodynamic was driven only by sea water tidal oscillation at the offshore boundary condition (condition in Fig. 6A ). The resulting surface current field during the experiment condition is shown in Fig. 7B, the current is perpendicular to the shoreline.

The field velocity appears comparable to the drifters' paths, both in direction and magnitude, so the model looks well calibrated.

### 5.2 Experiment II September 1, 2009

During the experiment carried out on September 1, 2009, the drifters were launched in the water in a plume of sewage water disposal from the canal of Cesenatico harbour. Two drifters were deployed in the plume centre and two at the plume front. The two drifters deployed at the plume front followed the plume front evolution during the experiment lasting 4 hours. Wind speed was approx. 30 m/s, significant wave height 0.5 m (Fig. 8A) and the tide descending. The plume and the drifters moved in the wind direction at an average speed of 0.2 m/s (Fig. 8B).

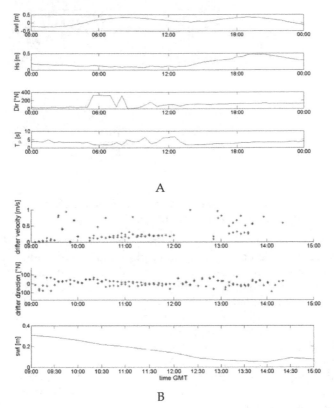

A

B

Fig. 8. A) Measured swl (top panel), significant wave height ($H_S$), direction and period ($T_P$). B) Drifters' velocity (top panel), direction (central panel) and contemporary swl (bottom panel).

A

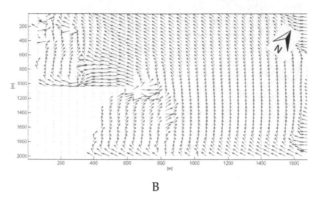

B

Fig. 9. A) Satellite view of the study area and pattern of three drifters launched on September 1, 2009. B). Surface currents' field for experiment II.

Differently from the previous examined condition, we observe here that the drifters' paths are north deviated by the action of the wind on the surface layer with higher velocity (Fig. 9A). The reorientation of the trajectory increases when the drifters approach the coast. Similar behaviour is observed in the hydrodynamic simulation results (Fig. 9B).

The observed and simulated effect is the result of the composition of the marine current driven by tidal oscillation, together with surface wind effect. The described condition is typical in summer in the final hours of the morning.

A model validation was also carried out by comparing simulated and observed salinity vertical profiles into the plume at section N3 during experiment II. The comparison (Fig. 10) shows a good agreement between observed and simulated values also in the vertical profiles. A more extensive comparison of vertical profiles with other parameters and at other sections will be performed in the future.

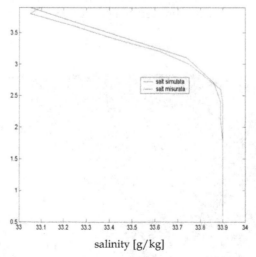

salinity [g/kg]

Fig. 10. Vertical salinity behaviour: observed in point N3 (red) and simulated by the model (blu).

During the experiments the presence of biological aggregates and foams was observed on the sea surface interested by the plume (Fig. 11). The presence of biological traces in sea areas interested by freshwater dispersion is a well known phenomenon. In a few cases bacterial and dead algae aggregate come directly from internal channels where variation in water depth provides alternance of photosynthetic and bacterial activity. Here, high aerobic biomass levels are produced by bacterial synthesis sustained by the production of photosynthetic oxygen of high growing algae populations. When oxygen, dissolved during light hours, cannot supply nightly bacteria/algal demand, the water column is interested by the presence of many species of died organic substances with the associated settling and floating phenomena. Production of biological foams can occur also when variations in salinity concentrations increase the mortality of a phytoplankton population growth in a low salinity environment. In these cases, foam presence is often registered in the last part of the harbour canal, near the sea mouth, and upon the plume boundary of the sea outfalled plume.

Two vertical profiles of temperature (Fig. 12A), dissolved oxygen, pH, (Fig. 12B) redox potential and salinity concentrations (Fig. 12A) were registered and analysed "on site" in order to check the main plume direction. Fixed investigated points are N1 and S1 focused as representing the north and south near the sea mouth area (see reference map in Fig. 2). Parameters are traced with reference to profile P6 at fixed points located on the east boundary in front of the harbour canal and chosen as indicators of offshore sea conditions.

No appreciable variations on salinity vertical distribution are registered in the south zone, where measured values appear very similar in S1 (south near mouth) and P6 (offshore sea). On the contrary, N1 vertical profile presents a salinity distribution which reveals the arrival in the surface layers of volumes coming from the mouth section enriched by internal freshwater. A difference of 2 g/l between bottom and surface layers with thermocline from depth of 60 to 120 cm is registered. Similarly, temperature does not show vertical variations in the south zone, even if media values appear lower in coastal rather than offshore sea water (26.5 °C) according with the cooling effects produced in September by internal water volumes. This is confirmed by the N1 temperature profile which presents lower values in surface layers (25.6°C) than in the underlying thermocline (26.4 °C) but inversion does not interrupt stratification which is maintained by variation in density. Similar temperature values in N1 and S1 points are registered within the thermocline thickness. At thermocline depths a temperature decrease is appreciable due to the colder masses stored at the bottom of the harbour canal.

N1, N2, N3 points, interested by the dispersion plume, show a pH vertical profile similar to temperature profile. Low pH values usually indicate biological organic substance degradation or nitrification phenomena typically active in waters of internal channels receiving wastewater. In N1 near the mouth point, higher values are confined in a 1 metre thickness layer, sited at a 1 metre depth. On this layer, lower pH values confirm the presence of a plume conditioned by freshwater also indicated by lower temperature.

Fig. 13 and Fig. 14 show the sequence of profiles obtained following the plume trajectory starting from P1 (internal point corresponding to the slipway) towards to N5 external point placed on the north boundary investigation area. As expected, freshwater volumes are progressively mixed with external high salinity volumes proceeding from internal to external sections. Vertical profiles of salinity behaviour at P1, P2, P3 internal points show that freshwater plume interests a 2 metre depth surface layer. At the last internal section (Gambero rosso), turbulence realizes a linear decrease on salinity concentration from 34 g/l at 2 m depth to 31 g/l at the surface. This layer overflows upon an almost static high salinity volume placed at the bottom channel. Both P4 and N1 external profiles indicate clear stratification conditions with a 60 cm floating layer. Here, wastewater presence is appreciable and thermocline is located into the underlying 60 cm. Measured salinity surface values together with behaviour of vertical profiles allows the identification of an area interested by plume dispersion limited to a northerly direction by N3 fixed investigation point. Similar profiles at points N4 and N5 reveal that in experiment tidal and currents conditions are typical of offshore sea water volumes.

Fig. 11. View of the floating biological foams observed on the north plume boundary during the September 1, 2009 experiments. Photo taken from the N3 position (see Fig. 2) beach oriented.

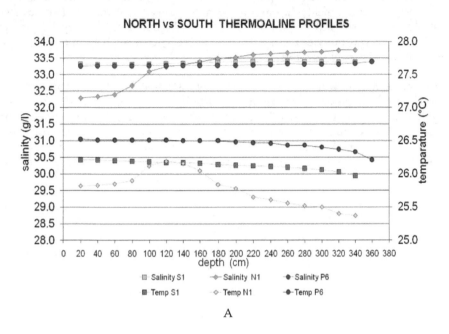

A

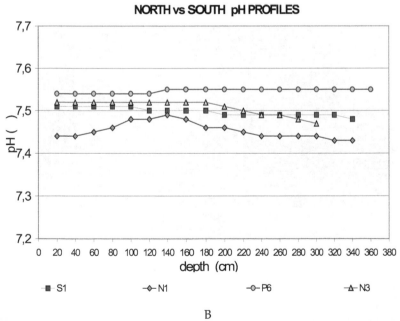

B

Fig. 12. A) Thermoaline and B) pH profiles at the beginning of the experiment at sections S1, N1, N3, P6 (see Fig.2).

The sequence of temperature profiles (Fig. 14) reveals very similar vertical trends and values among all profile sections inside the harbour canal (sections P1, P2 and P3). Perhaps a small effect of the external sea water's warmer mass could be noted in the deeper layers at P3 section sited in the proximity of the mouth. Excluding a 40cm sea bottom layer, all points' indicators of dispersion plume area present temperature values lower at surface (N1). As just reported in Fig. 12's comments on comparison of N1 and S1 thermoaline profiles, this initial thermal inversion which does not yet allow a stratification break, confirms salinity indications about plume areal extension. N5 profile, located at the northern boundary investigation area and not interested by colder freshwater coming from the internal basin, maintains a classic summer temperature profile for Adriatic coastal sea. In this case we observe a 26.4 °C constant temperature in a 120 cm depth surface layer, a thermocline to a depth of 240 cm and another 1 metre bottom layer with a constant temperature of 25.2 °C.

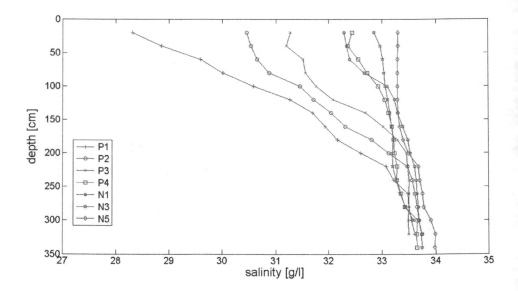

Fig. 13. Vertical profiles of salinity measured at the profile points during the experiment conducted on 1 September, 2009.

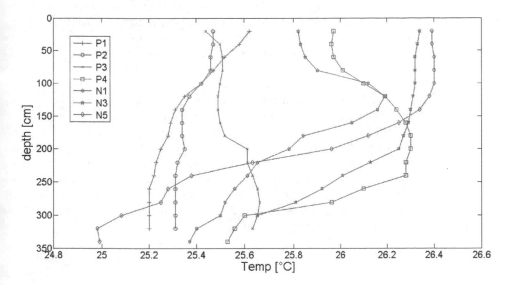

Fig. 14. Vertical profiles of temperature measured at the profile points during experiment conducted on 1 September, 2009.

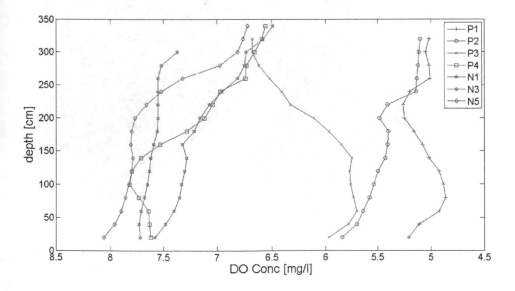

Fig. 15. Vertical profiles of dissolved oxygen at the profile points during experiment conducted on 1 September, 2009.

As expected, oxygen values averaged at each section (Fig. 15) increase, proceeding from internal to external points. At P1 and P2 profiles, photosynthesis produces maximum values in a 60 cm surface layer. At the P3 point (internal but near the mouth), a strong influence of external sea water on bottom layers is confirmed, which shows the same oxygen value, while at surface layers values are typical of internal waters. No information about plume dispersion could be obtained at external points where oxygen distribution is characterised by classic coastal sea profiles with oxygen decreasing values in the direction of deeper layers where photosynthesis is low and bacterial consumption increases.

Results of simulated salinity concentration (Fig. 16), similar to those presented in Fig. 4B, indicate a northerly oriented freshwater dispersion, different from the case analysed in Fig. 4B, which presents in the first phases a less oriented dispersion plume and during the following times (hour 15 – 18) a prevailing orientation to the southern coastal zone. In the actual case, the plume is west bounded by the continuous breakwaters, this means that the geometry is well reproduced in the model, and is dispersed to the north, for the effect of the wind, which was negligible in the previous examined condition.

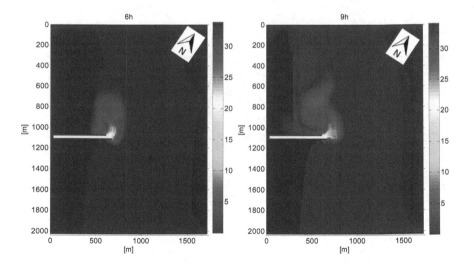

Fig. 16. Simulation of the freshwater plume dispersion during experiment II.

## 6. Conclusions

A freshwater dispersion plume in the sea has been described in depth in the present paper with the aim of producing a 3D numerical model and with the validation of two field campaigns carried out in different conditions. The investigated area concerns the coastal zone near Cesenatico (Adriatic Sea, Italy). The fresh water is dispersed by the canal harbour mouth into the open sea.

The model shows good performance in the application here presented, which is characterised by the presence of complex sea structures, requiring a very detailed and small mesh dimension in the geometry description.

Field data were acquired during two field campaigns and are of different typology: surface lagrangian paths, acquired by innovative properly designed drifters (in both campaigns); vertical profiles of temperature and salinity and dissolved oxygen acquired by a multiparameter probe in properly defined fixed points (in the second campaign). During the first campaign the hydrodynamic was driven only by the tidal oscillation and during the second also by surface wind, the tested conditions were therefore different and interesting for understanding the complex dynamics.

Comparison between model results and measurements are good for the surface hydrodynamic description and for the areal and vertical distribution of concentration, in particular, the resulting salinity values compared with experimental data have shown a surprisingly good agreement.

During the second experiment the presence of biological aggregates and foams was observed on the sea surface interested by the plume. The presence of biological traces in sea areas interested by freshwater dispersion is a well known phenomenon.

Vertical measurement of thermoaline parameters shows appreciable variations on salinity vertical distribution in the southern zone, where measured values appear very similar in the south near mouth and offshore sea. On the contrary, at the northern zone the vertical profiles present a salinitydistribution which reveals the arrival in the surface layers of volumes coming from the mouth section enriched by internal freshwater. A difference of 2 g/1 between bottom and surface layers with thermocline from depth of 60 to 120 cm is registered. Similar behaviour was observed for temperature. In fact in the north the temperature profile presents lower values in surface layers (25.6°C) than in the underlying thermocline (26.4 °C), but inversion does not interrupt stratification which is maintained by variation in density. At thermocline depths a temperature decrease is appreciable due to the colder masses stored at the bottom of the harbour canal.

The points, interested by the dispersion plume, showed a pH vertical profile similar to temperature profile. Low pH values usually indicate biological organic substance degradation or nitrification phenomena typically active in waters of internal channels receiving wastewater. In N1 near the mouth point, higher values are confined in a 1 metre thickness layer, sited at a 1 metre depth. On this layer lower pH values confirm the presence of a plume conditioned by freshwater also indicated by a lower temperature.

The methodology proposed in this paper appears to be useful and accurate enough to simulate the dynamics of the freshwater dispersion at the investigated scale.

The results here presented are original and have allowed a general comprehension of the thermoaline and hydrodynamic assessment of the dispersion area.

The model now validated can in the future be applied to investigate the dispersion in other meteo climatic conditions, tides and other canal mouth geometries.

## 7. Acknowledgements

Authors are grateful to CIRI Edilizia e Costruzioni, UO Fluidodinamica for the financial support.

## 8. References

Archetti R. (2009). Design of surface drifter for the oil spill monitoring. *REVUE PARALIA. Coastal and Maritime Mediterranean Conference.* Hammamet, Tunisie (2009). 2-5 Dec. 2009. vol. 1, pp. 231 - 234 http://www.paralia.fr/cmcm/hammamet-2009.htm.

Bragadin G.L., Mancini M.L., Turchetto A. (2009). Wastewater discharge by estuarine transition flow and thermoaline conditioning in shore habitat near Cesenatico breakwaters. *Proceeding of the Fifth International Conference on Coastal Structures. Coastal Structures 2007*-Venice.July 2-4, 2007 (vol.II, pp.1101-1112).

Broche P. Devenon J.L., Forget P., Maistre C., Naudin J. and Cauwet G. (1998). Experimental study of the Rhone plume. Part I. Physics and dynamics. *Oceanol Acta. 21, 725-738. ISSN: 0399-1784.*

Burrage D., J. Wesson, C. Martinez, T. Pérez, O. Möller Jr., A. Piola. (2008). Patos Lagoon outflow within the Río de la Plata plume using an airborne salinity mapper: Observing an embedded plume. *Continental Shelf Research*, 28 (13), 1625-1638. ISSN: 0278-4343.

Di Giacomo P. M., Washburn L., Holt B. and Jones B. H. (2004). Coastal pollution hazards in southern California observed by SAR imagery: stormwater plumes, wastewater plumes and natural hydrocarbon seeps. *Marine Pollution Bulletin* 49 (2004) 1013–1024. ISSN 0025-326X.

Duran N., Fiandrino A., Fraunié P., Ouillon S., Forget P. and Naudin J. (2002). Suspended matter dispersion in the Ebro ROFI: an integrated approach. *Continental Shelf Research* 22, 267-284. ISSN 0278-4343.

Fichez R., Jickells T. D. and Edmunds H. M. (1992). Algal blooms in high turbidity, a result of the conflicting consequences of turbulence on nutrient cycling in a shallow water estuary. *Estuary Coast Shelf Sci.* 35. 577 – 593. ISSN 0272-7714.

Figueiredo da Silva F., Duck R.W., Hopkins T.S. and Anderson J.M. (2002). Nearshore circulation revealed by wastewater discharge from a submarine outfall, Aveiro Coast, Portugal. *Hydrology and Earth System Sciences* 6(6), 983–989 (2002). ISSN: 1027-5606.

Froidefond J.M., Jegou A.M., Hermida J., Lazure P., Castaing P. (1998). Variability of the Gironde turbid plume by remote sensing. Effect of climate factor. *Oceanol Acta.* 21: 191-207. ISSN: 0399-1784.

Garvine R.W. (1987). Estuarine plumes and fronts in shelf waters: a layer model. *Journal of Physical Oceanography* 17 (1987), 1877–1896. ISSN: 0022-3670.

Garvine, R.W. (1995). A dynamical system for classifying buoyant coastal discharges. *Continental Shelf Research* 15 (13) (1995), pp. 1585–1596. ISSN: 0278-4343.

Grimes C. and Kingford M. (1996). How do riverine plumes of different sizes influence fish larvae: do they enhance recruitment? *Marine Freshwater Research.* 47, 191-208. ISSN: 1323-1650.

Jouanneau J. M. and Latouche C. (1982). Estimation of fluxes to the ocean from mega tidal estuaries under moderate climates and the problems they present. *Hydrobiologia*, 91. 23:29. ISSN: 0018-8158.

Kourafalou V.H., Lee T.N., Oey L.-Y. and Wang J.D. (1996). The fate of river discharge on the continental shelf, 2. Transport of coastal low-salinity waters under realistic wind and tidal forcing. *Journal of Geophysical Research* 101 (1996), 3435–3455. ISSN 0148-0227.

Lamberti A., Archetti R., Kramer M., Paphitis D., Mosso C., Di Risio M. (2005). European experience of low crested structures for coastal management. *Coastal Engineering*. Vol. 52.(10-11), 841 - 866 ISSN 0378-3839.

Liu S.K. and Leendertse J.J. (1978). Multidimensional numerical modelling of estuaries and coastal seas, *Advances in Hydroscience*, Vol. 11, Academic Press, New York (USA), 1978. ISSN: 0065-2768.

Mestres M., Sierra J.P., Sánchez-Arcilla A. (2007). Factors influencing the spreading of a low-discharge river plume. *Continental Shelf Research*, 27, (16-15), 2116-2134. ISSN: 0278-4343.

Mancini M.L. (2008). Wastewater finishing by combined algal and bacterial biomass in a tidal flow channel. Modeling and field experiences in Cesenatico. *International Symposium on Sanitary and Environmental Engineering-SIDISA 08 -Proceedings, ROMA, ANDIS*, 2008, pp. 50/1 - 50/8

Mancini M.L. (2009). Wastewater discharge and thermoaline conditioning in south Cesenatico (I) coastal area near breakwaters. *Proceedings of the Ninth International Conference on the Mediterranean Coastal Environment.-MEDCOAST 09*. Sochi-Russia. 10-14 November 2009. vol. 1, 143/1 - 143/7.

Mestres M., Sierra J.P., Sanchez Arcilla, A., Del Rio, J.G., Wolf T., Rodriguez A. and Ouillon S. (2003). Modelling of the Ebro River plume. Validation with field observations. *Scientia Marina*. 67 (4). 379 – 391. ISSN 0214-8358.

Molleri G. S. F., De M. Novo E. M. L., Kampel M. (2010). Space-time variability of the Amazon River plume based on satellite ocean color. Continental Shelf Research 30 (3-4). 342-352. ISSN: 0278-4343.

Nezlin G.P and DiGiacomo P.M. Satellite ocean color observations of stormwater runoff plumes along the San Pedro Shelf (southern California) during 1997–2003. (2005). *Continental Shelf Research*, 25,(14), 1692-1711. ISSN: 0278-4343.

Ogston A. S., Cacchione D. A., Sternberg R. W., Kineke G. C. (2000). Observations of storm and river flood-driven sediment transport on the northern California continental shelf. *Continental Shelf Research*, 20, (16), 2141-2162.

O'Donnell J. (1990). The formation and fate of a river plume: a numerical model. *J. Phys. Oceanogr*. 20, 551-569. ISSN: 1520-0485.

Sherwin T. J., Jonas P. J. C. and Sharp C. Subduction and dispersion of a buoyant effluent plume in a stratified English bay. *Marine Pollution Bulletin*, Vol. 34, No. 10, 827-839, 1997. ISSN 0025-326X.

Siegel H., Gerth M. and Mutze A. (1999). Dynamics of the Oder River plume in the southern Baltic Sea: Satellite data and numerical modelling. *Continental Shelf Research* 19 (1143 – 1159). ISSN: 0278-4343.

Stumpf R.P., Gefelbaum G. and Pennock J.R. (1993). Wind and tidal forcing of a buoyant plume, Mobile Bay, Alabama. *Continental Shelf Research* 13, 1281-1301. ISSN: 0278-4343.

Yankovsky E. and Chapman D.C. (1997). A simple theory for the fate of buoyant coastal discharges. *Journal of Physical Oceanography* 27 (1997), 1386–1401 ISSN: 1520-0485.

Warrick J. A. and Stevens A. W. (2011). A buoyant plume adjacent to a headland — observations of the Elwha River plume. *Continental Shelf Research*, 31,85-97. ISSN: 0278-4343.

# 7

# Numerical Modeling of the Ocean Circulation: From Process Studies to Operational Forecasting – The Mediterranean Example

Steve Brenner

*Department of Geography and Environment, Bar Ilan University*
*Israel*

## 1. Introduction

The Earth is often referred to as the water planet, although water accounts for only 0.023% of the mass of the planet. Nevertheless, water is found mainly at or near the surface and in the atmosphere and therefore is a very prominent planetary feature when viewed from space. Water as a substance appears in all three physical phases – solid, liquid, and gas. Under the present day climatic conditions, ice is found mainly in the polar regions, at latitudes north of 60°N and south of 60°S. Liquid water is found in the hydrosphere which includes the oceans, marginal seas, lakes, and rivers. The oceans cover nearly 70% of the surface of the Earth, with an average depth of ~ 4000 m. Water vapor, the gaseous phase, appears in the atmosphere and accounts for up to 4% of the mass. The hydrologic cycle describes the continuing transfer of water among these three components. All three forms of water also play important roles in the climate system. Water vapor is the main absorber of infrared radiation and therefore is a major contributor to the greenhouse effect. Clouds and ice are the major factors that determine the albedo of the Earth and therefore are mostly responsible for the reflection of approximately 30% of the incoming solar radiation. The specific heat capacity of water is nearly four times that of air and therefore the oceans serve as a major heat reservoir and regulator of the climate system. Furthermore ocean currents are responsible for more than one third of the heat transport from the equator to the poles and therefore affect the horizontal temperature gradients in the atmosphere which are closely linked to the development of major weather systems on various temporal and spatial scales.

The oceans also serve as a major source of food and natural resources and are important for commerce and transportation. For hundreds and perhaps even thousands of years, mariners intuitively understood some of the salient features of the surface circulation in the most highly traversed parts of the ocean. In 1770 Benjamin Franklin and Timothy Folger published the first map of the Gulf Stream, the major ocean current that flows northward along the east coast of North America and then turns northeastward and flows across the North Atlantic Ocean. The purpose of this map was to help mail ships sailing from Europe to North America to avoid this current and thereby shorten the duration of their trip. Yet despite the interest in and the importance of the oceans, oceanography as a formal science is relatively young, being only slightly more than a century old. In the early years, it was primarily a descriptive science based on sparse and scattered observations. The quantitative aspects of physical and dynamical oceanography saw a major breakthrough with the publication of Henry Stommel's seminal work on the North Atlantic circulation (Stommel, 1948). With a simple mathematical model of the wind driven circulation he was able to elegantly explain the phenomenon of westward intensification (i.e., the formation of strong western boundary currents such as the Gulf Stream) as a result of the meridional variation of the Coriolis force.

The idea of using numerical models to further expand the understanding of the intricacies and complexities of the ocean circulation was introduced nearly twenty years later in the pioneering work of Bryan & Cox (1967). As with Stommel's research, they too investigated the circulation of the North Atlantic Ocean which at the time was the most highly observed ocean basin. The purpose of the model was to solve an initial value problem based on a simplified version of the Navier-Stokes equations. Through their model they were able to study the interaction between the wind driven and the thermohaline components of the circulation. Their work drew heavily from the experience of numerical weather prediction which took nearly thirty years to develop the capability of producing skillful forecasts beginning with Richardson's (1922) original concept but unsuccessful attempt and continuing to Charney et al. (1950) producing the first successful 24 hr forecast. As computational capabilities have increased exponentially over the past thirty years, so too has ocean modeling developed from a tool for simplified and focused process studies to fully operational forecasting systems. In this sense, the distinction between process studies (or simulations) and a forecasting system can be explained as follows. In the former, the goal is to understand the physical basis of the process without regard to reproducing specific details at any particular instant in time. In the latter, attention is focused on being able to produce the most accurate simulation of a particular realization of the flow at a specific time. The development of models for process studies and simulations was a necessary step in the development of forecasting systems. Furthermore, the useful range of a forecast, which is closely related to the limit of predictability, is limited by the chaotic behavior of the fluid flow. One the other hand, longer term simulations for the projection of future climate change is perhaps the most common example today of a process study. In both modern process studies and forecasting systems, the initial focus of model development has been the circulation, but today major progress has been made in developing components for simulating and predicting the fundamental biogeochemical processes of the oceanic ecosystem as well.

The goal of this chapter is to present an overview of modern ocean modeling as a tool for basic research as well as for operational forecasting. Considering the rapid developments and extensive experience of the Mediterranean oceanographic research community from recent years, we will use the Mediterranean as the prototype to explain and demonstrate these capabilities and successes in ocean modeling.

## 2. The governing equations and basic ocean dynamics

In order to fully appreciate the role and importance of numerical ocean models, it is helpful to first understand some of the basic dynamics of the ocean circulation. Mathematically, investigating the ocean circulation can be considered as solving an initial boundary value problem described by the Navier Stokes equations. These form a set of nonlinear, partial differential equations which describes the motion of any Newtonian fluid. The core of this is

Numerical Modeling of the Ocean Circulation: From Process Studies to Operational Forecasting – The
Mediterranean Example

137

the three dimensional equation for the conservation of momentum which is essentially an expression of Newton's second law of motion. The two fundamental forces that must be considered are the pressure gradient force and gravity. For geophysical fluids, rotation of the Earth is also important and therefore Coriolis force must also be added to the equations. To complete the description of the motion equations for mass conservation (continuity) and for the conservation of internal energy must also be added. The latter can be expressed in terms of density or in terms of temperature and salinity. To make these equations more tractable and directly applicable to the ocean circulation, various simplifications and approximations are applied. These simplifications are usually based on a scale analysis of the various terms in the equations. The two most common approximations are: (1) the vertical extent or depth of the fluid layer is much smaller than the horizontal scale of motion, and (2) the Boussinesq approximation in which the density variations are assumed to be small compared to the mean value and are therefore neglected except in the buoyancy term of the equation. As a result of the first approximation, the vertical component of the conservation of momentum can be reduced to a diagnostic equation for hydrostatic balance (i.e., the vertical component of the pressure gradient force exactly balances gravity or the weight of the fluid). The second approximation, which is roughly equivalent to assuming that seawater is incompressible, means that mass continuity can be reduced to a diagnostic equation for the conservation of volume (i.e., three dimensional nondivergence). The final set of the governing equations (usually referred to as the primitive equations) in Cartesian coordinates (x, y, z), includes seven equations as follows:

*Horizontal momentum*

$$\frac{\partial u}{\partial t} + u\frac{\partial u}{\partial x} + v\frac{\partial u}{\partial y} + w\frac{\partial u}{\partial z} - fv = -\frac{1}{\rho_0}\frac{\partial p}{\partial x} + DIFF(u) \tag{1}$$

$$\frac{\partial v}{\partial t} + u\frac{\partial v}{\partial x} + v\frac{\partial v}{\partial y} + w\frac{\partial v}{\partial z} + fu = -\frac{1}{\rho_0}\frac{\partial p}{\partial y} + DIFF(v) \tag{2}$$

Where u, v, w are the velocity components in the x, y, z directions, t is time, $\rho$ is the density (the subscript 0 indicates the mean value), $f = 2\Omega sin\varphi$ is the Coriolis parameter ($\Omega$ is the rotation rate of the Earth and $\varphi$ is the latitude), p is the pressure, and $DIFF(\psi)$ is the diffusion given by $DIFF(\psi) = \frac{\partial}{\partial x}\left(A_h\frac{\partial\psi}{\partial x}\right) + \frac{\partial}{\partial y}\left(A_h\frac{\partial\psi}{\partial y}\right) + \frac{\partial}{\partial z}\left(A_z\frac{\partial\psi}{\partial z}\right)$, where $A_h$ and $A_z$ are the horizontal and vertical diffusion coefficients, respectively;

*Vertical momentum* (hydrostatic equation)

$$\frac{\partial p}{\partial z} = -\rho g \tag{3}$$

Where g is gravity;

*Mass continuity*

$$\frac{\partial u}{\partial x} + \frac{\partial v}{\partial y} + \frac{\partial w}{\partial z} = 0 \tag{4}$$

**Conservation of internal energy** (can be written in terms of density or temperature and salinity)

$$\frac{\partial T}{\partial t} + u\frac{\partial T}{\partial x} + v\frac{\partial T}{\partial y} + w\frac{\partial T}{\partial z} = DIFF(T) \tag{5}$$

$$\frac{\partial S}{\partial t} + u\frac{\partial S}{\partial x} + v\frac{\partial S}{\partial y} + w\frac{\partial S}{\partial z} = DIFF(S) \tag{6}$$

Where $T$ and $S$ are the temperature and salinity, respectively;

*Equation of state*

$$\rho = \rho(S, T, p) \tag{7}.$$

Details of the derivation of the governing equations can be found in any text book on geophysical fluid dynamics such as Cushman-Roisin (1994). In order to solve the equations it is necessary to specify appropriate spatial boundary conditions and the top, the bottom and the sides of the domain as well as initial conditions. There is no general formulation of the boundary conditions since they depend upon the particular problem being addressed. Examples of boundary conditions at the top include wind stress for the momentum equations, or heat and mass fluxes for the internal energy equations. The bottom boundary conditions usually consist of frictional drag and no vertical mass flux. Lateral boundary conditions may be as simple as no flow at the coastline or some type of wave radiation condition at an open lateral boundary which allows waves to escape with no reflection (e.g., Orlanski, 1976).

The equations as they appear above describe a wide range of atmospheric and oceanic motions (except sound waves which are filtered out by the Boussinesq approximation). To study particular phenomena or processes, they can be further simplified, usually through additional scale analysis which leads to neglecting other terms. In some cases analytical solutions can be found, but in most cases numerical approaches are necessary. A very powerful and widely used simplification of Eq. (1) and (2) is geostrophic flow in which the local time derivative, the nonlinear advections terms, and diffusion are neglected. The remaining leading order terms, which roughly balance each other, are the Coriolis force and the horizontal pressure gradient force (last term on the left hand side and first term of the right hand side, respectively). The immediate implication is that the currents must flow parallel to the isobars (lines of constant pressure) rather than from high pressure to low pressure zones as in non-rotating fluids. This also means that the currents can be diagnosed directly from the pressure or mass field. The practical importance of this is that it is much easier and cheaper to measure the mass field variables (i.e., temperature, salinity, and pressure) than to measure the motion field (currents). Consequently the vast majority of physical oceanographic measurements consists of the three dimensional distribution of temperature and salinity. Combining the geostrophic equations with the hydrostatic equation allows us to compute the vertical shear of the currents due to horizontal pressure or density gradients. The two main weaknesses of the geostrophic approximation are that it breaks down in tropical areas where the Coriolis force is very weak, and it does not allow for temporal changes.

Another common method for simplifying the equations is to reduce their spatial dimensionality. For example, the primary external forcing of the ocean originates in the atmosphere and is applied from above (winds and heat flux). Consequently, the vertical gradients of the primary dependent variables in Eq. (1), (2), (5), and (6) are much larger than the horizontal gradients. It is therefore quite common to study the importance of this stratification through the use of one-dimensional water column models. A classic example is the study of the wind forced surface boundary layer by Ekman (1905) in which Eqs. (1) and (2) are reduced to steady state equations balancing the Coriolis force with the vertical component of diffusion. The solution is the so called Ekman spiral for the surface layer in which the current magnitude decays with depth and the current vector rotates clockwise with depth in the northern hemisphere. Another example is the investigation of vertical convective mixing and its role in the deepening of the surface mixed layer through the use of the one-dimensional version of Eq. (5) and (6) in which the local time derivative is balanced by the vertical component diffusion (e.g., Martin, 1985).

In contrast to the vertical column models, other processes in which the horizontal variations are important or of interest can be investigated using two dimensional, depth integrated versions of the equations. To study the wind driven gyres in the upper ocean, (Stommel (1948) and Munk (1950) both started with the geostrophic form of Eqs. (1) and (2) with the addition of a frictional drag term as an alternative to horizontal diffusion. They took advantage of the non-divergence of the geostrophic flow to recast the equations as a single equation for vorticity. Through the solutions of the equations they were able to explain the underlying dynamics of the observed circulation in the North Atlantic Ocean. The general anticylonic (clockwise) gyre was driven by the curl of the wind stress (i.e., change in the direction from the easterly Trade Winds in the tropics and subtropics to the Westerlies in the mid-latitudes). The appearance of the intense western boundary current (i.e., the Gulf Stream) was a result of intensification due the accumulation of anticylonic relative vorticity by the wind stress and anticylonic planetary vorticity due to meridional variations of the Coriolis parameter but bounded by the damping effect of friction with the east coast of North America.

The various examples presented above are meant to demonstrate some of the basic and salient features of ocean processes which have been investigated over the past 100 years through the use of various simplified versions of the governing equations for geophysical fluid dynamics. It does not even scratch the surface of the vast body of scientific literature in this rapidly expanding and exciting field of study. An in depth survey of these processes can be found in many of the excellent modern textbooks published in recent years such as Vallis (2006).

## 3. Numerical ocean modeling

As noted in the introduction, the rapid development of computer technology over the past few decades has encouraged the massive development and advancement of numerical ocean models since the original effort of Bryan & Cox (1967). The main advantage of numerical modeling as compared to simplified process studies is that the numerical models are based on the more complete form of the governing equations presented in the previous section. This allows us to investigate more complex flow regimes and processes than in the past. In fact some of the simplifications such as hydrostatic balance in the governing equations are also being removed in recent models, thereby restoring the full time dependent equation for the vertical component of velocity. This is driven by the interest in and capabilities to investigate and simulate smaller scale processes. A model has the potential to fill in the many gaps left by limited in situ observations, subject of course to the computational and mathematical limitations of any model. One disadvantage of using more and more complex models is that it becomes more difficult to isolate and understand specific dynamical processes and thereby we develop the tendency to use a model as a black box. Even when running the most complex models we must never lose sight of what exactly the model is doing. Successful completion of a simulation does not guarantee proper results.

We must always critically examine the results to be sure the model is doing what it should. With this in mind we present a very brief survey of the most commonly used numerical methods used in ocean modeling today.

The governing equations presented in the previous section form a set of time dependent, hyperbolic partial differential equations. Numerical methods for solving such equations have been developed and have appeared in the mathematical literature over many years. As noted in the introduction, the approach to constructing numerical ocean models and the choice of particular methods has benefited greatly from and closely followed the development of atmospheric models and numerical weather prediction, which preceded ocean modeling by 10-15 years. The most common method used today in ocean models is finite differencing. In recent years finite elements have also become popular. For various reasons, spectral methods have not been widely used, perhaps due to the associated difficulties of dealing with irregular boundaries (i.e., coastlines). Without loss of generality, we will use the finite difference method to illustrate the fundamental principles of a typical approach to ocean modeling. A detailed presentation of ocean modeling methodology and applications can be found in the recent books of Kampf (2009, 2010).

The first step in developing a model is to define the domain of interest and divide it into a discrete set of grid points in space. The goal of the model is to approximate the continuous equations with a compatible set of algebraic equations which are solved at the grid points. This requires accurate methods for representing the first and second order spatial derivatives based on the values of the relevant variables at the grid points. If we consider a dependent variable $y$ as a function of the spatial coordinate $x$, say $y(x)$, then the gradient or first derivative $\partial y / \partial x$ can be approximated from a Taylor series expansion around the $i$-th grid point $x_i$ as

$$\frac{\partial y}{\partial x} \approx \frac{y_{i+1} - y_i}{\Delta x} + O(\Delta x) \tag{8}$$

or

$$\frac{\partial y}{\partial x} \approx \frac{y_i - y_{i-1}}{\Delta x} + O(\Delta x) \tag{9}$$

which are referred to as the forward and backward differencing schemes, respectively, and where $\Delta x$ is the grid spacing. These schemes are first order accurate as indicated by the truncation error $O(\Delta x)$. By averaging these two schemes we obtain the more accurate centered differencing scheme

$$\frac{\partial y}{\partial x} \approx \frac{y_{i+1} - y_{i-1}}{2\Delta x} + O(\Delta x^2) \tag{10}$$

which is second order accurate. Higher order schemes that are even more accurate can be formed by various weighted combinations of the respective Taylor series expansions, although most models use centered differencing. The second derivative is approximated by the three point stencil

$$\frac{\partial^2 y}{\partial x^2} \approx \frac{y_{i+1} - 2y_i + y_{i-1}}{\Delta x^2} \tag{11}.$$

The location of the dependent variables is a matter of choice. They can all be co-located at the grid points, or can be located on a staggered grid in which certain variables are shifted by half of a grid point. A commonly staggered grid is the Arakawa C-grid in which the velocity components are shifted one half of a grid point in their respective directions relative to the mass variables (Arakawa & Lamb, 1977). This arrangement ensures that the numerical equations will preserve certain integral properties of the continuous equations such as energy conservation.

In the time integration of the equations, the computational stability of the differencing scheme must also be considered. This means that the time step and spatial grid spacing must be chosen in such a way as to properly resolve the motion of the fastest moving waves that can be simulated by the model (usually the free surface or external mode gravity wave). Most models use some type of centered explicit or split explicit scheme that is also second order accurate. In explicit schemes all variables can be advanced to the forward time step based on the values known at the present and/or backward time step. In split explicit schemes, the terms in the momentum equations that are identified with the fastest moving waves are integrated separately with a shorter time step. This is done to improve the numerical efficiency and execution time of the model. An alternative is to use a fully implicit time scheme in which advancing the model to the forward step involves simultaneously knowing the values of the variables at the forward, present, and backward time steps. This has the advantage of the scheme being absolutely stable (i.e., no numerical amplification) but the disadvantage of being computationally cumbersome and slow due to the need to invert a large tridiagonal matrix at every time step.

The final point to consider in the construction of an ocean model is how to account for the unresolved scales of motion. The grid spacing or resolution of a model limits the explicitly resolved processes. Strictly speaking, from a mathematical perspective the shortest length scale that can be explicitly resolved by a model is $2\Delta x$. Practically however, scales shorter than ~4-6 $\Delta x$ are often misrepresented due to numerical damping or phase speed errors which may be an inherent characteristic of certain differencing schemes. However there are also important processes which may occur on scales smaller than the grid spacing which affect the larger scale flow. The classic example of this is vertical convective mixing forced by the wind or by induced by static instability when the water is cooled from above. This mixing will transport various properties of the water but is accomplished by small scale turbulent eddies which are not usually explicitly resolved. Such processes are accounted for by adding sub grid scale parameterizations, often in the form of a diffusion term but with a diffusion coefficient that is several orders of magnitude larger than the value for molecular diffusion. The quasi-empirical method for computing these eddy diffusion coefficients is usually referred to as a turbulence close scheme (e.g., Mellor & Yamada, 1982; Pacanowski & Philander, 1981). An analogous term is usually included for horizontal mixing. Finally, ecosystem models, which are becoming more common components of ocean models, require the addition of advection-diffusion equations, similar to Eqs. (5) and (6), for the relevant biogeochemical variables in addition to all of the biogeochemical processes which are treated computationally the same as sub grid scale processes.

## 4. The Mediterranean Sea – a laboratory ocean basin

Since the early 20th century the Mediterranean Sea was known to be a concentration basin where excess evaporation drives a basin wide thermohaline cell in which less saline water enters from the Atlantic Ocean through the Strait of Gibraltar (Nielsen, 1912). This surface water becomes more saline and denser. It sinks to a depth of ~ 250 m and then returns to the strait where it is carried by a subsurface outflow back to the Atlantic. During the past 25-30 years scientific interest in the oceanography of the Mediterranean Sea was renewed for various reasons. As a result of intensive field campaigns, it became clear that the circulation is far more complex than originally envisioned. It is now known that the Mediterranean Sea functions as a mini-ocean with dynamical processes occurring over a broad spectrum of spatial and temporal scales ranging from the basin wide thermohaline cell, driven by deep water formation, with a time scale of tens of years to energetic mesoscale eddies varying over a period of several weeks to months (e.g., Millot, 1999; Robinson & Golnaraghi, 1994).

Following the new description of the circulation that emerged from these programs, various numerical models were applied to the Mediterranean to further investigate the processes that drive the circulation. Initially, low resolution, basin wide models were used to study the climatological mean circulation of the entire Mediterranean (e.g., Roussenov et al., 1995; Zavatarelli & Mellor, 1995). Other models focused on particular process studies such as deep water formation (e.g., Wu et al., 2000) and/or the sub-basin circulation, and were used to study the response of the general circulation to interannual atmospheric variability (e.g., Korres et al., 2000). Most recently, a rather unique and fascinating phenomenon that occurred in the Eastern Mediterranean involved an abrupt shift in the source region of deep water formation from the Adriatic Sea to the Aegean Sea during the 1990's. This has been called the Eastern Mediterranean Transient (Roether et al., 2007). Several models have been used to simulate the evolution of this process (e.g., Lascaratos et al., 1999; Samuel et al., 1999) in response to changes in atmospheric forcing. As the data and research models provided new understanding of the circulation, and as observational systems and computer technology advanced, by the late 1990's it was decided to apply this new knowledge to the problem of operational ocean forecasting. An up to date review of the present understanding of the Mediterranean circulation can be found in Brenner (2011).

In the next two sections we will present some examples of both process studies and ocean forecasting taken from some of our most recent research efforts. This represents only a small fraction of many of the ongoing investigations being conducted by many scientists around the Mediterranean. In no way is this intended to be an exhaustive survey. It is simply a small sample meant to demonstrate the state-of-the-art of applications of numerical ocean models. It is mainly out of convenience that we take examples from our own personal experience of research in the Mediterranean.

## 5. Process studies and simulations

In this section we present an example from some of our recent and ongoing Mediterranean modeling research. It is a one dimensional (vertical) coupled hydrodynamics-ecosystem model for a typical point located in the Eastern Mediterranean Sea. The goal of this model is to investigate the fundamental biogeochemical processes and the influence of the annual cycle of vertical mixing upon them. The hydrodynamic part of the model is a one dimensional version of the Princeton Ocean Model (POM) originally described by (Blumberg & Mellor, 1987). POM is a three dimensional, time dependent model based on the primitive equations with the Boussinesq and hydrostatic approximations as described above in Section 2. It also included a free surface, which turns the continuity equation, Eq. (4), into a time dependent equation for the height of the free surface. POM contains full thermodynamics as well as the turbulence closure sub-model of (Mellor & Yamada, 1982). It is forced at the surface through the boundary conditions which specify the wind stress, heat flux components, and fresh water flux. In the vertical column version all horizontal

Numerical Modeling of the Ocean Circulation: From Process Studies to Operational Forecasting – The Mediterranean Example

143

advection and diffusion are neglected and the focus is on the role of vertical mixing only. Complex ecosystem or biogeochemical models are young relative to hydrodynamic models and are therefore in a stage of rapid development. For this particular study we have used the Biogeochemical Flux Model (BFM) described by (Triantafyllou et al., 2003 and Vichi et al., 2003). The model simulates several classes of phytoplankton, zooplankton, the carbon cycle, and the nitrogen cycle. The specific coupling of the models and implementation for the southeastern Mediterranean Sea presented here is based on the work of Suari (2011). In terms of the hydrodynamics, the main challenge in running the one dimensional model for the eastern Mediterranean was to account for the inflow of relatively fresh Atlantic Water which prevents unrealistic increases in salinity, which would cause the model to eventually become unstable. This was solved by adding a relaxation term in which the simulated salinity profile was nudged towards monthly mean climatological profiles. The model was configured with 40 unevenly spaced layer from the surface to a depth of 600 m and was forced at the surface with a repeating annual cycle that consisted of daily mean winds and heat fluxes that were computed from the multiyear average of the data taken from the NCEP/NCAR reanalysis covering the period from 1950-2006 (Kalnay et al., 1996). The model was run for 50 years with this perpetual year forcing. The purpose of such experiments is to assess the long term behavior and stability of the system without regard to the high frequency or inter annual variability.

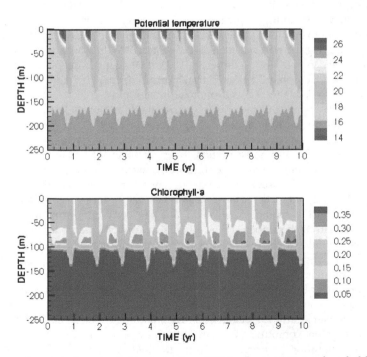

Fig. 1. Time series plot of: potential temperature (°C) in the upper panel and chlorophyll-a (µg L$^{-1}$) in the lower panel from the last 10 years of a 50 year simulation of a one dimensional coupled hydrodynamic-ecosystem model for the eastern Mediterranean.

The results presented in Fig. 1 show the potential temperature (upper panel) and the chlorophyll-a (lower panel) from the last ten years of the fifty year simulation. By this point the model has passed through the spin-up phase and produces a relatively stable repeating annual cycle. The surface temperature varies between a maximum of approximately 26°C in summer and a winter minimum of 16.6°C. The shallow surface mixed layer, in which the temperature is relatively high and uniform, is clearly visible in summer when it extends from the surface to a depth of 30 m. It is driven mainly by wind mixing, which generates enough turbulent kinetic energy to mix the water against the density gradient. By autumn the surface begins to cool and as a result the water column begins to mix vertically due to free convection, as indicated by the deepening of the green shaded contour. The free convection is driven by gravitational instability of the water column due to cooling from above. By late winter (early to mid March) the mixed layer has deepened to it maximum extent of 190 m as indicated by the uniform cyan contour extending from the surface from late February through mid March. This cycle and the values of the temperature and mixed layer depth are consistent with the observations from this region (e.g., Hecht, et al., 1988; Manca et al., 2004; Ozsoy et al., 1993).

The lower panel of Fig. 1 shows that chlorophyll-a (concentration in $\mu M\ L^{-1}$), which is the proxy for phytoplankton biomass, is confined to the upper part of the water column where there is sufficient light for photosynthesis. Nutrients (mainly nitrate, phosphate, and silicate) are also necessary for the cells to function. The nutrients are injected into the upper layers from below the nutricline (begins at ~150-200 m and extends to ~ 600 m) during deep winter mixing or during wind induced upwelling events. They are rapidly depleted form the photic zone when photosynthesis commences. Chlorophyll-a in the figure exhibits a pattern that is typical for an oligotrophic sea such as the eastern Mediterranean. During spring and summer it is confined mainly to the upper 90-100 m. During the deep mixing in the latter part of winter, the phytoplankton are transported deeper by convective mixing. A combination of factors leads to reduced photosynthesis and biomass concentration during this period. Sun light is less intense and the phytoplankton spend less time in the photic zone. Also due to the deeper mixing they are distributed over a larger volume and therefore the concentration is lower as indicated by the cyan contour. The warmer colors indicate two important features on the marine ecosystem. In early spring the yellow contours show a layer of relatively high chlorophyll concentration extending from the surface to a depth of ~80 m and which lasts for 2-3 weeks. This phenomenon is referred to as the spring bloom. It occurs shortly after the end of the winter (i.e., end of net surface cooling) and the onset of net surface heating in the spring. As a result the free convective mixing ceases and the phytoplankton remain in the upper layers. At this time nutrients are abundant due to the import of high nutrient waters from the deeper layers during winter. These two factors combined with the increasing intensity of the sunlight lead to a rapid increase in photosynthesis and therefore a substantial increase in chlorophyll-a concentration. The nutrients are consumed by the photosynthetic activity of the cells. Since the nutrient source in deep water has been cut off by the cessation of free convection, the nutrients in the photic zone are rapidly depleted and the bloom ends within a few weeks. This is indicated by the transition to the green contours. Later in the summer a subsurface layer with high chlorophyll-a concentration appears at a depth of 70-90 m (yellow and orange contours). This phenomenon referred to as the deep chlorophyll maximum, DCM, is due to the complex interaction between light intensity, leakage of nutrients from the nutricline, and the density stratification. Its occurrence is quite common in the oligotrophic Mediterranean Sea

Numerical Modeling of the Ocean Circulation: From Process Studies to Operational Forecasting – The
Mediterranean Example

145

(e.g., Estrada et al., 1993; Yacobi et al., 1995). The simulated pattern and values of chlorophyll-a concentration are consistent with observed values for this region reported by (Manca et al., 2004 and Yacobi et al., 1995).

## 6. Ocean forecasting

In this section we present another example of the powerful use and application of numerical ocean models as part of an operational forecasting system. In contrast to process studies or simulations which are designed to help us understand the particular dynamical process of interest, the goal of a forecasting system is to provide the most accurate prediction of the circulation at a particular instant in time, but within the constraint of producing the forecast in reasonably short period of time so that it considered to be useful. Clearly a 24 hour forecast that requires 24 hours of computer time has no value. A balance must therefore be reached between the acceptable level of forecast error and the time required to produce the forecast. Furthermore in a forecast system, in addition to the model itself, the specification of the initial conditions is a central consideration. Experience from numerical weather prediction has shown that during the first few days the forecast errors depend mainly on errors in the initial conditions, whereas at longer forecast lead times model errors and uncertainties have a larger impact on forecast errors. In addition to collecting data, accurate mathematical methods are necessary for interpolating the observations to the model grid while creating a minimal amount of numerical noise. This entire procedure, referred to as data assimilation (e.g., Kalnay, 2003), will not be discussed here. Our focus will be on the numerical model itself.

The development of the Mediterranean Forecasting System, MFS, began in 1998 as a cooperative effort of nearly 30 institutions with the goal of producing a prototype operational forecasting system and to demonstrate its feasibility. The project included components of in situ and remotely sensed data collection, data assimilation and model development. The model development component was structured to include a hierarchy of nested models with increasing resolution. The overall system was driven by the coarse resolution, full Mediterranean model. At the next level, sub-basin scale models, which covered large sections of the western, central, and eastern Mediterranean with a threefold increase in resolution, were nested in the full basin model. Nesting is the procedure through which the initial conditions were interpolated to the higher resolution grid, and the time dependent lateral boundary conditions were extracted from the coarser grid model. Finally, very high resolution local models for specific regions were nested in the sub-basin models with an additional two to threefold increase in resolution. An overall description of the prototype system and its implementation can be found in (Pinardi et al., 2003). While the initial model development focused on mainly climatological simulations with the nested model, the next phase led to the pre-operational implementation of short term forecasting with all three levels of models. This system has evolved into Mediterranean Operational Ocean Network, which is perhaps one of the most advanced operational ocean forecasting systems today (MOON, 2011). It routinely provides daily forecasts for the circulation at all scales and the ecosystem at the larger scales.

One component of MOON is a high resolution local model for the southeastern continental shelf zone of the eastern Mediterranean. The model was developed initially within MFS (Brenner, 2003) and has subsequently gone through a number of improvements and refinements. The version presented here is described in detail by (Brenner et al., 2007). It is based on the full three dimensional, primitive equations Princeton Ocean Model which was described above in Section 3. The horizontal grid spacing is 1.25 km and there are 30 vertical levels distributed on a terrain-following vertical coordinate. Data for the lateral boundary conditions are extracted from a sub-basin, regional model which covers most of the Levantine, Ionian, and Aegean basins. The domain and bathymetry of the model are shown in Fig. 2. The mathematical formulation of the boundary conditions along the two open boundaries consists of specifying the normal and tangential components of the horizontal velocity at all boundary grid points and the tracers (temperature and salinity) at inflow points. At outflow points the boundary values are extrapolated from the first interior grid point using a linearized advection equation.

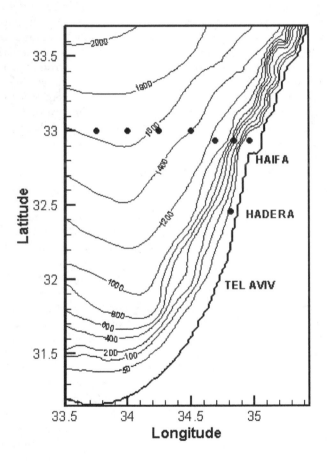

Fig. 2. Domain and bathymetry of the high resolution southeastern Levantine model. Dots indicate locations were observations were available for verification.

This model runs daily and produces forecasts of the temperature, salinity, free surface, and currents out to four days. As noted previously, the primary goal of a forecasting system is to produce the best possible prediction of the circulation at a specific instant in time. Thus forecast verification is an important aspect of assessing the usefulness of the system. In comparison to the atmosphere, ocean observations are extremely limited. The best spatial and temporal data are provided by satellites but are generally limited to sea surface temperature (SST) and sea surface height. The former are usually available several times daily while the latter are limited to approximately weekly, depending upon the path of the satellite. Other measurements are available from ships of opportunity or from fixed buoys but these are sporadic and limited in both time and space. With all of these reservations in mind, in the next few figures we present some examples of the verification of the forecasts produced by this model. In Fig. 3 we show the forecast skill of SST for a one year period as a function of forecast lead time. The skill scores used are the domain averaged root mean square error (RMSE) and the anomaly correlation coefficient. The former measures the magnitude of the forecast error while the latter provides a measure of the pattern error. The figure shows the forecast skill for the high resolution shelf model (red line) and for the coarser resolution regional model (green line). We also include the error for a persistence forecast (i.e., no change from the initial conditions), which is considered to be the minimum skill forecast. From both the RMSE and the anomaly correlation it is clear that the forecast skill degrades as the forecast length increases. The value added to the forecast by the high resolution model is substantial as it outperforms the regional model in both scores (i.e., lower error magnitude and higher pattern correlation). Both models also manage to significantly beat persistence for RMSE, but the regional model is only marginally better in the pattern correlation.

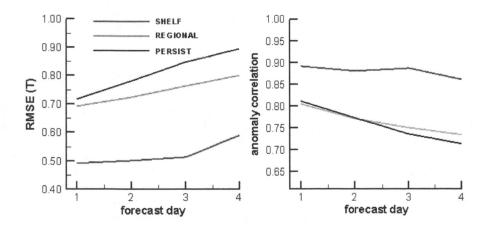

Fig. 3. Forecast skill for one year of forecasts in terms of root mean square error and anomaly correlation coefficient.

While the skill of the SST forecast is impressive, it is also important to validate the ability of the model to predict the subsurface fields. Unfortunately here the data are much more limited in space and time. In Fig. 4 we show a scatter plot of the predicted versus the observed temperature taken from a sea level measurement station located offshore near Hadera (see map in Fig. 2 for location). The instrument was located at a depth of ~15 m below the surface and the bottom depth is ~27 m. The comparison shown here also covers a one year period. Overall the comparison is excellent with a correlation coefficient of nearly 0.97. During winter (low temperatures) and summer (high temperatures) the points tend to be roughly evenly scatter above and below the regression line thus indicating that there is no clear bias in the forecasts. During the transition seasons of spring and autumn (mid range temperatures), there is a strong tendency for the model to under predict the temperature and therefore develop a cold bias. This is most likely due to the more rapid temperature changes during the transitions seasons as compared to summer or winter.

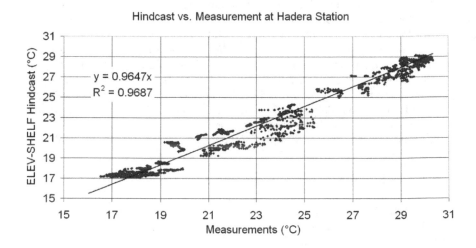

Fig. 4. Scatter plot of the predicted versus measured temperature at a depth of 15 m at an offshore station.

Finally, as a measure of the spatial distribution of the prediction of the subsurface fields, a comparison was made between all measurements collected during a single, one day cruise in the late summer along a transect of points that extend westward from Haifa (see Fig. 2 for location). The measurements were obtained from an instrument that measures nearly continuous profiles of temperature and salinity from the surface to the bottom or to a depth of 1000 m, whichever is deeper. From below the surface mixed layer, the model did an excellent job of predicting the temperature and salinity at all depths and stations along the transect. In the mixed layer the model showed a warm bias with simulated temperatures that were too high by 1-2°C. This error is probably due to the specification of surface heat fluxes that were too high and/or winds that were too weak which prevented the model from creating a deep enough mixed layer. The high resolution forecast was significantly better than the regional model forecast in this area which again demonstrates the value added by a high resolution model. It should be noted however that this comparison was conducted for a single forecast only.

# 7. Conclusion

In this chapter we have presented a concise overview of more than 40 years of research and development of numerical ocean circulation models. The pioneering work of Bryan & Cox (1967) set the stage for subsequent model development. The rapid development of computer technology of the past two decades has been a major factor allowing for the design of increasingly more complex and realistic models. By complementing field data and the associated gaps, numerical ocean models have proven to be an indispensible tool for enhancing our understanding of a wide range and variety of processes in oceanic hydrodynamics. Consequently, most modern oceanographic studies will almost always include a highly developed modeling component. Models are routinely used for processes studies and as the central component operational ocean forecasting systems as demonstrated by the examples presented.

# 8. Acknowledgement

The modeling results presented in this chapter were supported by the European Commission through the Sixth Framework Program European Coastal Sea Operational Observing and Forecasting System (ECOOP) Contract Number 36355, and Mediterranean Forecasting System Towards Environmental Prediction (MFSTEP) Contract Number EVKT3-CT-2002-00075.

# 9. References

Arakawa and Lamb (1977). Computational design of the basic dynamical processes of the UCLA general circulation model. In: *Methods in Computational Physics*, Vol. 17, pp. 174-267, Academic Press, London.

Blumberg A. and Mellor, G.L. (1987). A description of a three-dimensional coastal ocean circulation model. In: *Three-Dimensional Coastal Ocean Models*, N. Heaps, editor, pp. 1-16, American Geophysical Union, Washington, DC.

Brenner, S. (2003). High-resolution nested model simulations of the climatological circulation in the southeastern Mediterranean Sea. *Annales Geophysicae*, Vol. 21, pp. 267-280.

Brenner, S. (2011). Circulation in the Mediterranean Sea, In: *Life in the Mediterranean Sea: A Look at Habitat Changes*, N. Stambler, editor, chapter 4, Nova Science Publishers, ISBN 9781-61209-644-5.

Brenner, S., Gertman, I., and Murashkovsky, A. (2007). Pre-operational ocean forecasting in the southeastern Mediterranean: Model implementation, evaluation, and the selection of atmospheric forcing. *Journal of Marine Systems*, Vol. 65, pp. 268-287.

Bryan, K. and Cox, M.D. (1967). A numerical investigation of the oceanic general circulation. *Tellus*, Vol. 19, pp. 54-80.

Charney, J.G., Fjortoft, R. and von Neuman, J. (1950). Numerical integration of the barotropic vorticity equation. *Tellus*, Vol. 2, pp. 237-254.

Cushman-Roisin, B. (1994). *Introduction to Geophysical Fluid Dynamics*, Prentice Hall, ISBN 0-13-353301-8, Englewood Cliffs, New Jersey.

Estrada, M., Marrase, C., Latasa, M., Berdalet, E., Delgado, M., and Riera, T. (1993). Variability of deep chlorophyll maximum characteristics in the Northwestern Mediterranean. *Marine Ecology Progress Series,* Vol. 92, pp. 289-300.

Ekman, V.W. (1905). On the influence of the Earth's rotation on ocean currents. *Arkiv for Matematik, Astronomi och Fysik,* Vol. 2, No. 11, pp. 1-52.

Kalnay, E. (2003). *Atmospheric Modeling, Data Assimilation, and Predictability.* Cambridge University Press, ISBN 13-978-0-521-79629-3, Cambridge.

Kalnay, E., Kanamitsu, M., Kistler, R., Collins, W., Deaven, D., Gandin, L., Iredell, M., Saha, S. White, G., Woollen,J., Zhu, Leetmaa, A., Reynolds, R., Chelliah, M., Ebisuzaki, W., W.Higgins, Janowiak, J., Mo, K.C., Ropelewski, C., Wang, J., Jenne, R., and Joseph, D. (1996). The NCEP/NCAR 40-year reanalysis project. *Bulletin of the American Meteorological Society,* Vol. 77, pp. 437-471.

Hecht, A., Pinardi, N., and Robinson, A. (1988): Currents, water masses, eddies and jets in the Mediterranean Levantine Basin, *Journal of Physical Oceanography,* Vol. 18, pp. 1320-1353.

Kampf, J. (2009). *Ocean Modeling for Beginners Using Open Source Software.* Springer, ISBN 978-3-642-00819-1, Heidelberg.

Kampf, J. (2010). *Advanced Ocean Modeling Using Open Source Software.* Springer, ISBN 978-3-642-10609-5, Heidelberg.

Korres, G., N. Pinardi, and A. Lascaratos, 2000. The ocean response to low-frequency interannual atmospheric variability in the Mediterranean Sea, Part I: Sensitivity experiments and energy analysis. *Journal of Climate,* Vol. 13, pp. 705-731.

Lascaratos, A., Roether, W., Nittis, K., and Klein, B. (1999). Recent changes in deep water formation and spreading in the Mediterranean Sea: a review. *Progress in Oceanography,* Vol. 44, 1-36.

Manca, B.B., Burca, M., Giorgetti, A., Coatanoan, C., Garcia, M.-J., and Iona, A. (2004). Physical and biochemical averaged profiles in the Mediterranean regions: an important tool to trace the climatology of water masses and to validate incoming data from operational oceanography. *Journal of Marine Systems,* Vol. 48, pp. 83-116.

Martin, P. (1985). Simulation of the mixed-layer at OWS November and Papa with several models. *Journal of Geophysical Research,* Vol. 90, pp. 903-916.

Mellor, G.L. and Yamada, T. (1982). Development of a turbulence closure scheme for geophysical fluid problems. *Reviews of of Geohpysics and Space Physics,* Vol. 20, pp. 851-875.

Millot, C. (1999). Circulation in the Western Mediterranean Sea. *Journal of Marine Systems,* Vol. 20, pp. 423-440.

MOON. (2011). *Mediterranean Operational Oceanography Network.* Available from: http://www.moon-oceanforecasting.eu.

Nielsen, J.N. (1912). Hydrography of the Mediterranean and adjacent seas. Danish Oceanographic Expeditions 1908-10, Report I, pp. 72-191.

Orlanski, I. (1976). A simple boundary condition for unbounded hyperbolic flows. *Journal of Computational Physics,* Vol. 21, pp. 251-259.

Numerical Modeling of the Ocean Circulation: From Process Studies to Operational Forecasting – The
Mediterranean Example

151

Ozsoy, E., Hecht,A., Unluata, U.,  Brenner, S., Sur, H., Bishop, J., Oguz, T. , Rozentraub, Z., and Latif., M.A. (1993). A synthesis of the Levantine Basin circulation and hydrography, 1985-1990. *Deep Sea Res. II*, Vol. 40, pp. 1075-1120.

Pacanowski, R. and Philander, S.G.H. (1981). Parameterization of vertical mixing in numerical models of tropical oceans. *Journal of Physical Oceanography*, Vol. 11, pp. 1443-1451.

Pinardi, N., Allen, I., and Demirov, E. (2003). The Mediterranean Ocean Forecasting System: the first phase of implementation. *Annales Geophysicae*, Vol. 21, pp. 3-20.

Richardson, L.F. (1922). *Weather Prediction by Numerical Process*. Cambridge University Press, London.

Robinson, A.R., and Golnaraghi, M. (1994). The physical and dynamical oceanography of the Mediterranean Sea. In *Ocean Processes in Climate Dynamics: Global and Mediterranean Examples*. Malanotte-Rizzoli, P. and Robinson, A.R. (Eds.), pp. 255-306, Kluwer Academic Publishers, Dordrecht.

Roether, W., Klein, B., Manca, B.B., Theocharis, A., and Kioroglou, S. (2007). Transient Eastern Mediterranean deep waters in response to the massive dense-water output of the Aegean Sea in the 1900's. *Progress in Oceanography*, Vol. 74, pp. 540-571.

Roussenov, V., Stanev, E., Artale,V.,  and Pinardi, N. (1995). A seasonal model of the Mediterranean Sea general circulation. *Journal of Geophysical Research*, Vol. 100, pp. 13515-13538.

Samuel, S.L., Haines, K., Josey, S.A., and Myers, P.G. (1999). Response of the Mediterranean Sea thermohaline circulation to observed changes in the winter wind stress field in  the period 1980-93. *Journal of Geophysical Research*, Vol. 104, pp. 5191-5210.

Stommel, H. (1948). The westward intensification of wind-driven ocean currents. *Transactions, American Geophysical Union*, Vol. 29, pp. 202-206.

Suari, Y. (2011). *Modeling the impact of riverine nutrients input on the southeastern Levantine planktonic ecology*. Ph.D. Thesis, Bar Ilan University, under review.

Triantafyllou, G., Petihakis, G., and Allen, I.J. (2003). Assessing the performance of the Cretan Sea ecosystem model with the use of high frequency M3A buoy data set. *Annales Geophysicae*, Vol. 21, pp. 365-375.

Vallis, G.K. (2006). *Atmospehric and Oceanic Fluid Dynamics*. Cambridge University Press, ISBN 978-0-521-84969-2, Cambridge.

Vichi,M., Oddo, P., Zavatarelli, M., Colucelli, A., Coppini, G., Celio, M., Umani, S.F., and Pinardi, N. (2003). Calibration and validation of a one-dimensional complex marine biogeochemical flux model in different areas of the northern Adriatic shelf. *Annales Geophysicae*, Vol. 21, pp., pp. 413-436.

Wu, P., Haines, K. and Pinardi, N. (2000). Toward an understanding of deep-water renewal in the Eastern Mediterranean. *Journal of Physical Oceanography*, Vol. 33, pp.  443-458.

Yacobi, Y.Z., Zohary, T., Kress, N., Hecht, A., Robarts, R.D., Waiser, M., Wood, A.M., and Li, W.K.W.  (1995) Chlorophyll distribution throughout the southeastern

Mediterranean in relation to the physical structure of the water mass. *Journal of Marine Systems*, Vol. 3, pp. 179-190.

Zavatarelli, M. and Mellor, G.L. (1995). A numerical study of the Mediterranean Sea circulation. *Journal of Physical Oceanography*, Vol. 25, pp. 1384-1414.

# Part 3

## Tidal and Wave Dynamics: Estuaries and Bays

# 8

# Astronomical Tide and Typhoon-Induced Storm Surge in Hangzhou Bay, China

Jisheng Zhang[1], Chi Zhang[1], Xiuguang Wu[2]
and Yakun Guo[3]
[1]*State Key Laboratory of Hydrology-Water Resources and Hydraulic Engineering,*
*Hohai University, Nanjing, 210098*
[2]*Zhejiang Institute of Hydraulics and Estuary, Hangzhou, 310020*
[3]*School of Engineering, University of Aberdeen,*
*Aberdeen, AB24 3UE*
[1,2]*China*
[3]*United Kingdom*

## 1. Introduction

The Hangzhou Bay, located at the East of China, is widely known for having one of the world's largest tidal bores. It is connected with the Qiantang River and the Eastern China Sea, and contains lots of small islands collectively referred as Zhoushan Islands (see Figure 1). The estuary mouth of the Hangzhou Bay is about 100 km wide; however, the head of bay (Ganpu) which is 86 km away from estuary mouth is significantly narrowed to only 21 km wide. The tide in the Hangzhou Bay is an anomalistic semidiurnal tide due to the irregular geometrical shape and shallow depth and is mainly controlled by the $M_2$ harmonic constituent. The $M_2$ tidal constituent has a period about 12 hours and 25.2 minutes, exactly half a tidal lunar day. The Hangzhou Bay faces frequent threats from tropical cyclones and suffers a massive damage from its resulting strong wind, storm surge and inland flooding. According to the 1949-2008 statistics, about 3.5 typhoons occur in this area every year. When typhoon generated in tropic open sea moves towards the estuary mouth, lower atmospheric pressure in the typhoon center causes a relatively high water elevation in adjacent area and strong surface wind pushes huge volume of seawater into the estuary, making water elevation in the Hangzhou Bay significantly increase. As a result, the typhoon-induced external forces (wind stress and pressure deficit) above sea surface make the tidal hydrodynamics in the Hangzhou Bay further complicated.

In the recent years, some researches have been done to study the tidal hydrodynamics in the Hangzhou Bay and its adjacent areas. For example, Hu et al. (2000) simulated the current field in the Hangzhou Bay based on a 2D model, and their simulated surface elevation and current field preferably compared with the field observations. Su et al. (2001), Pan et al. (2007) and Wang (2009) numerically investigate the formulation, propagation and dissipation of the tidal bore at the head of Hangzhou Bay. Also, Cao & Zhu (2000), Xie et al. (2007), Hu et al. (2007) and Guo et al. (2009) performed numerical simulation to study the typhoon-induced

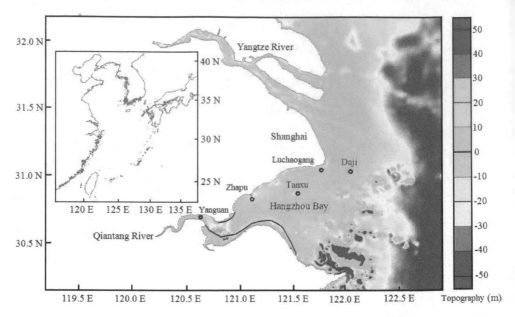

Fig. 1. Global location and 2005's bathymetry of the Hangzhou Bay and its adjacent shelf region

storm surge. However, most of them mainly focused on the 2D mathematical model. The main objective of this study is to understand the characteristics of (i) astronomical tide and (ii) typhoon-induced storm surge in the Hangzhou Bay based on the field observation and 3D numerical simulation.

## 2. Field observation

To understand the astronomical tides in the Hangzhou Bay, a five-month in situ measurement was carried out by the Zhejiang Institute of Hydraulic and Estuary from 01 April 2005 to 31 August 2005. There were eight fixed stations (T1-T8) along the banks of the Hangzhou Bay, at which long-term tidal elevations were measured every 30 minutes using ship-mounted WSH meter with the accuracy of $\pm 0.03$ m. The tidal current velocity was recorded every 30 minutes at four stations H1-H4 using SLC9-2 meter, manufactured by Qiandao Guoke Ocean Environment and Technology Ltd, with precisions of $\pm 4°$ in direction and $\pm 1.5\%$ in magnitude. The topography investigation in the Hangzhou Bay was also carried out in the early April 2005. Figure 2 shows the tidal gauge positions and velocity measurement points, together with the measured topography using different colors.

On 27/08/1981, a tropical depression named Agnes was initially formed about 600 km west-northwest of Guam in the early morning and it rapidly developed as a tropical storm moving west-northwestward (towards to Zhejiang Province) in the evening. Agnes became a typhoon in the morning of 29/08/1981, 165 km southwest of Okinawa next day. Agnes started to weaken after entering a region of hostile northerly vertical wind shear. The cyclonic center was almost completely disappeared by the morning of 02/09/1981. During Typhoon Agnes

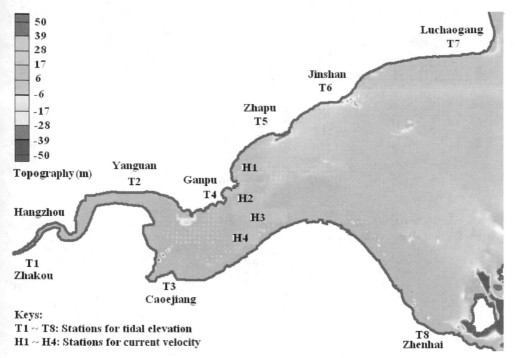

Fig. 2. A sketch of measurement stations and topography

(No.8114), which resulted in one of extremely recorded high water levels in the Hangzhou Bay, wind fields were observed every hour and storm tides were recorded every three hours at Daji station and Tanxu station (see Figure 1). Only the surge elevations were recorded and no current velocity was measured.

## 3. Numerical simulation

### 3.1 Governing equations
A 3D mathematical model based on FVCOM (an unstructured grid, Finite-Volume Coastal Ocean Model) (Chen et al., 2003) is developed for this study. The model uses an unstructured triangular grid in horizontal plane and a terrain-following $\sigma$-coordinate in vertical plane (see Figure 3), having a great ability to capture irregular shoreline and uneven seabed. The most sophisticated turbulence closure sub-model, Mellor-Yamada 2.5 turbulence model (Mellor & Yamada, 1982), is applied to compute the vertical mixing coefficients. More details of FVCOM can be found in Chen et al. (2003). Only the governing equations of the model are given here for completeness and convenience.

$$\frac{\partial \zeta}{\partial t} + \frac{\partial Du}{\partial x} + \frac{\partial Dv}{\partial y} + \frac{\partial \omega}{\partial \sigma} = 0 \qquad (1)$$

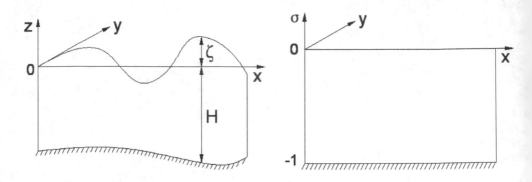

Fig. 3. Coordinate transformation of the vertical computational domain. Left: $z$-coordinate system; Right: $\sigma$-coordinate system

$$\frac{\partial uD}{\partial t} + \frac{\partial u^2 D}{\partial x} + \frac{\partial uvD}{\partial y} + \frac{\partial uw}{\partial \sigma} = fvD - \frac{D}{\rho_o}\frac{\partial P_{atm}}{\partial x} - gD\frac{\partial \zeta}{\partial x}$$

$$- \frac{gD}{\rho_o}[\frac{\partial}{\partial x}(D\int_\sigma^0 \rho d\sigma') + \sigma\rho\frac{\partial D}{\partial x}] + \frac{\partial}{\partial \sigma}(\frac{K_m}{D}\frac{\partial u}{\partial \sigma}) + DF_u \tag{2}$$

$$\frac{\partial vD}{\partial t} + \frac{\partial uvD}{\partial x} + \frac{\partial v^2 D}{\partial y} + \frac{\partial vw}{\partial \sigma} = -fuD - \frac{D}{\rho_o}\frac{\partial P_{atm}}{\partial y} - gD\frac{\partial \zeta}{\partial y}$$

$$- \frac{gD}{\rho_o}[\frac{\partial}{\partial y}(D\int_\sigma^0 \rho d\sigma') + \sigma\rho\frac{\partial D}{\partial y}] + \frac{\partial}{\partial \sigma}(\frac{K_m}{D}\frac{\partial v}{\partial \sigma}) + DF_v \tag{3}$$

$$\frac{\partial TD}{\partial t} + \frac{\partial TuD}{\partial x} + \frac{\partial TvD}{\partial y} + \frac{\partial T\omega}{\partial \sigma} = \frac{\partial}{\partial \sigma}(\frac{K_h}{D}\frac{\partial T}{\partial \sigma}) + DF_T \tag{4}$$

$$\frac{\partial SD}{\partial t} + \frac{\partial SuD}{\partial x} + \frac{\partial SvD}{\partial y} + \frac{\partial S\omega}{\partial \sigma} = \frac{\partial}{\partial \sigma}(\frac{K_h}{D}\frac{\partial S}{\partial \sigma}) + DF_S \tag{5}$$

$$\rho = \rho(T, S) \tag{6}$$

$$\frac{\partial q^2 D}{\partial t} + \frac{\partial uq^2 D}{\partial x} + \frac{\partial vq^2 D}{\partial y} + \frac{\partial \omega q^2}{\partial \sigma} = \frac{2K_m}{D}[(\frac{\partial u}{\partial \sigma})^2 + (\frac{\partial v}{\partial \sigma})^2] + \frac{2g}{\rho_o}K_h\frac{\partial \rho}{\partial \sigma}$$

$$- \frac{2Dq^3}{B_1 l} + \frac{\partial}{\partial \sigma}(\frac{K_{q^2}}{D}\frac{\partial q^2}{\partial \sigma}) + DF_{q^2} \tag{7}$$

$$\frac{\partial q^2 lD}{\partial t} + \frac{\partial uq^2 lD}{\partial x} + \frac{\partial vq^2 lD}{\partial y} + \frac{\partial \omega q^2 l}{\partial \sigma} = \frac{lE_1 K_m}{D}[(\frac{\partial u}{\partial \sigma})^2 + (\frac{\partial v}{\partial \sigma})^2] + \frac{lE_1 g}{\rho_o}K_h\frac{\partial \rho}{\partial \sigma}$$

$$- \frac{Dq^3}{B_1}\widetilde{W} + \frac{\partial}{\partial \sigma}(\frac{K_{q^2}}{D}\frac{\partial q^2 l}{\partial \sigma}) + DF_{q^2 l} \tag{8}$$

where x, y and $\sigma$ are the east, north and upward axes of the $\sigma$-coordinate system; u, v and w are the x, y and $\sigma$ velocity components, respectively; t is the time; $\zeta$ is the water elevation; D is the total water depth (=H+$\zeta$, in which H is the bottom depth); $P_{atm}$ is the atmospheric pressure; $\rho$ is the seawater density being a polynomial function of temperature T and salinity S (Millero & Poisson, 1981); f is the local Coriolis parameter (dependent on local latitude and the angular speed of the Earth's rotation); g is the acceleration due to gravity (=9.81 m/s$^2$); $\rho_0$ is the mean seawater density (=1025 kg/m$^3$); $K_m$ and $K_h$ are the vertical eddy viscosity coefficient and thermal vertical eddy diffusion coefficient; $F_u$, $F_v$, $F_T$ and $F_S$ are the horizontal u-momentum, v-momentum, thermal and salt diffusion terms, respectively; q$^2$ is the turbulent kinetic energy; l is the turbulent macroscale; $K_{q^2}$ is the vertical eddy diffusion coefficient of the turbulent kinetic energy; $\widetilde{W}$ is a wall proximity function (=1+E$_2$$\left(\frac{l}{\kappa L}\right)^2$, where the parameter $L^{-1}$=($\zeta$-z)$^{-1}$+(H+z)$^{-1}$); $F_{q^2}$ and $F_{q^2 l}$ represent the horizontal diffusion terms of turbulent kinetic energy and turbulent macroscale; and B$_1$, E$_1$ and E$_2$ are the empirical constants assigned as 16.6, 1.8 and 1.33, respectively.

Mode splitting technique is applied to permit the separation of 2D depth-averaged external mode and 3D internal mode, allowing the use of large time step. 3D internal mode is numerically integrated using a second-order Runge-Kutta time-stepping scheme, while 2D external mode is integrated using a modified fourth-order Runge-Kutta time-stepping scheme. A schematic solution procedure of this 3D model is illustrated in Figure 4. The point wetting/drying treatment technique is included to predict the water covering and uncovering process in the inter-tide zone. In the case of typhoon, the accuracies of the atmospheric pressure and wind fields are crucial to the simulation of storm surge. In this study, an analytical cyclone model developed by Jakobsen & Madsen (2004) is applied to predict pressure gradient and wind stress. The shape parameter and cyclonic regression parameter are determined by the formula suggested by Hubbert et al. (1991) and the available field observations in the Hangzhou Bay (Chang & Pon, 2001), respectively. Please refer to Guo et al. (2009) for more information.

### 3.2 Boundary conditions

The moisture flux and net heat flux can be imposed on the sea surface and bottom boundaries, but are not considered in this study. The method of Kou et al. (1996) is used to estimate the bottom shear stress induced by bottom boundary friction, accounting for the impact of flow acceleration and non-constant stress in tidal estuary. A river runoff (Q=1050 m$^3$/s) from the Qiantang River according to long-term field observation is applied on the land side of the model domain. The elevation clamped open boundary condition and atmospheric force (wind stress and pressure gradient) above sea surface are the main driving forces of numerical simulation. In modeling astronomical tide, the time-dependent open-sea elevations are from field observation at stations T7-T8 and zero atmospheric force is given. In modeling typhoon-induced storm surge, the time-dependent open-sea elevations are from FES2004 model (Lyard et al., 2006) and typhoon-generated water surface variations and atmospheric force is estimated by the analytical cyclone model. In this study, the external time step is $\Delta t_E$=2 sec and the ratio of internal time step to external time step is I$_S$=5.

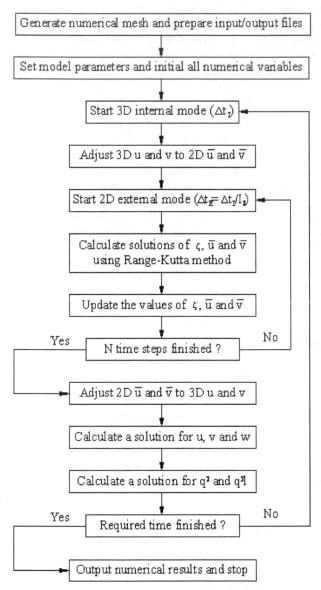

Fig. 4. A schematic solution procedure of 3D estuarine modeling

## 3.3 Mesh generation

As shown in Figures 1 and 2, the Hangzhou Bay has a very irregular shoreline. Therefore, to accurately represent the computational domain of the Hangzhou Bay, unstructured triangular meshes with arbitrarily spatially-dependent size were generated. The area of the whole solution domain defined for astronomical tide modeling is about 5360 km$^2$. The computational meshes were carefully adapted and refined, especially in the inter-tide zone, until no significant changes in the solution were achieved. The final unstructured grid having 90767 nodes and 176973 elements in the horizontal plane (each $\sigma$-level) was used with minimal distance of 20 m in the cells (see Figure 5). In the vertical direction, 11 $\sigma$-levels (10 $\sigma$-layers) compressing the $\sigma$ mesh near the water surface and sea bottom symmetrically about the mid-depth are applied.

In modeling typhoon-induced storm surge, a large domain-localized grid resolution strategy is applied in mesh generation, defining very large computational domain covering the main area of typhoon and locally refining the concerned regions with very small triangular meshes. The whole computational domain covers an extensive range of 116-138$^0$E in longitude and 21-41$^0$N in latitude. The final unstructured grid having 111364 nodes and 217619 elements in the horizontal plane (each $\sigma$-level) was used with the minimal 100 m grid size near shoreline and the maximal 10000 m grid size in open-sea boundary (see Figure 6). In the vertical direction, 6 $\sigma$-levels (5 $\sigma$-layers) is uniformly applied.

## 4. Results and discussion

The results from field observation and numerical simulation are compared and further used to investigate the characteristics of tidal hydrodynamics in the Hangzhou Bay with/without the presence of typhoon.

### 4.1 Astronomical tide
### 4.1.1 Tidal elevation

Figures 7 and 8 are the comparison of simulated and observed tidal elevations at 5 stations (T2, T3, T4, T5 and T6) in spring tide and neap tide, respectively. The x-coordinate of these figures is in the unit of day, and, for example, the label '21.25 August 2005' indicates '6:00am of 21/08/2005'. Both the numerical simulation and field observation for spring and neap tides show that the tidal range increases significantly as it travels from the lower estuary (about 6.2 m in spring tide and 3.1 m in neap tide at T6) towards the middle estuary (about 8.1 m in spring tide and 3.7 m in neap tide at T4), mainly due to rapid narrowing of the estuary. The tidal range reaches the maximum at Ganpu station (T4) and decreases as it continues traveling towards the upper estuary (about 4.4 m in spring tide and 2.5 m in neap tide at T2). In general, very good agreement between the simulation and observation is obtained. There exists, however, a slight discrepancy between the computed and observed tidal elevations at T2 (Yanguan). The reason for this may be ascribed to that the numerical model does not consider the tidal bore, which may have significant effect on the tidal elevations at the upper reach. Such impact on tidal elevations, however, decreases and becomes negligible at the lower reach of the estuary.

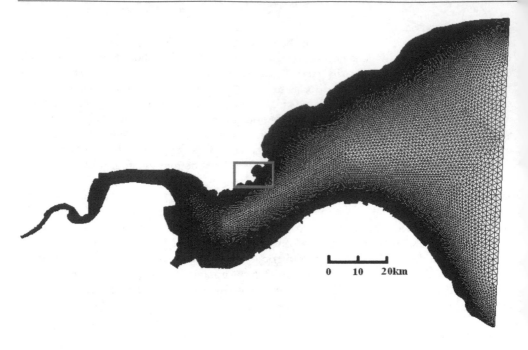

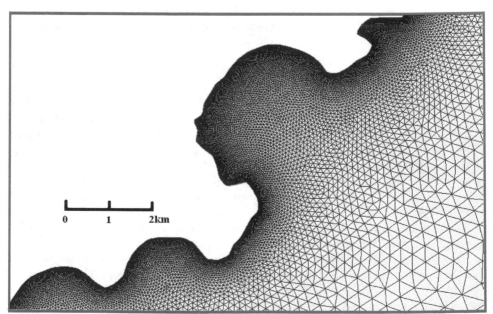

Fig. 5. A sketch of triangular grid (upper) and locally zoomed in mesh near Ganpu station (lower) for modeling astronomical tide

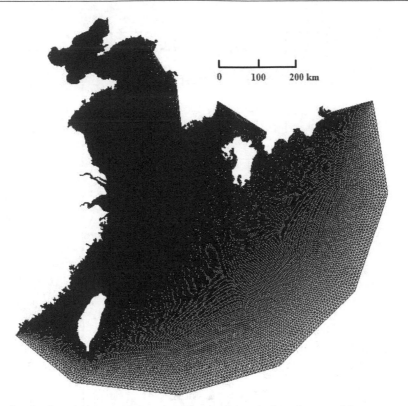

Fig. 6. A sketch of triangular grid for modeling typhoon-induced storm tide

### 4.1.2 Current velocity

It is clearly seen from Figures 7 and 8 that the maximum tidal ranges occur at the Ganpu station (T4). Thus, it is expected that the maximum tidal current may occur near this region. The tidal currents were measured at four locations H1-H4 across the estuary near Ganpu. These measurements are used to verify the numerical model. Figures 9 and 10 are the comparison between simulated and measured depth-averaged velocity magnitude and direction for the spring and neap tidal currents, respectively. It is seen that the flood tidal velocity is clearly greater than the ebb flow velocity for both the spring and neap tides. The maximum flood velocity occurs at H2 with the value of about 3.8 m/s, while the maximum ebb flow velocity is about 3.1 m/s during the spring tide. During the neap tide, the maximum velocities of both the flood and ebb are much less than those in the spring tide with the value of 1.5 m/s for flood and 1.2 m/s for ebb observed at H2. The maximum relative error for the ebb flow is about 17%, occurring at H2 during the spring tide. For the flood flow the maximal relative error occurs at H3 and H4 for both the spring and neap tides with values being about 20%. In general, the depth-averaged simulated velocity magnitude and current direction agree well with the measurements, and the maximal error percentage in tidal current is similar as that encountered in modeling the Mahakam Estuary (Mandang & Yanagi, 2008).

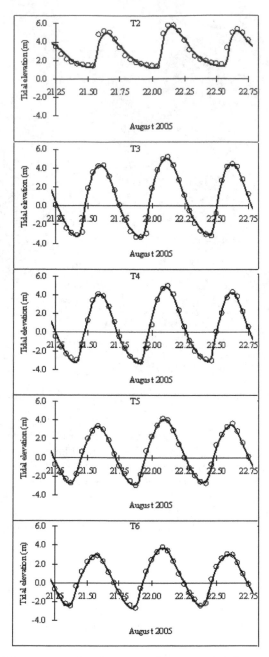

Fig. 7. Comparison of the computed and measured spring tidal elevations at stations T2-T6.
—: computed; o: measured

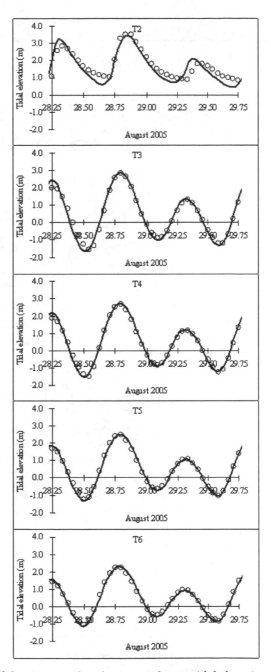

Fig. 8. Comparison of the computed and measured neap tidal elevations at stations T2-T6. —: computed; ○: measured

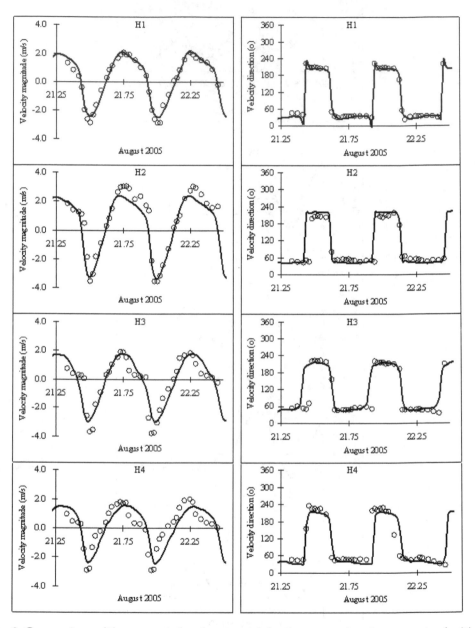

Fig. 9. Comparison of the computed and measured depth-averaged spring current velocities at stations H1-H4. −: computed; o: measured

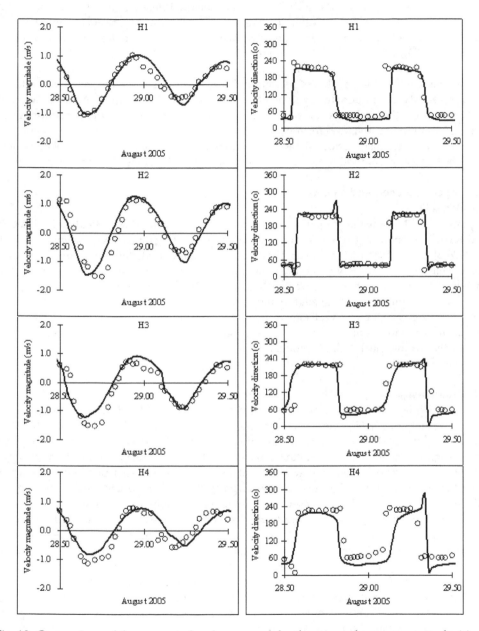

Fig. 10. Comparison of the computed and measured depth-averaged neap current velocities at stations H1-H4. —: computed; o: measured

The vertical distributions of current velocities during spring tide are also compared at stations H1 and H4. The measured and simulated flow velocities in different depths (sea surface, 0.2D, 0.4D, 0.6D and 0.8D, where D is water depth) at these two stations are shown in Figures 11 and 12. It is noted that the current magnitude obviously decreases with a deeper depth (from sea surface to 0.8D), while the flow direction remains the same. The numerical model generally provides accurate current velocity along vertical direction, except that the simulated current magnitude is not as high as that of measured during the flood tide. The maximum relative error in velocity magnitude during spring tide is about 32% at H4 station. Analysis suggests that the errors in the tidal currents estimation are mainly due to the calculation of bottom shear stress. Although the advanced formulation accounts for the impacts of flow acceleration and non-constant stress distribution on the calculation of bottom shear stress, it can not accurately describe the changeable bed roughness that depends on the bed material and topography.

## 4.2 Typhoon-induced storm surge
### 4.2.1 Wind field
Figures 13 and 14 show the comparisons of calculated and measured wind fields at Daji station and Tanxu station during Typhoon Agnes, in which the starting times of x-coordinate are both at 18:00 of 29/08/1981 (Beijing Mean Time). In general, the predicted wind directions agree fairly well with the available measurement. However, it can be seen that calculated wind speeds at these two stations are obviously smaller than observations in the early stage of cyclonic development and then slightly higher than observations in later development. The averaged differences between calculated and observed wind speeds are 2.6 m/s at Daji station and 2.1 m/s at Tanxu station during Typhoon Agnes. This discrepancy in wind speed is due to that the symmetrical cyclonic model applied does not reflect the asymmetrical shape of near-shore typhoon.

### 4.2.2 Storm surge
Figure 15 displays the comparison of simulated and measured tidal elevations at Daji station and Tanxu station, in which the starting times of x-coordinate are both at 18:00 on 29/08/1981 (Beijing Mean Time). It can be seen from Figure 15 that simulated tidal elevation of high tide is slightly smaller than measurement, which can be directly related to the discrepancy of calculated wind field (shown in Figures 13 and 14). A series of time-dependent surge setup, the difference of tidal elevations in the storm surge modeling and those in purely astronomical tide simulation, are used to represent the impact of typhoon-generated storm. Figure 16 having a same starting time in x-coordinate displays simulated surge setup in Daji station and Tanxu station. There is a similar trend in surge setup development at these two stations. The surge setup steadily increases in the early stage (0-50 hour) of typhoon development, and then it reaches a peak (about 1.0 m higher than astronomical tide) on 52nd hour (at 22:00 on 31/08/1981). The surge setup quickly decreases when the wind direction changes from north-east to north-west after 54 hour. In general, the north-east wind pushing water into the Hangzhou Bay significantly leads to higher tidal elevation, and the north-west wind dragging water out of the Hangzhou Bay clearly results in lower tidal elevation. The results indicate that the typhoon-induced external forcing, especially wind stress, has a significant impact on the local hydrodynamics.

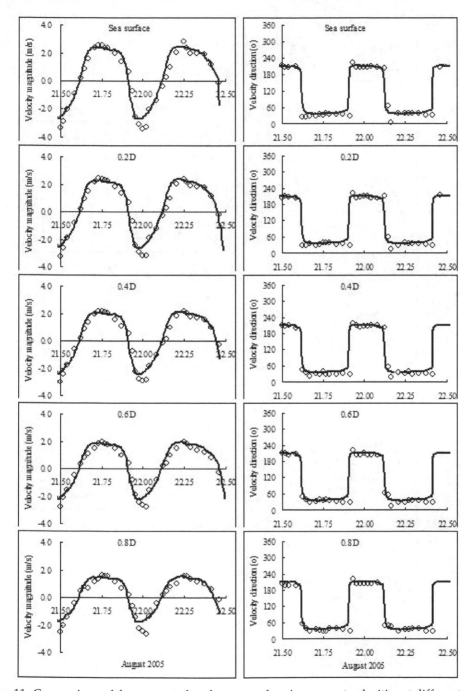

Fig. 11. Comparison of the computed and measured spring current velocities at different depths at station H1. —: computed; o: measured

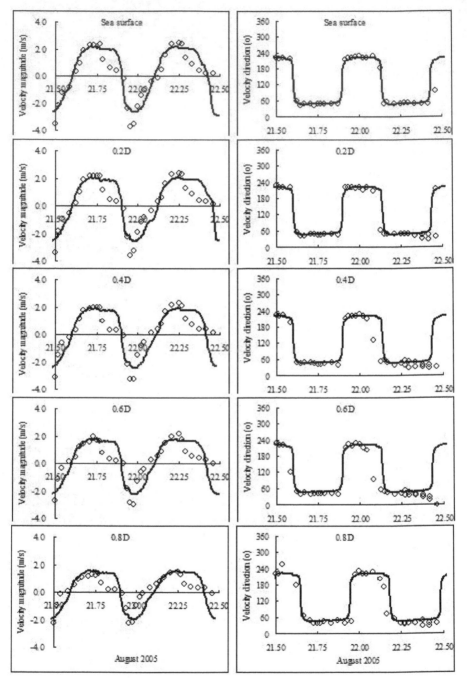

Fig. 12. Comparison of the computed and measured spring current velocities at different depths at station H4. −: computed; ○: measured

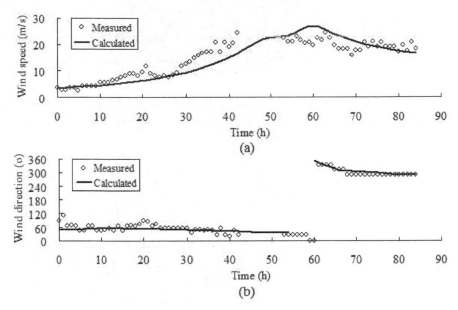

Fig. 13. Comparison of calculated and measured wind fields at Daji station during Typhoon Agnes. (a): wind speed; (b): wind direction. Starting time 0 is at 18:00 of 29/08/1981

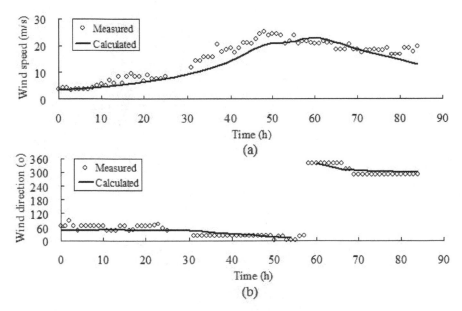

Fig. 14. Comparison of calculated and measured wind fields at Tanxu station during Typhoon Agnes. (a): wind speed; (b): wind direction. Starting time 0 is at 18:00 of 29/08/1981

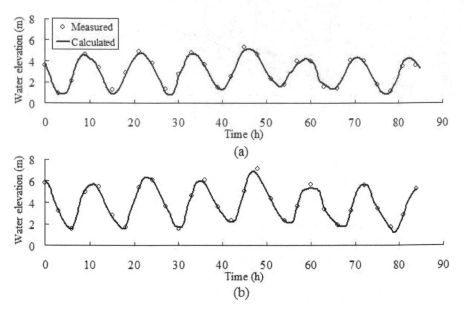

Fig. 15. Comparison of calculated and measured water elevations during Typhoon Agnes. (a): Daji station; (b): Tanxu station. Starting time 0 is at 18:00 of 29/08/1981

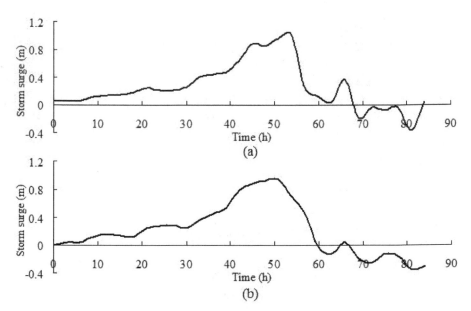

Fig. 16. The simulated surge setup at two stations during Typhoon Agnes. (a) Daji station; (b) Tanxu station. Starting time 0 is at 18:00 of 29/08/1981

## 5. Conclusions

In this study, the results from field observation and 3D numerical simulation are used to investigate the characteristics of astronomical tide and typhoon-induced storm surge in the Hangzhou Bay. Some conclusions can be drawn as below:

1. Tidal hydrodynamics in the Hangzhou Bay is significantly affected by the irregular geometrical shape and shallow depth and is mainly controlled by the $M_2$ harmonic constituent. The presence of tropical typhoon makes the tidal hydrodynamics in the Hangzhou Bay further complicated.

2. The tidal range increases significantly as it travels from the lower estuary towards the middle estuary, mainly due to rapid narrowing of the estuary. The tidal range reaches the maximum at Ganpu station (T4) and decreases as it continues traveling towards the upper estuary.

3. The flood tidal velocity is clearly greater than the ebb flow velocity for both the spring and neap tides. The maximum flood velocity occurs at H2 with the value of about 3.8 m/s, while the maximum ebb flow velocity is about 3.1 m/s during the spring tide. During the neap tide, the maximum velocities of both the flood and ebb are much less than those in the spring tide with the value of 1.5 m/s for flood and 1.2 m/s for ebb observed at H2.

4. The vertical distributions of current velocity at stations H1 and H4 show that the current magnitude obviously decreases with a deeper depth (from sea surface to 0.8D), while the flow direction remains the same.

5. Tropical cyclone, in terms of wind stress and pressure gradient, has a significant impact on its induced storm surge. In general, the north-east wind pushing water into the Hangzhou Bay significantly leads to higher tidal elevation, and the north-west wind dragging water out of the Hangzhou Bay clearly results in lower tidal elevation.

## 6. References

Cao, Y. & Zhu, J. "Numerical simulation of effects on storm-induced water level after contraction in Qiantang estuary," *Journal of Hangzhou Institute of Applied Engineering*, vol. 12, pp. 24-29, 2000.

Chang, H. & Pon, Y. "Extreme statistics for minimum central pressure and maximal wind velocity of typhoons passing around Taiwan," *Ocean Engineering*, vol. 1, pp. 55-70, 2001.

Chen, C., Liu, H. & Beardsley, R. "An unstructured, finite-volume, three-dimensional, primitive equation ocean model: application to coastal ocean and estuaries," *Journal of Atmospheric and Oceanic Technology*, vol. 20, pp. 159-186, 2003.

Guo, Y., Zhang, J., Zhang, L. & Shen, Y. "Computational investigation of typhoon-induced storm surge in Hangzhou Bay, China," *Estuarine, Coastal and Shelf Science*, vol. 85, pp. 530-536, 2009.

Hu, K., Ding, P., Zhu, S. & Cao, Z. "2-D current field numerical simulation integrating Yangtze Estuary with Hangzhou Bay," *China Ocean Engineering*, vol. 14(1), pp. 89-102, 2000.

Hu, K., Ding, P. & Ge, J. "Modeling of storm surge in the coastal water of Yangtze Estuary and Hangzhou Bay, China," *Journal of Coastal Research*, vol. 51, pp. 961-965, 2007.

Hubbert, G., Holland, G., Leslie, L. & Manton, M. "A real-time system for forecasting tropical cyclone storm surges," *Weather Forecast*, vol. 6, pp. 86-97, 1991.

Jakobsen, F. & Madsen, H. "Comparison and further development of parametric tropic cyclone models for storm surge modeling," *Journal of Wind Engineering and Industrial Aerodynamics*, vol. 92, pp. 375-391, 2004.

Kou, A., Shen, J. & Hamrick, J. "Effect of acceleration on bottom shear stress in tidal estuaries," *Journal of Waterway, Port, Coastal and Ocean Engineering*, vol. 122, pp. 75-83, 1996.

Lyard, F., Lefevre, F., Letellier, T. & Francis, O. "Modelling the global ocean tides: modern insights from FES2004," *Ocean Dynamics*, vol. 56, pp. 394-415, 2006.

Mandang, I. & Yanagi, T. "Tide and tidal current in the Mahakam estuary, east Kalimantan, Indonesia," *Coastal Marine Science*, vol. 32, pp. 1-8, 2008.

Mellor, G. & Yamada, T. "Development of a turbulence closure model for geophysical fluid problems," *Reviews of Geophysics and Space Physics*, vol. 20, pp. 851-875, 1982.

Millero, F. & Poisson, A. "International one-atmosphere equation of seawater," *Deep Sea Research Part A*, vol. 28, pp. 625-629, 1981.

Pan, C., Lin, B. & Mao, X. "Case study: Numerical modeling of the tidal bore on the Qiantang River, China," *Journal of Hydraulic Engineering*, vol. 113(2), pp. 130-138, 2007.

Su, M., Xu, X., Zhu, J. & Hon, Y. "Numerical simulation of tidal bore in Hangzhou Gulf and Qiantangjiang," *International Journal for Numerical Methods in Fluids*, vol. 36(2), pp. 205-247, 2001.

Wang, C. "Real-time modeling and rendering of tidal in Qiantang Estuary," *International Journal of CAD/CAM*, vol. 9, pp. 79-83, 2009.

Xie, Y., Huang, S., Wang, R. & Zhao, X. "Numerical simulation of effects of reclamation in Qiantang Estuary on storm surge at Hangzhou Bay," *The Ocean Engineering*, vol. 25(3), pp. 61-67, 2007.

# 9

# The Hydrodynamic Modelling of Reefal Bays – Placing Coral Reefs at the Center of Bay Circulation

Ava Maxam and Dale Webber
*University of the West Indies*
*Jamaica*

## 1. Introduction

Reefal bays are a common type of bay system found along most Caribbean coasts including the Jamaican coastline. These bay systems are associated with and delimited by arching headland with sub-tending reef arms broken by a prominent channel. Traditionally, these bays are termed "semi-enclosed" as their limits are defined by the sand bar or reef partially cutting off waters behind them from open sea (Nybakken, 1997). Yet, it has been shown that circulatory patterns emanating from the lee of reef structures can persist beyond the fore-reef (Prager, 1991; Gunaratna et al., 1997). This raises the possibility of re-characterizing these systems where the reef is defined as the centre of a dynamic bay, inducing a continuous re-circulation of the inside waters beyond the traditional limit (Figure 1). In this study, hydrodynamic modelling, particle tracking and a novel gyre analysis method were used to assess the reefal bay's signature spatial and temporal patterns in circulation, with the goal of characterizing the reefal bay as unique in its function. This was carried out on the Hellshire southeast coast of Jamaica where four of seven bays are typical reefal bays.

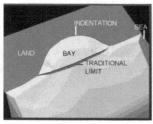
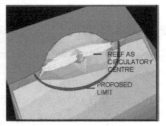

(A) open bay         (B) traditional reefal bay         (C) proposed reefal bay

Fig. 1. A number of hypothetical bays are presented where A represents the open bay, B the traditional definition of the reefal bay, and C the reef proposed as circulatory centre of the reefal bay system.

Reef systems often function to reduce the shoreline wave action and influence sediment dynamics. They therefore provide the ecological link between land and sea, as nurseries offering protection for marine life, as recreational sites, and as receiving sites for industrial and biological effluent. Their distinctive circulatory patterns have, however, been understudied and not fully characterized. This research aims to describe the signature circulatory patterns of the subtending reef bay system, including the effects of bathymetry, wind, tides and over-the-reef flow on this circulatory emanation. Hydrodynamic modelling, particle tracking and a novel gyre analysis method were utilized to characterize the reefal bay circulation and determine those features that make this reef-centered bay system unique.

Reefal bays carry unique patterns of circulation, however, very few reef hydrodynamic studies have focused on the particular circulation associated with fringing Caribbean reef systems. One study on a shallow, well-mixed Caribbean type back-reef lagoon in St. Croix documents that circulation was dominated by wind and over-the-reef flow (Prager, 1991). Another study on the Grand Cayman Island reefs documented that the outer reef tended to be dominated by wind-driven currents and the inner by high frequency waves. Deep water waves and tides, winds and over-the-reef flow controlled the hydrodynamic sub-system found in the lagoon (Roberts et al., 1988). At the reef crest, wave breaking and rapid energy transfers resulted in a sea level set-up which drove strong reef-normal surge currents (Roberts et al., 1992). In both the Grand Cayman and St. Croix reef systems, flow over the reef was often the dominant forcing mechanism driving lagoon circulation (Roberts, 1980; Roberts & Suhayda, 1983; Roberts et al., 1988). Whereas previous studies have contributed to Caribbean reefal hydrodynamics, their application to the reefal bay systems in particular falls short in a number of ways. The reefal bay dynamics has never been distinguished from other reef systems as a unique coastal system. It is instead often broadly categorized under the larger fringing reef system or as a fully enclosed lagoon system. Also, the contribution of reef-induced eddies to the hydrodynamic make-up is understated. Smaller-scale eddy features were not examined in these Caribbean studies. These are important features to note, whether transient or permanent in nature (Sammarco & Andrews, 1989) because of their ability to trap water, sediments, larvae and plankton around reefs. Sammarco & Andrews (1989) showed that attenuation of tidal effects within lagoons and tidal anomalies generated by the reef were responsible for creating or maintaining eddies on isolated systems. More comprehensive research is now necessary to determine the characteristic circulatory dynamics and responsible forcing functions.

## 2. Numerical modelling development and challenges for reef systems

The lagoons formed by coral reefs exhibit some of the most variable bathymetry of coastal oceanography and present a challenge to understanding their dynamics (Hearn, 2001). The ideal model must be able to account for all the forcing factors and conditions typical of the coral reef environment including wave and current propagation and interaction, density flows, channel exchange, reef topology and reef morphology. The modelling becomes even more complex when attempts are made to process spatial scales ranging from tens of kilometers down to sub-meter at the same time. These difficulties continue to confound localized studies of reef phenomena.

Several numerical models have been applied to lagoon hydrodynamics using one-dimensional (Smith, 1985), two-dimensional (Prager, 1991; Kraines et al., 1998) and three-dimensional models (Tartinville et al., 1997; Douillet et al., 2001). Wave breaking and overtopping remain phenomena that are difficult to describe mathematically because the physics is not completely understood (Feddersen & Trowbridge, 2005; Pequignet, 2008). The large range of combinations of reef types, shapes, tidal environments and wave climates makes all existing analyses of wave-generated flow on coral reefs limited in their applications (Gourlay & Colleter, 2005). Instrument-measured field data, however, confirm that the wave dynamics is responsible for a significant proportion of the reefal lagoon/bay hydrodynamics (Symonds et al., 1995; Hearn, 1999, 2001). As the waves break, a maximum set-up occurs near the reef edge. The maximum set-up on the reef top is proportional to the excess wave height (Hearn, 2001). The set-up creates the pressure gradient required to drive the wave-generated flow across the reef (Gourlay & Colleter, 2005). Friction coefficients are also important to consider and so these are presented as large values in recognition of the great roughness of reefs (Symonds et al., 1995). In consideration, however, of reefs with steep faces where waves break to the reef edge, wave set-up is reduced by the velocity head of the wave generated current. In this case, influence of bottom friction in the surf zone is ignored. Wave overtopping has been developed and described as two linked functions by Van der Meer (2002):- one for breaking waves applicable to more intense wave conditions (here, wave overtopping increases for an increasing breaker parameter), and the other for the maximum achieved for non-breaking waves applicable to significantly reduced wave conditions where waves no longer break over the reef.

Three-dimensional models continue to evolve in simulating wave-driven flow across a reef. An attempt is made in this chapter to simulate the three-dimensional flow associated with reefal bays by incorporating equations for wave breaking and overtopping at the reef into a finite element-based model for stratified flow.

## 3. Reefal bay sites

Southeast of Jamaica, a 15 km stretch of coastline, the Hellshire east sector (Figure 2), consists of seven bays - four of which are reefal. Three bays were compared for their circulatory signatures – Wreck Bay, Engine Head Bay and Sand Hills Bay. Two of the three, Wreck Bay and Sand Hills Bay, have prominent reef parabola stretching between headlands with a central, narrow channel breaking the reef continuum. Wreck Bay, with its narrower channel, is more enclosed than Sand Hills Bay. Associated reefs are emergent and exposed, more so at low tide. Both reefal bays are separated along the coastline by Engine Head Bay, an open bay with no development of reef arms. Engine Head Bay was therefore considered as a control given it is non-reefal and its position exposes it to the same conditions as the two reefal bays.

A diurnal variation in the wind records is typical of the southeast coast of Jamaica (Hendry, 1983) due to the influence of the sea-land regime. The tidal range is microtidal ranging from 0.3 - 0.5 m with an annual mean of 0.23 m (Hendry, 1983) and demonstrating a mixed tidal regime. Tidally generated currents are therefore small in amplitude compared to wind-driven currents. The wave climate of the southeast coast is influenced mainly by trade wind-generated waves that approach Jamaica from the northeast. Offshore waves impact the shelf edge off Hellshire from a predominantly east-south-easterly direction after undergoing southeast coast refraction. Swell waves approach the coast at a typical period range of 6-9 seconds, but these are soon affected by complex bathymetry. Wave decay occurs when the land-breeze emanates along the coast. The shelf along which these bays fringe are made up of basement rock composed of Pliestocene limestone eroded during low sea levels in the Pliestocene epoch. As a result, bathymetric highs are now shoals, banks, reefs and cays, and on the inshore, karst limestone relief facilitates freshwater sub-marine seeps into the bays (Goodbody et al., 1989).

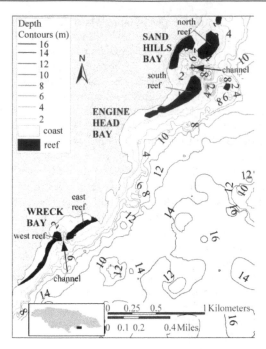

Fig. 2. Map showing the study site of three bays located on the Hellshire South East Coast of Jamaica. Wreck Bay and Sand Hills Bay are the two reefal bays under investigation, along with the open bay Engine Head Bay located between the other two.

Environmental stress studies conducted inshore and offshore these bays used plankton population size and species composition as indicators. Lowest values in biomass, primary production and density were recorded in the southernmost bays. These bays were therefore considered generally removed from the effects of the highly productive Kingston Harbour and Great Salt Pond waters to the north, with the exception of during flood occasions when elevated levels were recorded in the southernmost bay, Wreck Bay. The authors suggested the possibility of long retention times due to localized circulation (White, 1982; Webber, 1990). These results were of great interest given the implications presented for the protective role played by reefal bays as nurseries for the early aquatic stages of marine and terrestrial species; for the significance of its distance down-shore from the main harbor not inhibiting its eutrophication; and for sediment transport and exchange along the shoreline. In fact, physicochemical variables were also robust in characterizing the persistence of bay waters beyond the reef (Maxam & Webber, 2009). This indicated the need for appropriate numerical simulations to adequately describe the circulatory patterns in these bays - the findings of which are presented in this chapter.

## 4. Methods for Simulating the reefal bay system

Oceanographic and meteorological data were collected for the Hellshire coast and served as inputs into the hydrodynamic model. Field data were also used for model verification after executing model simulations under various meteorological conditions. This was followed by an analysis of bay contraction and expansion due to circulation induced by the presence of the subtending reef, and ultimately the development of particular circulatory signatures defining the reefal bay.

## 4.1 Oceanographic and meteorological data collection

Bathymetric depth points were digitized from Admiralty bathymetric charts for the Hellshire coastline area and the entire South-East Shelf. For the finer-scale bathymetry required of the reef and bay areas, water depth (± 0.1 m) was measured to supplement the Admiralty data using an echo-sounder with Trimble Garmin GPS and post processed to account for tidal elevation differences from mean sea level. Wind speed (± 0.1 m s$^{-1}$) and wind direction (± 0.1°) data were collected from the nearby Normal Manley International Airport weather center as continuous two-minute averages over the entire sampling period (1999 to 2003). Long-term current measurements for speed (± 0.10 cm s$^{-1}$) and direction (± 0.1°) were recorded continuously by Inter-Ocean S4 current meters at four sites inside (Table 1) and outside of Wreck Bay.

| | Mooring Location | Depth (m) | Deployment Dates | Duration (wks) | On / every |
|---|---|---|---|---|---|
| 1 | Channel | 4.0 | 24 May – 13 Jun 2000 | 3 | 5 min / 1 hr |
| 2 | Channel | 4.0 | 11 Jul – 03 Sep 2000 | 7 | 5 min / 1 hr |
| 3 | Channel | 4.0 | 20 Dec 2002 - 10 Jan 2003 | 3 | 1 min / 10 min |
| 4 | Channel | 4.0 | 14 Mar – 28 Mar 2003 | 1 | 1 min / 10 min |
| 5 | West Back-reef | 2.0 | 11 Jul – 03 Sep 2000 | 7 | 5 min / 1 hr |
| 6 | East Back-reef | 0.7 | 20 Jul – 01 Sep 2000 | 1 | 5 min / 1 hr |

Table 1. Deployment specifications for long-term field current data collection in Wreck Bay. Hydrodynamic model outputs were compared with these measurements for verification.

Hourly tidal amplitudes (± 1 mm) were calculated using Foreman's Tidal Analysis (Foreman, 1977) and Prediction Program, incorporating mean sea-level and tidal amplitude data over a 40-year period from Port Royal, a nearby tide station. Hourly incident wave height values (± 1 cm) used in the over-the-reef flow calculations were taken from Refraction-Diffraction (REFDIF) wave models (Kirby & Dalrymple, 1991) of the shoreline (Burgess et al., 2005). The deepwater wave climate obtained from JONSWAP (Hasselmann et al., 1973) analysis was used to run the REFDIF models in order to carry the deepwater waves from the continental shelf to the shoreline. Near-shore conditions were simulated at 50% occurrence (average conditions) and used as input into the hydrodynamic model.

## 4.2 Hydrodynamic modelling

A hydrodynamic model, RMA-10, was utilized to simulate the depth-averaged velocity field of the fore-reef and back-reef along with the shoreline flow under wind and tidal conditions typical of the Jamaican south-east coastal area. RMA-10 is a three-dimensional finite element model for simulation of stratified flow in bays and streams (King, 2005). The primary features of RMA-10 are the solution of the Navier-Stokes equations in three-dimensions; the use of the shallow-water and hydrostatic assumptions; coupling of advection and diffusion of temperature, salinity and sediment to the hydrodynamics; the inclusion of turbulence in Reynolds stress form; horizontal components of the non-linear terms; and vertical turbulence quantities are estimated by either a quadratic parameterisation of turbulent exchange or a Mellor-Yamada Level 2 turbulence sub-model (Mellor & Yamada, 1982). Computations in the model are based on the Reynolds form of the Navier-Stokes equations for turbulent flows and employ an iterative process that solves simultaneous equations for conservation of mass and momentum. RMA-10 requires the input of nodal x, y and z data depicting sea floor bathymetry, parameters for roughness and eddy viscosity, and boundary conditions of flow discharge. The iterative process computes nodal values of water surface elevation, flow, depth and layered horizontal velocity components or vertically averaged velocity components if this option is used.

Two-dimensional depth-averaged approximations were used for the Hellshire bays' simulations. Depth-averaged results are appropriate given the shallowness of the reefal bay and the knowledge that this usually presents a well-mixed system. Boundary conditions were entered into RMA-10 using a list of nodes defined as flow continuity checks simulating flow over the reefs and also used to specify initial values of salinity concentration (36.0 ppt), temperature (28.0 °C) and suspended sediment concentration (2.0 $\mu gL^{-1}$) conditions along the model east and west open boundaries. Boundary conditions were also read from a wind velocity and direction file derived from wind data. This was input as hourly averaged wind velocity and direction and allowed the model to read dynamic wind conditions useful in examining the influence of a diurnal wind regime. Boundaries were also subject to a tidal-graph of hourly tidal elevation data for interpolation.

Reef parabola were represented by continuity lines where hydrograph data of dynamic flow over the reef were interpolated. Flow over the reef was calculated as hourly-averaged values using the wave run-up and overtopping Van der Meer equations (Van der Meer, 2002) as a base. Wave overtopping is the average discharge per linear meter of width, $q$, and is calculated in relation to the height of the reef crest line. The final flow value Q used in the hydrograph file is given as the length of the reef parabola long axis multiplied by the average discharge $q$. For breaking waves ($\gamma_b \xi_0 \leq 2$), wave overtopping increases for increasing breaker parameter $\xi_0$. Assumptions are made of a fully developed wave at the reef crest and so the incident wave height is used. Determination of correct wave period for heavy wave-breaking on a shallow fore-shore is neglected here as this requires complex wave transformation Boussinesq models (Nwogu et al., 2008) and lies beyond the objectives of this study. Instead, an average value for the wave period is used. Other influences are included in the general formula such as roughness on the reef slope and the reef slope itself (considered here to be equal to or steeper than 1:8 close to the reef crest). The wave overtopping formula is given as exponential functions with the general form:

$$q = a \cdot \exp(b.R_c) \tag{1}$$

The coefficients $a$ and $b$ are still functions of the wave height, slope angle, breaker parameter and the influence factors of reef roughness and slope; $R_c$ being the free crest height above still water line. Wave heights used varied around the predicted value of 0.48 m but were not simulated for extreme events (<1-year event occurrence). A set of turbulent exchange, turbulent diffusion and Chezy coefficients was applied at all nodes. The turbulent exchange coefficient associated with the x and y direction shear of the x and y direction flow was set as -5.4 Pa s. The turbulent exchange coefficient of the z direction shear of the x and y direction flow was set at 0.44 Pa s. The turbulent diffusion coefficient associated with the x and y directions were set at 2.11 m$^2$ s$^{-1}$ and that associated with the z direction set at 0.21 m$^2$ s$^{-1}$. The Chezy coefficient of 0.029 m$^{0.5}$ s$^{-1}$ was used for all nodes except at the shoreline where it was reduced to 0.0015.

Particular conditions at the Hellshire coastline led to adding a third variable, Y, to account for the diurnal effect of the wind regime. It was found that emanation of the land-breeze significantly reduced wave heights and caused more variation in the flow over the reef than predicted by the Van de Meer calculations. This variable Y is a function of the southward wind flow and leads to a large reduction in the $q$ value once the land-breeze emanates. At a maximum the final overtopping formula becomes:

$$\frac{q}{\sqrt{gH_{m0}^3}} = 0.2 \cdot \exp(Y) \cdot \exp\left[-2.3 \frac{R_c}{H_{m0}} \frac{1}{(\gamma_f)(\gamma_b)}\right] \tag{2}$$

where:

| | | | |
|---|---|---|---|
| $q$ | = | average wave overtopping discharge | (m$^3$ s$^{-1}$ m$^{-1}$) |
| $g$ | = | acceleration due to gravity | (m s$^{-2}$) |
| $H_{m0}$ | = | significant wave height | (m) |
| $R_c$ | = | free crest height above still water line | (m) |
| $Y$ | = | wind y component | |
| $\gamma_f$ | = | influence factor for roughness | |
| $\gamma_b$ | = | influence factor for slope | |

Comparisons between RMA Model results and field-collected current measurements were tested for significance using the t-test.

The model mesh was built using assemblages of two-dimensional triangular and quadrilateral elements. The software RAMGEN (King, 2003), a graphics based pre-processor for RMA-10, was used to form the grid and create the interface file that the RMA software utilized. The regional mesh covered the entire south-east coastal shelf including Kingston Harbour to the north, and had two open boundaries - one at the east side and the other south-west. Courser elements (>1 km$^2$) were created for the offshore shelf areas. Elements were more refined (<100 m$^2$) closer to the shoreline or in areas where there were expectations of large changes.

Individual particles were tracked based on the velocity distribution used by the RMATRK software (King, 2005). This application is designed to track particles released into a surface water system that have been simulated with the RMA-10 model. It transports discrete objects through a surface water system defined by the RMA-10 finite element grid. Time steps were set up so that track increments were drawn for every six minutes in a one-hour or three-hour time block.

## 4.3 Gyre analysis

The horizontal expansion and contraction of gyres were measured to quantify the extent of bay fluctuation. Tracks produced by the RMATRK model were of three categories:

- Hourly plotted tracks: where new particles were introduced in the same positions at the beginning of each hour for as long as the duration of one and a half tidal cycles,
- Three-hourly tracks: where new particles were introduced in the same positions at the beginning of each three-hour time block for as long as the duration of one and a half tidal cycles, and

- Indefinite tracks: where one set of particles was tracked for as long as they remained in the reefal bay system - their paths plotted for every three hours they remained.

Hourly plotted tracks were used to predict the duration of a reef gyre. Three-hourly tracks were plotted to capture the full horizontal extent of the circulation.

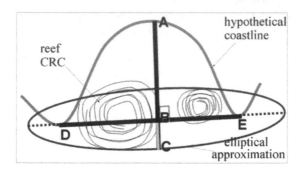

Fig. 3. Diagram of the reefal bay dimensions used in calculating the circulatory extent of the bay. The extent (BC) is calculated as a fraction of the AC distance normal to a line (DE) joining the land projections. AC is derived from an elliptical approximation of the outer, seaward curve of the looping currents.

Extents were measured from these plots as a proportion of the linear distance, from the shore to the elliptical arc, normal to a straight line joining the land projections at the ends of the bay indentation (Figure 3). The ellipse best approximates the seaward edge of the gyre. The elliptical major axis is always equal to or greater than the length of the straight line joining the land projections. Therefore, the reef circulation lateral extension, $L_c$, is given as the percentage:

$$L_c = \frac{BC}{AB} \times 100 \tag{3}$$

Indefinite tracks allowed predictions of the retention ability of gyres. The number of particles remaining around the reef was counted after each 3-hr track run.

## 5. Results

### 5.1 General current flow description based on field measurements

Results from fixed S4 current measurements in Wreck Bay (Figure 4) showed water flowing through the channel generally exited in a south-south-eastward direction, with a deflection southwards when current speeds were high.

Mean speed values for channel currents peaked at 22 cm s[-1], and flow directions were southward from 173° to 181°. On the western arm speeds averaged 28 cm s[-1] with a mean flow direction of 102°, and on the eastern arm mean speed was 22 cm s[-1] with a flow direction of 290°. Flow persisted southwards out through the channel from the back-reef currents continuously, except during very rare occasions of in-flow at mid-depth when velocities were at their lowest (mean of 2.9 cm s[-1]). Channel currents in Wreck Bay were greatly influenced by the back-reef feeder currents, more than the direct influence of wind and tides. Correlations of channel and back reef flow components showed that the western arm current magnitude was almost five times more strongly correlated (cross correlation r =

0.62) than the eastern arm currents (cross correlation r = 0.18) with channel currents. The west reef feeder currents therefore contributed much more to channel flow than the east reef. Multiple regression values showed that the back-reef currents combined accounted for 47% of the variability in the channel currents, compared to wind and tides accounting for 29%.

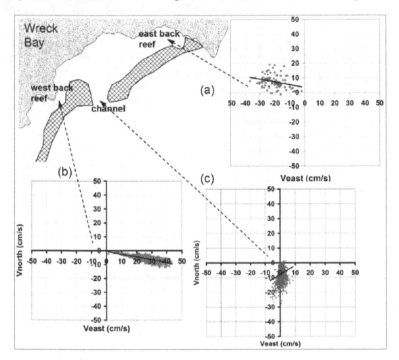

Fig. 4. Current component plots are shown for the east back-reef (a), the west back-reef (b) and channel (c) of Wreck Bay, collected from long-term deployment of S4 current meters moored at all three sites at the same time. This field data compared favorably with RMA model results.

Accountability by winds and tides of the overall variability in the current magnitude decreased from highs of 55-56% for the spring and winter data to 29% for the summer currents. During this summer period the lowest recorded mean channel current speed (7.7 cm s$^{-1}$) was observed as well as an equality of the relative contributions of tides and winds to the overall variability.

## 5.2 RMA model simulations
### 5.2.1 Current flow
Velocity results from S4 current meters compared well with RMA model results (depth averaged) for the dominant north (Y) component of the channel site at Wreck Bay (Figure 5), giving no significance for difference by t-test. For the month of August (2000) , S4 north component readings averaged -7.8 cm s$^{-1}$ while the RMA model averaged slightly lower at - 8.2 cm s$^{-1}$ (Table 2). The north component was used to represent the channel flow given its high cross-correlation value of -0.99 with the channel flow magnitude.

Depth-averaged velocity results from hydrodynamic modelling showed that currents circulated the reef arms constantly. This circling of the reef was strongest during the combined condition of a rising tide with prevalent sea-breeze (Figure 6). This particular condition generated some of the strongest currents on the west reef of Wreck Bay (the 28 to 32 cm s[-1] category) and the corresponding south reef of Sand Hills Bay. Back reef current highs by the model, however, were less than measured in the field. Field-measured monthly average for the Wreck Bay east back reef current magnitude was 22 cm s[-1] and agrees with model averages, however, the variation in flow is not replicated and spikes in back reef speeds (up to 38 cm s[-1]) not captured. In Sand Hills Bay, model currents strongly circulated the south reef at up to 28 cm s[-1] on the southern curve of the gyre. Engine Head Bay showed no formation of looping currents. The combination of a prevalent sea-breeze with falling tide strengthened the east reef circulation in Wreck Bay (Figure 7). Horizontal current fields depicted velocities of up to 32 cm s[-1] in this gyre, the fastest speeds occurring on the western side of the gyre. For Sand Hills Bay, the north reef gyre was pronounced with a central inner gyre showing closed circulation. Horizontal current fields depicted velocities of the 18 to 20 cm s[-1] category around the north reef. Engine Head Bay again showed no formation of horizontal circulatory currents.

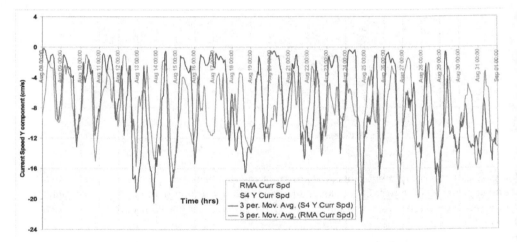

Fig. 5. RMA model and S4 field north component current data comparisons for the Wreck Bay Channel area. A t-test reported no significance for difference when both current data sets were input as independent samples (t = 1.46; p = 0.15).

| | Y-COMP VELOCITY RESULTS (cm s[-1]) | |
| --- | :---: | :---: |
| | RMA Model Data | S4 Field Data |
| Average: | -8.2 | -7.8 |
| Maximum: | -0.7 | 0.3 |
| Minimum: | -24.5 | -24.7 |
| Range: | 23.8 | 25.0 |

Table 2. RMA model and S4 field north component current data statistics and comparisons for Wreck Bay Channel.

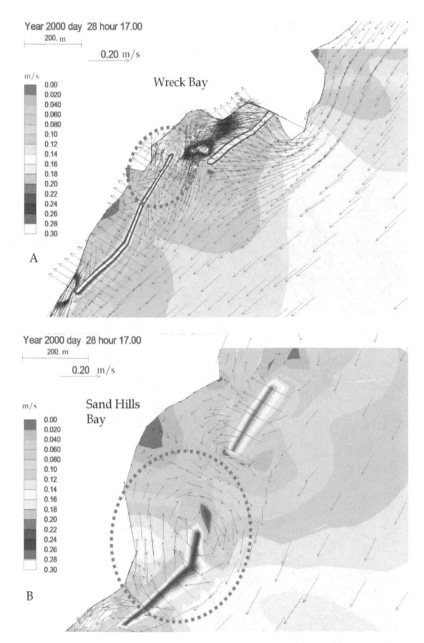

Fig. 6. Depth-averaged current field maps for (a) Wreck Bay and (b) Sand Hills Bay during a dominant rising tide combined with sea-breeze regime. Current vectors depict well-formed, closed looping circulation on the down-shore reef arm (circled), causing both bays to be expanded beyond the reef.

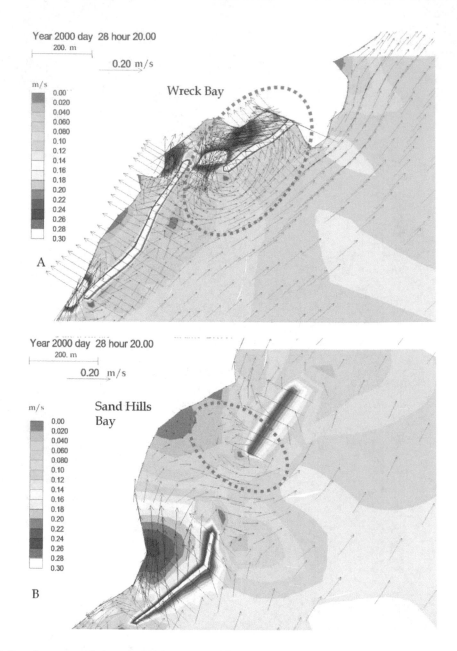

Fig. 7. Depth-averaged current field maps for (a) Wreck Bay and (b) Sand Hills Bay during a dominant falling tide combined with sea-breeze regime. Current vectors depict well-formed, closed looping circulation on the up-shore reef arm (circled), causing both bays to be expanded beyond the reef.

## 5.2.2 Particle tracking and retention

Under only the rising tide regime, 19 % particles remained in Sand Hills Bay after 9 hrs. The rising tide combined with land-breeze regime increased the remaining particles to 22 % after 9 hrs. When the sea-breeze dominated, however, combined with the rising tide the retention dropped to 2 % in 9 hrs. Therefore particles were likely to remain trapped in Sand Hills Bay the longest when introduced at the beginning of the rising tide cycle during a land-breeze regime and were likely to be flushed out the quickest if introduced during the sea-breeze with mid-falling tide.

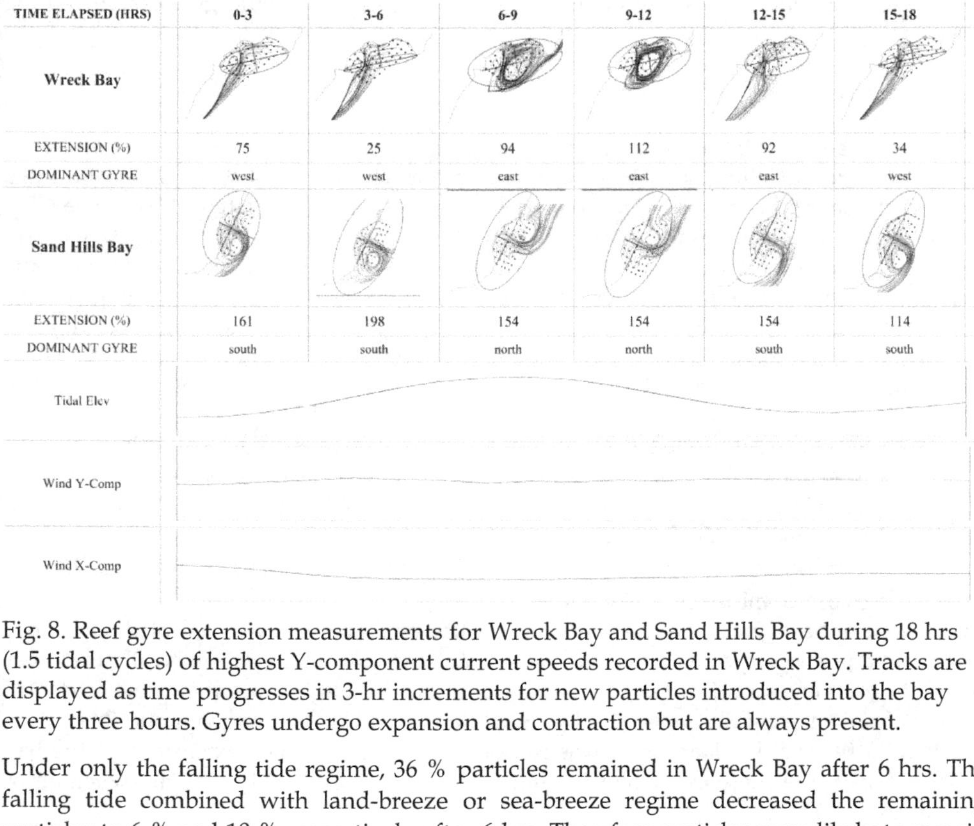

| TIME ELAPSED (HRS) | 0-3 | 3-6 | 6-9 | 9-12 | 12-15 | 15-18 |
|---|---|---|---|---|---|---|
| Wreck Bay | | | | | | |
| EXTENSION (%) | 75 | 25 | 94 | 112 | 92 | 34 |
| DOMINANT GYRE | west | west | east | east | east | west |
| Sand Hills Bay | | | | | | |
| EXTENSION (%) | 161 | 198 | 154 | 154 | 154 | 114 |
| DOMINANT GYRE | south | south | north | north | south | south |
| Tidal Elev | | | | | | |
| Wind Y-Comp | | | | | | |
| Wind X-Comp | | | | | | |

Fig. 8. Reef gyre extension measurements for Wreck Bay and Sand Hills Bay during 18 hrs (1.5 tidal cycles) of highest Y-component current speeds recorded in Wreck Bay. Tracks are displayed as time progresses in 3-hr increments for new particles introduced into the bay every three hours. Gyres undergo expansion and contraction but are always present.

Under only the falling tide regime, 36 % particles remained in Wreck Bay after 6 hrs. The falling tide combined with land-breeze or sea-breeze regime decreased the remaining particles to 6 % and 10 % respectively after 6 hrs. Therefore particles were likely to remain trapped in Wreck Bay the longest if introduced at the beginning of the falling tide cycle and were likely to be flushed out the quickest if introduced at the beginning of the rising tide.

### 5.3 Gyre extension assessment

Gyres expanded and contracted around reefs as the forcing conditions changed (Figure 8). As the gyre on one reef arm strengthened the other weakened. Wreck Bay had its largest extension ($L_c$ = 112 %) during the falling tide phase and when the sea-breeze emanated. The largest extensions were produced by the east reef circulation and coincided with the greatest current component speeds flowing out of the channel. This channel current formed the western edge of the east gyre. When the west reef circulation emanated, gyre extensions were smaller and did not exceed 75 %. West reef gyres were most developed at low-to-rising tide during land-breeze emanation and coincided with the lowest current component speeds recorded in the channel at that time. The longest duration of this closed western gyre was observed during 15 hrs of some of the smallest tidal changes recorded.

Sand Hills Bay had its largest extension at 198 % during the combination of a rising tide and when the sea-breeze emanated. This was due to the south reef gyre that also tended to be more closed than the north reef's. The north reef gyre was most developed at the rising-to-high tide (also when the sea-breeze emanated) and had its largest extension at 154 %. In the absence of large tidal changes and developed wind regimes, the south gyre dominated the extension.

## 6. Discussion

### 6.1 Circum-reef circulation defining the reefal bay

Hydrodynamic modelling showed that circulation around the Wreck Bay and Sand Hills Bay reef parabola continuously looped the reef as circum-reef circulation (CRC). The CRC was considered "closed" when fore-reef currents fed water back into the back-reef and "open" when main fore-reef flow continued along-shore (Figure 9). Channel surge currents were responsible for the propagation of inner bay waters seawards, and encouraged open CRC. Tracking models revealed the longevity and spatial spread of this flow, simulating the patterns first observed in these bays by field drogues and fixed measurements that depicted continuous current flow around reef arms at surface and depth (Maxam & Webber, 2010). The presence of the reef induced this persistence and localized the (CRC). The lack of reefs in Engine head bay supported this premise as gyre formation and localization was not evident in the non-reefal bay. This was confirmation that open bays did not facilitate recycling of their inside waters from the outside as reefal bays do. In the absence of prominent reef arms, the CRC cannot exist.

### 6.2 Reef arm crc dominance and cycling

Simulations of new particles introduced into the bay on an hourly basis revealed that under particular tide and wind regimes, one reef's circulation was strengthened while the other abated in the same bay (Figures 10, 11). This simulated the dynamics that prevented field drogues from entering the weaker reef gyre while trapped in the dominant one (Maxam & Webber, 2010). The dominant gyre was responsible for the greatest extensions of the bay system, and so the presence of two prominent reef arms resulted in regular switching of dominance.

Full development of both reef arm gyres occurred in one tidal cycle. The reef gyre downstream the main long-shore flow appeared strengthened on the rising tide while the adjacent reef up-shore was strengthened by the falling tide. It is important to note that these simulations accurately portray the importance of the tidal influence in a micro-tidal environment where it was otherwise expected to be overwhelmed by wind- and wave-induced stresses. In the absence of large tidal fluctuations, as during a neap tide, the up-shore gyre was too weak to be developed and the down-shore gyre dominated. Up-shore reef arms were more reliant on tidal changes to effect gyre formation than down-shore reefs. The sea-breeze aided in strengthening both gyres during simulation, agreeing with long-term field data that showed this correlation (Maxam & Webber, 2010). This wind regime

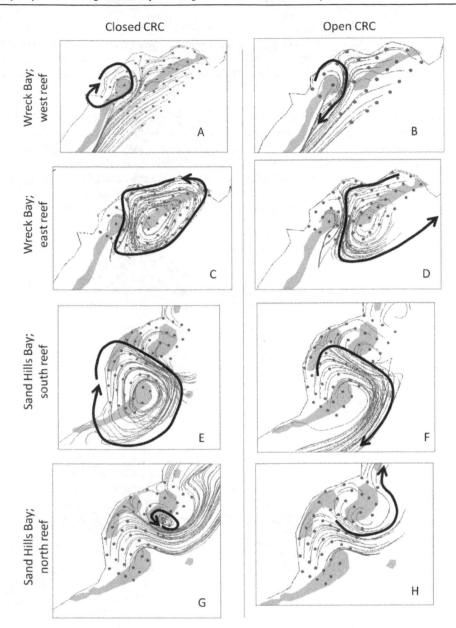

Fig. 9. Diagrams depicting closed and open circum-reef circulation (CRC) simulated from RMATRK discrete particle tracking modelling. The closed CRC displaying recirculation were evident for Wreck Bay west reef (A) and east reef (B) arms, as well as Sand Hills Bay south reef (C) and north reef (D) arms particularly during wind calms. Open CRC is also displayed in Wreck Bay west reef (E) and east reef (F) arms, and again around Sand Hills Bay south reef (G) and north reef (H) arms particularly during increased channel flow.

induced more flow over the reef due to increased heights of waves impinging on the reef and at higher frequencies (Roberts et al., 1992). Breaking would occur and the rapid energy transferred caused an increase in water level, driving strong back-reef surge currents and increasing current speeds in the northern part of the gyre. These surges, however, reduced the retention times of these gyres.

This cycle of emanation and contraction is characteristic of the reefal bay system, giving the reefal bay a spatial pulse that is dependent on prevailing wind and tidal regimes. The reefal bay does not have a static bay area but instead will be at a minimum when the CRC is most contracted and at a maximum when the CRC is most extended. At its minimal spatial extent, the horizontal area of the hydrodynamic reefal bay is dependent on the size of the reef. The larger reef in Wreck Bay, the east reef arm, gave the lager dominant gyre resulting in the greater seaward extensions of the bay. The same was observed in Sand Hills Bay where the south reef was the larger reef and therefore gave the greater extensions (Figure 12).

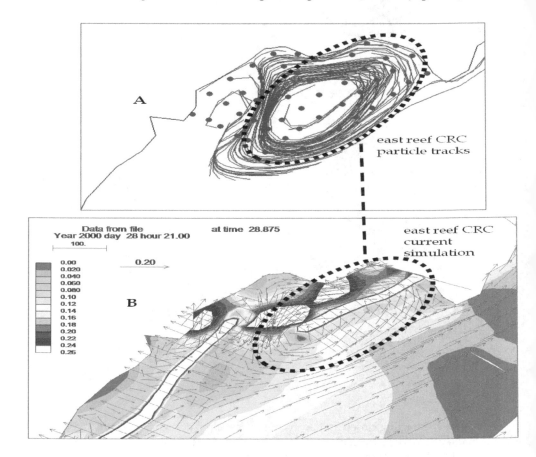

Fig. 10. Dominant east reef CRC in Wreck Bay due to large falling tide range is displayed in A and B as circled area in model particle tracks (A) and model vectors (B). CRC formation on the opposing reef arm is weakened during dominance of the other.

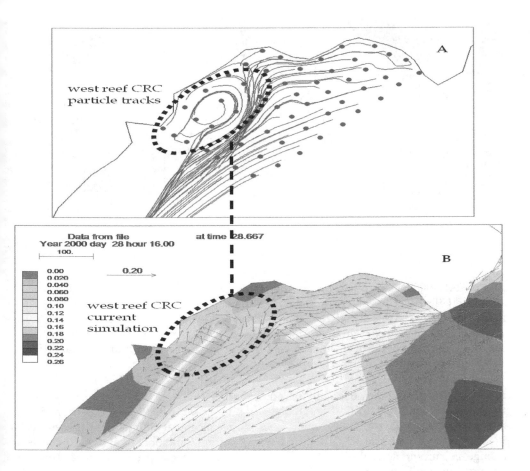

Fig. 11. Dominant west reef CRC in Wreck Bay is shown here typically occurring during neap periods when bay extension was due primarily to wind and over-the-reef forcing. CRC is displayed as circled area in model particle tracks (A) and model vectors (B). CRC formation on the opposing reef arm is weakened during dominance of the other.

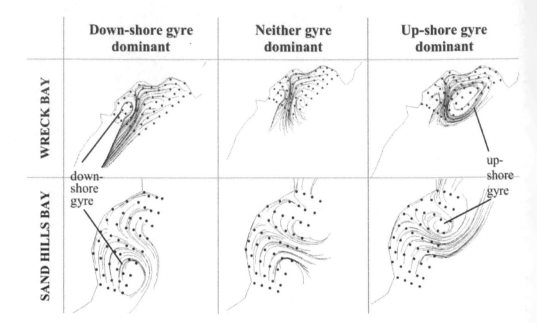

Fig. 12. RMTRK Tracking model outputs depicting gyre dominance cycling in Wreck Bay and Sand Hills Bay. Closed gyre formation is dominant on the down-shore reef during rising tide regimes, abate at high tide, then re-form on the up-shore reef during falling tide. The larger reef in both bays produced the larger dominant gyre resulting in the greater seaward extensions of the bay. The east reef for Wreck Bay and the south reef at Sand Hills Bay therefore expanded the bays the most.

### 6.3 Reef CRC persistence between paired reef arms

Persistence of one reef CRC over another was observed with the reef pairs and was characteristic of one reef only, unlike reef dominance that alternated between reefs. Persistence of a reef arm CRC occurred when, during conditions that caused the least change in current flow, the CRC was continuously propagated on that reef. This was observed during a combination of decreased over-the-reef flow and small changes in tidal amplitude, when the Wreck Bay west reef arm and the Sand Hills Bay south reef arm displayed continuous CRC while the other reef arms in the pair showed none, even during changing tidal cycles. This persistence, along with the larger west reef flow, has led to the west reef contributing more than the east reef overall to the channel flow in Wreck Bay.

### 6.4 Variability in retention

Sand Hills Bay retained particles longer than Wreck Bay in model simulations, with retention controlled mostly by the dominant reef gyre. The dominant reef gyre is maintained in Sand Hills Bay during the rising tide, while that of Wreck Bay is well-formed during the falling tide. This presents the likelihood that waterbourne particles flushed out of

Kingston Harbour to the north during a flood event undergo retention along the Hellshire shoreline all through the tidal cycle, particularly during wind clams, but alternating in these reefal bays depending on the stage of the cycle prevailing. The longest retention time derived from field data was 9 hrs (Maxam & Webber, 2010) and compares well with model results that showed the longer retention of particles ranging from 6 to 9 hrs.

Simulations also show that CRC presence is characterized by increased fluctuations in the retention of particles. Model simulations depicted that after 6 hrs, Wreck Bay and Sand Hills Bay showed the greatest variation in number of particles remaining and Engine Head Bay the least variation across all conditions. Therefore, those conditions that facilitated greater particle retention in the reefal bays, particularly wind calms (Maxam & Webber, 2010), significantly increased retention times over that of the open Engine Head Bay. The same is true for those conditions that facilitated decreased particle retention in the reefal bays where these were significantly lower than in the non-reefal bay.

Provided wind conditions did not dominate, Engine Head Bay produced similar retention times as particles oscillated back and forth inside the bay arc with the change of the tidal regime. This oscillation, however, did not extend outside the bay arc, unlike with the reefal bays. Reefal bays therefore display the ability to not only trap particles throughout tidal cycles, but also create a wider trapping area (extended seawards) than open bays along the Hellshire shoreline.

CRC strengthening was therefore evident
i.     with the closure of the looping circulation;
ii.    in the increased recirculation rate of particles resulting from increased gyre current speeds; and
iii.   in the broad spatial extent of the CRC occupying a greater portion of the bay area.

Hence, the CRC is considered persistent because it continuously loops the reef, and is strengthened when gyres are closed and it broadens horizontally. This closing re-circulation demonstrates very well the connectivity and continuity of the channel outflow re-entering the bay over the reef, and therefore best confirms the reef as the circulatory centre of the bay. Model simulations did not produce a reversal in back-reef currents at any time, evidence that the CRC is never completely reversed but instead may become severely weakened, usually coinciding with very rare events of channel reversal at depth (recorded by field instruments in Maxam & Webber, 2010). The functional bay is therefore seen to exist around the reef such that the reef parabola are the center of the system. Increased flow over the reef, especially during the sea-breeze regime, caused surges in channel currents that would increase the speed of the current loop and result in faster flushing times. Reefal bay flushing and retention regimes have direct implications on the dynamics of vulnerable planktonic species important to reef establishment (Wolanksi & Sarsenski, 1997), and the ability of these bays to draw in, retain and flush pollutants (Black et al., 1990; Lasker & Kapela, 1997).

## 6.5 Bathymetric characteristics necessary for promoting CRC

The topography of the reefal bay allows it to produce signature dynamics driven primarily by over-the-reef flow, wind and tidal forcings. Waves break over the reef and the generated flow feed reef-parallel currents that in turn supply a major channel outflow. The channel (Figure 13) features significantly in this system and its prominence is the main bathymetric difference from other more popularly studied reef systems such as atolls, platform and ribbon reef. The channel in the reefal bay is the main conduit of back-reef water exiting to the sea, and therefore sets up the hydrodynamics to produce jet currents that help complete the circum-reef current. This CRC has been shown to either close in on itself , which is when gyres are formed that cause particles to re-circulate on the reef, or to flow along the fore-reef and join the general long-shore flow, causing particles to leave the system.

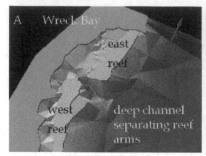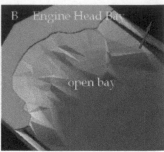

Fig. 13. Spatial 3D model of Wreck Bay (A) and Engine Head Bay (B) revealing their differences in topography. In Wreck Bay, reef arms are emergent at high tide and the deepest part of the system is its prominent channel. This is topographically more complex than the open, non-reefal Engine Head Bay (spatial 3D Models are exaggerated vertically).

Bathymetric characterization includes the reef arms, where their presence localizes the CRC and relative size becomes an important factor. The larger reef arm generates the more expansive gyres and therefore greatest emanations of the bay. This geomorphology is typical of many Caribbean reefal bays. By over-generalization, however, bays have been classified geomorphologically by variations in their coastlines' indented shape (Rea & Komar, 1975; Silvester et al., 1980). This has been applied to systems for which the circulation can be persistent or temporary. Gently-sloping shorelines, for example, exposed to wave action may contain gyre circulations, similar to the CRC, that comprise a seaward rip current diffusing beyond the breaker zone and returning landward as slow mass drift under wave action (Carter 1988). Unlike the reefal-bay system, however, the stability of these gyres is heavily dependent on high energy wave action and so rip features are hardly permanent or in the same location. Ultimately, the bathymetry unique to these reefal-bay systems is principal in forming and maintaining the CRC, as seen in the simulation of the longer-lasting gyres when both the wind and tidal contributions are reduced.

### 6.6 Reliability of the hydrodynamic model

The hydrodynamic model used flow over-the-reef along a boundary line to simulate wave breaking and captured the effects of shorter period wind-wave driven flow important in driving channel currents. Current simulations in the channel were therefore in good agreement with S4 field data and are considered most important in these models since they form the main link in the CRC formation, in addition to being the direct driver of CRC emanation and contraction. Simulations, however, fell short in capturing some effects caused by the reef flat (Cetina-Heredia et al., 2008), and the contribution of longer period swell, seiching and infra-gravity waves (Lugo-Fernandez et al., 1998; Pequignet, 2008). This affected the back-reef outputs where currents were faster and less variable than simulated by the RMA model. Results from the model, however, were sufficient for simulating the bay circulation around the reef, revealing signature patterns, and deriving the contributions of wind and tide regimes to driving gyre emanations.

## 7. Conclusion

The hydrodynamic modelling and tracking simulations were able to reproduce field observations, allowing the following signatures to be developed for characterizing the reefal bay system:

- A characteristic bathymetry comprised of reef arms broken by prominent channel, giving rise to a persistent circulation;
- A reef-centered circulation driven by wind, tides and over-the-reef flow;
- A reef-centered circulation that continuously looped the reef (circum-reef circulation or CRC) to form either a closed gyre (closed CRC) or to flow along the fore-reef as open loop (open CRC);
- A CRC that was persistent because it is always present and localized;
- A CRC with a spatial pulse indicated by cycles of expansion and contraction;
- The dominance of the CRC alternating between reef arms and dictating which reef arm was primarily responsible for bay extension;
- The persistence of particularly one reef arm's CRC regardless of the wind or tidal regime.

These signatures are now identified with the reefal bay system, where the reef is shown to be central to inducing the circum-reef circulation or CRC that encourages re-circulation of inner bay waters, and that this CRC formation is not found in non-reefal bays, where there is an absence of emergent reef between headlands. Driving forces such as wind, over-the-reef flow and tidal changes were responsible for maintaining the CRC including its contractions and emanations. These findings are important in their implications for stabilizing and protecting these systems as well as the shoreline of which they are a part. Incorporating the reef-parabola geomorphology as the centre of circulation gives predictability to other bay features such as the physicochemical, geo-physical and biological dynamics, which are all affected in greater part by local circulation. Many of these bays, for example, function as nurseries for marine and terrestrial species where their planktonic stages are directly influenced by current patterns and regimes. Identifying the CRC will aid in locating and protecting habitats conducive to plankton viability and survival, including reef growth and expansion.

## 8. Summary

Research on reefal bays revealed that inner bay waters exiting the channel between reefs re-circulated into the back-reef, and that this circulation was localized and permanent around reefs as the signature circulation. The distinctive topology of reef arms subtending the headland and separated by a prominent channel induced particles to circulate the reef in expanding and contracting gyres. Gyres expanded by as much as 98% of the horizontal distribution, with expansion and contraction linked to cyclical wind and tidal regimes, giving the reefal system a signature pulse in circulation. Strengthening of the circulation around the reef resulted in closure of looping circulation, increased recirculation rate of particles, increased gyre current speeds and broadening of the circulation's spatial extent.

One reef arm's circulation would dominate over the other at the peaks of these cycles, exhibiting gyre dominance. Increased variability of particle retention was also characteristic. These signatures were not evident in an adjacent open, non-reefal bay used as a comparison. The stability, spatial spread and localization of the circulation therefore defined this circum-reef circulation and identifies its association with reefal bays in particular, where the reef functions as the centre of a dynamic bay.

## 9. Acknowledgements

The authors are grateful to the Port Royal Marine Lab, the Center for Marine Sciences, the Japan International Corporation Agency and the Mona Geoinformatics Institute for providing funding, technical support and equipment to carry out this study. The Environmental Foundation of Jamaica in partnership with the Life Sciences Department, University of the West Indies, was significant in providing funding for training in hydrodynamic modelling. We acknowledge Christopher Burgess for guidance in the oceanographic statistics and modelling. Acknowledgement also goes to the dedication of Sean Townsend and the many student volunteers from the Department of Life Sciences, University of the West Indies, in assisting with the field work.

## 10. References

Black, K.P.; Gay, S.L. & Andrews, J.C. (1990). Residence times of neutrally-buoyant matter such as larvae, sewage or nutrients on coral reefs. *Coral Reefs* 9: 105-114.

Burgess, P.; Irwin, M.; Maxam, A. & Townsend, S. (2005). Oceanographic Study of Sand Hills Bay. Civil Environmental and Coastal Solutions Engineer's Report to the UDC, 79 pp, Kingston Jamaica.

Carter, R.W.G. (1988). Coastal Environments: an introduction to the physical, ecological and cultural systems of coastlines. Academic Press. Great Britain.

Cetina-Heredia, P.; Connolly, S. & Herzfeld, M. (2008). Modeling larval retention around reefs by local scale circulation features, *Proceedings of the 11th International Coral Reef Symposium*, July, 2008, Florida.

Douillet, P.; Ouillon, S. & Cordier, E. (2001). A numerical model for fine suspended sediment transport in the southwest lagoon of New Caledonia. *Coral Reefs* 20: 361-372.

Feddersen, F. & Trowbridge, J. H. (2005). The effect of wave breaking on surf-zone turbulence and alongshore currents: A modelling study. *Journal of Physical Oceanography*. 35: 2187 – 2203.

Foreman, M.G.G. (1977). Manual for Tidal Heights Analysis and Prediction. *Pacific Marine Science Report* 77-10, Institute of Ocean Sciences, Patricia Bay, Sidney, B.C., 58 pp.

Goodbody, I.; Bacon, P.; Greenaway, A.; Head, S.; Hendry, M. & Jupp, B. (1989). Caribbean Coastal Management Study: The Hellshire Coast, St. Catherine, Jamaica. *Marine Science Unit Research Report* no.1. Ed. I. Goodbody. University of the West Indies. pp 176.

Gourlay, M.R. & Colleter, G. (2005). Wave-generated flow on coral reef – an analysis for two-dimensional horizontal reef-tops with steep faces. *Coastal Engineering*. 52: 353 – 387.

Gunaratna, P.P.; Justesen, P. & Abeysirigunawardena, D.S. (1997). Mathematical modeling of hydrodynamics in a reef protected coastal stretch. 2nd DHI Software User Conference, Denmark. Paper C7-1.

Hasselmann, K.; Barnett, T. P.; Bouws, E.; Carlson, H.; Cart-wright, D. E.; Enke, K.; Ewing, J. A.; Gienapp, H.; Hasselmann, D. E.; Kruseman, P.; Meerburg, A.; Muller, P.; Olbers, D. J.; Richter, K.; Sell, W. & Walden, H. (1973). Measurements of wind-wave growth and swell decay during the Joint North Sea Wave Project (JONSWAP), Erganzungsheft zur Deutschen Hydrographischen Zeitschrift, Reihe A. 12. 95 pp, Deutsches Hydrographisches Institut, Hamburg, Germany.

Hearn, C. J. (1999). Wave-breaking hydrodynamics within coral reef systems and the effect of changing relative sea level. *Journal of Geophysical Research*, Series C 104, 30007 – 30019.

Hearn, C.J. (2001). Introduction to the special issue of *Coral Reefs* on "Coral Reef Hydrodynamics". *Coral Reefs* 20: 327-329.

Hendry, M.D. (1983). The influence of the sea-land breeze regime on beach erosion and accretion : an example from Jamaica. *Caribbean Geography, 1* (1), 13-23.

King, I.P. (2003). A finite element model for stratified flow; RMA-10 version 7.1B. 66 pp, Resource Modelling Associates, Sydney, Australia.

King, I.P. (2005). A particle tracking model compatible with the RMA series of finite element surface water models; RMATRK version 3.2, 19 pp, Resource Modelling Associates, Sydney, Australia.

Kirby, J.T. & Dalrymple, R.A. (1991). User's Manual -Combined Refraction/Diffraction Model: REF/DIF 1 Version 2.3. Center for Applied Coastal Research. University of Delaware. Newark, Delaware.

Kraines, S.B.; Yanagi, T.; Isobe, M. & Komiyama, H. (1998). Wind-wave driven circulation on the coral reef at Bora Bay, Miyako Island. *Coral Reefs.* 17: 133-143.

Lasker, H.R. & Kapela, W.J. Jr. (1997). Heterogeneous water flow and its effects on the mixing and transport of gametes. *Proceedings of the 8th International Coral Reef Symposium.* 2: 1109-1114.

Lugo-Fernandez, A.; Roberts, H.H. & Wiseman, W.J. (1998). Water level and currents of tidal and infragravity periods at Tague Reef, St. Croix (USVI). *Coral Reefs,* 17 (4), 343-349.

Maxam, A.M. & Webber, D.F. (2009). Using the distribution of physicochemical variables to portray reefal bay waters. *Journal of Coastal Research,* 25 (6), 1210-1221.

Maxam, A.M. & Webber, D.F. (2010). The influence of wind-driven currents on the circulation and bay dynamics of a semi-enclosed reefal bay, Jamaica. *Estuarine, Coastal and Shelf Science,* 87, 535-544.

Mellor, G.L. & Yamada, T. (1982). Development of a Turbulence Closure Model for Geophysical Fluid Problems. *Reviews of Geophysics and Space Physics,* 20 (4), pp 851-875.

Nwogu, O.; Demirbilek, Z. & Merrifield, M. (2008). Non-linear wave transformation over shallow fringing reefs, *Proceedings of the 11th International Coral Reef Symposium,* July, 2008. Florida.

Nybakken, J.W. (1997). Marine Biology: An Ecological Approach, 4th ed. Addison-Wesley Educational Publishers Inc.

Pequignet, A. (2008). Importance of infragravity band in the wave energy budget of a fringing reef, *Proceedings of the 11th International Coral Reef Symposium,* July, 2008. Florida.

Prager, E.J. (1991). Numerical simulation of circulation in a Caribbean-type back reef lagoon, *Coral Reefs,* 10, 177-182.

Rea, C.C. & Komar, P.D. (1975). Computer simulation models of a hooked beach's shoreline configuration. *Journal of Sedimentology and Petrology* 45: 866-872.

Roberts, H.H. (1980). Physical processes and sediment flux through reef-lagoon systems, *Proc 17th Int. Coastal Engineering Conf.* , ASCE, Sydney, Australia, pp 946-962.

Roberts, H.H. & Suhayda, J.N. (1983). Wave-current interactions on a shallow reef. *Coral Reefs*, 1, 209-214.

Roberts, H.H.; Lugo, A.; Carter, B. & Simms, M. (1988). Across reef flux and shallow subsurface hydrology in modern coral reefs, *Proceedings of the 6th International Coral Reef Symposium*, 2, 509-515. Townsville, Australia.

Roberts, H.H.; Wilson, P.A. & Lugo-Fernandez, A. (1992). Biologic and geologic responses to physical processes: examples from modern reef systems of the Caribbean-Atlantic region, *Continental Shelf Research*, 12 (7/8), 809-834.

Sammarco, P.W. & J.C. Andrews (1989). The Helix experiment: differential localised dispersal and recruitment patterns in Great Barrier Reef corals, *Limnolology and Oceanography*, 34, 898-914.

Silvester, R., Tsuchiya, Y., & Shibano, Y. (1980). Zeta bays, pocket beaches and headland control. *Proceedings in the 17th International Conference of Coastal Engineering.*, ASCE 2: 1306-1319.

Smith, N.P. (1985). The decomposition and simulation of the longitudinal circulation in a coastal lagoon. *Estuarine Coastal Shelf Science* 21: 623-632.

Symonds, G.; Black, K.P. & Young, I.R. (1995). Wave driven flow over shallow reefs. *Journal of Geophysical Research* 100 (C2): 2639-2648.

Tartinville, B.; Deleersnijder, E. & Rancher, J. (1997). The water residence time in the Mururoa atoll lagoon: sensitivity analysis of a three-dimensional model. *Coral Reefs* 16: 193-203.

Van der Meer, J.W. (2002). Wave run-up and wave overtopping on dikes. (in Dutch; original title: Golfoploop en golfoverslag bij dijken). WL Delft Hydraulics, *Report H2458/H3051*, June 1997.

Webber, D.F. (1990). Phytoplankton populations of the coastal zone and nearshore waters of Hellshire: St. Catherine, Jamaica. Ph.D. Thesis, 285 pp, University of the West Indies, Mona.

White, M. (1982). Ground water lenses in Hellshire Hills (East), a minor source of water for Hellshire Bay. Hydrology Consultants Ltd., Kingston MS Report. 34 pp.

Wolanski, E. & Sarsenski, J. (1997). Larvae dispersion in coral reefs and mangroves. *American Science*. 85: 236-243.

# 10

# Experimental Investigation on Motions of Immersing Tunnel Element under Irregular Wave Actions

Zhijie Chen[1], Yongxue Wang[2], Weiguang Zuo[2],
Binxin Zheng[1] and Zhi Zeng[1], Jia He[1]
[1]*Open Lab of Ocean & Coast Environmental Geology,*
*Third Institute of Oceanography, SOA*
[2]*State Key Laboratory of Coastal and Offshore Engineering,*
*Dalian University of Technology*
*China*

## 1. Introduction

An immersed tunnel is a kind of underwater transporting passage crossing a river, a canal, a gulf or a strait. It is built by dredging a trench on the river or sea bottom, transporting prefabricated tunnel elements, immersing the elements one by one to the trench, connecting the elements, backfilling the trench and installing equipments inside it (Gursoy et al., 1993). Compared with a bridge, an immersed tunnel has advantages of being little influenced by big smog and typhoon, stable operation and strong resistance against earthquakes. Due to the special economical and technological advantages of the immersed tunnel, more and more underwater immersed tunnels are built or are being built in the world.

Building an undersea immersed tunnel is generally a super-large and challenging project that involves many key engineering techniques (Ingerslev, 2005; Zhao, 2007), such as transporting and immersing, underwater linking, waterproofing and protecting against earthquakes. Some researches with respect to transportation, *in situ* stability and seismic response of tunnel elements are seen to be carried out (Anastasopoulos et al., 2007; Aono et al., 2003; Ding et al., 2006; Hakkaart, 1996; Kasper et al., 2008). The immersion of tunnel elements was also studied (Zhan et al., 2001a, 2001b; Chen et al., 2009a, 2009b, 2009c).

The immersion of a large-scale tunnel element is one of the most important procedures in the immersed tunnel construction, and its techniques involve barges immersing, pontoons immersing, platform immersing and lift immersing (Chen, 2002). In the sea environment, the motion responses of a tunnel element in the immersion have direct influences on its underwater positioning operation and immersing stability. So a study on the dynamic characteristics of the tunnel element during its interaction with waves in the immersion is desirable. Although, some researches on the immersion of tunnel elements were done in the past years, there is still much work remaining to study further. Also, the study on the immersion of tunnel elements under irregular wave actions is not seen as yet.

The aim of the present study is to investigate experimentally the motion dynamics of the tunnel element in the immersion under irregular wave actions based on barges immersing method. The motion responses of the tunnel element and the tensions acting on the controlling cables are tested.

The time series of the motion responses, i.e. sway, heave and roll of the tunnel element and the cable tensions are presented. The results of frequency spectra of tunnel element motion responses and cable tensions for irregular waves are given. The influences of the significant wave height and the peak frequency period of waves on the motions of the tunnel element and the cable tensions are analyzed. Finally, the relation between the tunnel element motions and the cable tensions is discussed.

## 2. Physical model test

### 2.1 Experimental installation and method

The experiments are carried out in a wave flume which is 50m long, 3.0m wide and 1.0m deep. The sketch of experimental setup is shown in Fig. 1. Assuming the movements of the barges on the water surface are small and can be ignored, the immersion of the tunnel element is directly done by the cables from the fixed trestle over the wave flume.

The immersed tunnel element considered in this study is 200cm long, 30cm wide and 20cm high, which is a hollow cuboid sealed at its two ends. The tunnel model is made of acrylic plate and concrete and the cables are modeled by springs and nylon strings that are made to lose their elasticity.

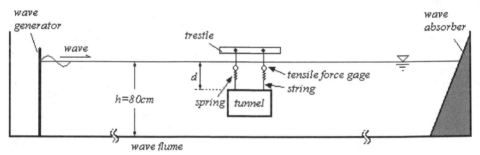

Fig. 1. Sketch of experimental setup

It is known that the immersion of the tunnel element in practical engineering is actually done by the ballast water, namely negative buoyancy, inside the tunnel element. The weight of the tunnel element model used in this experiment is measured as 1208.34N. When the model is completely submerged in the water, the buoyancy force acting on it is 1176.0N. So the negative buoyancy is equal to 32.34N, which is 2.75 percent of the buoyancy force of the tunnel element. The negative buoyancy makes the cables bear the initial tensions.

Water depth ($h$) in the wave flume is 80cm. The normal incident irregular waves are generated from the piston-type wave generator. The significant wave heights ($H_s$) are 3cm and 4cm, and the peak frequency period of waves ($T_p$) 0.85s, 1.1s and 1.4s, respectively. The experiments are conducted for the cases of three different immersing depths of the tunnel element, i.e., $d$=10cm, 30cm and 50cm, respectively. $d$ is defined as the distance from the water surface to the top surface of the tunnel element.

Corresponding to the three immersing depths of the tunnel element, three kinds of springs with different elastic constants are used in the experiment. According to the properties of cables using in practical engineering and the suitable scale of the model test, the appropriate springs are chosen. The relations between the elastic force and the spring extension are shown in Fig. 2. There are four strings that join four springs respectively to control the immersed tunnel element in the waves. Two strings are on the offshore side and the other two on the onshore side of the tunnel element. To measure the tensions acting on the strings, four tensile force gages are connected to the four strings respectively.

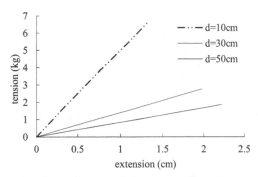

Fig. 2. Relations between the elastic force and the spring extension

The CCD (Charge Coupled Device) camera is utilized to record the motion displacements of the tunnel element during its interaction with waves. Two lights with a certain distance are installed at the front surface of the tunnel element, as shown in Fig. 3. When the tunnel element moves under irregular wave actions, the positions of the two lights are recorded by the CCD camera. Finally, the sway, heave and roll of the tunnel element are obtained from the CCD recorded images by the image analysing program.

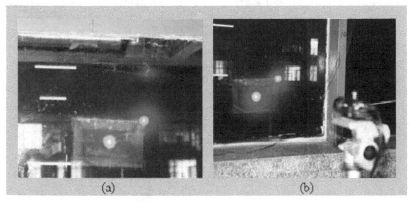

(a)                                                                  (b)

Fig. 3. Photo view of the tunnel element at the wave flume. (a) wave is propagating over the tunnel element; (b) the tunnel element and CCD

## 2.2 Simulation of wave spectra

In the experiment, Johnswap spectrum is chosen as the target spectrum to simulate the physical spectrum, and two significant wave heights, $H_s$=3.0cm and 4.0cm, and three peak frequency periods of waves, $T_p$=0.85s, 1.1s and 1.4s are considered. As examples, two groups of wave conditions, i.e. $H_s$=3.0cm, $T_p$=1.4s and $H_s$=4.0cm, $T_p$=1.1s, are taken to present the simulation of the physical wave spectra. Fig. 4 shows the results of the comparison between the target spectrum and physical spectrum. It is seen that they agree very well.

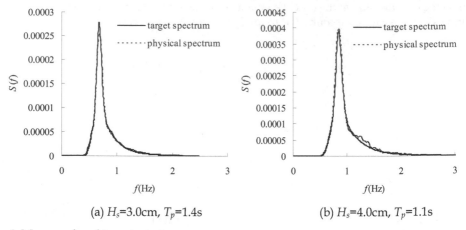

(a) $H_s$=3.0cm, $T_p$=1.4s                                    (b) $H_s$=4.0cm, $T_p$=1.1s

Fig. 4. Measured and target spectrum

## 3. Experimental results and discussion

### 3.1 Motion responses of the tunnel element

The significant wave height and the peak frequency period of waves are the main influencing factors on the motion responses of the tunnel element under irregular wave actions. Moreover, in the different immersing depth positions, the motions of the tunnel element make differences. In this experiment, the different immersing depths, significant wave heights and peak frequency periods are considered to explore their impacts on the motions of the tunnel element.

### 3.1.1 Time series of the tunnel element motion responses

As an example, the time series of the tunnel element motion responses in the wave conditions $H_s$=3.0cm, $T_p$=0.85s and $d$=10cm within the time 80s are shown in Fig. 5. Under the normal incident wave actions, the tunnel element makes two-dimensional motions, i.e. sway, heave and roll.

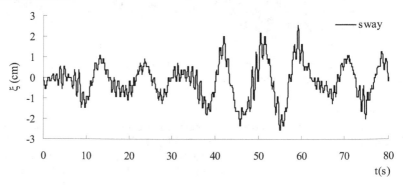

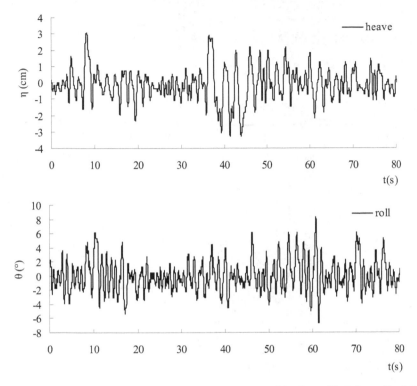

Fig. 5. Time series of the tunnel element motion responses ($d$=10cm, $H_s$=3.0cm, $T_p$=0.85s)

### 3.1.2 Motion responses of the tunnel element in the different immersing depth

Fig. 6 gives the results of the frequency spectra of the tunnel element motion responses in the wave conditions $H_s$=4.0cm and $T_p$=1.1s for different immersing depths of the tunnel element. From the peak values of the frequency spectra curves, it is obvious that the motion responses of the tunnel element are comparatively large for the comparatively small immersing depth. Comparing the motions of the tunnel element of sway, heave and roll, the area under the heave motion response spectrum is larger than that under the sway motion response spectrum when the immersing depths are 10cm and 30cm, as indicates that the motion of the tunnel element in the vertical direction is predominant. In addition, it can be observed that there are two peaks on the curves of the sway and heave motion responses spectra. This illuminates that the low-frequency motions occur in the tunnel element besides the wave-frequency motions. The low-frequency motions are caused by the actions of cables. For the sway, the low-frequency motion is dominant, while the wave-frequency motion is relatively small. From the figure, it can be seen that the low-frequency motion is always larger than the wave-frequency motion for the sway as the tunnel element is in the different immersing depths. It reveals that the low-frequency motion is the main of the tunnel element movement in the horizontal direction. This can also be obviously observed from the curve of time series of the sway in Fig. 5. However, for the heave, as the immersing depth increases, the motion turns gradually from that the low-frequency motion is dominant into that the wave-frequency motion is dominant.

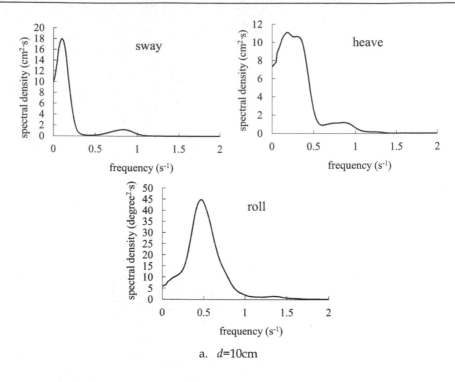

a.   *d*=10cm

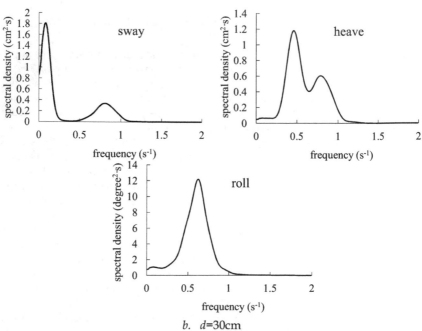

*b.*   *d*=30cm

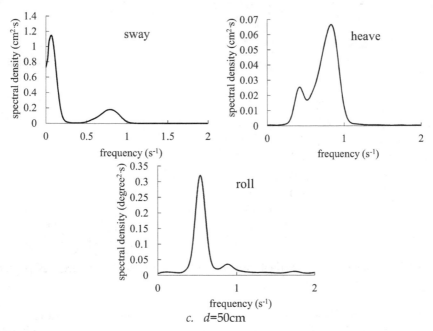

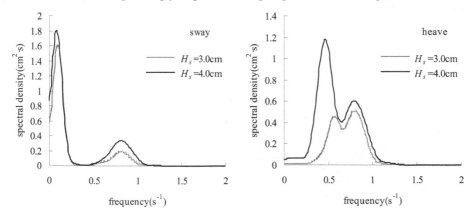

Fig. 6. Frequency spectra of the tunnel element motion responses for different immersing depths ($H_s$=4.0cm, $T_p$=1.1s)

### 3.1.3 Influence of the significant wave height on the tunnel element motions

The results of the frequency spectra of the tunnel element motion responses for different significant wave heights in the test conditions $d$=30cm and $T_p$=1.1s are shown in Fig. 7. From the figure, it is seen that the shapes of the frequency spectrum curves of the tunnel element motion responses are very similar for different significant wave heights, while just the peak values are different. Corresponding to the large significant wave height, the peak value is large, as well large is the area under the motion response spectrum. Apparently, the motion responses of the tunnel element are correspondingly large for the large significant wave height.

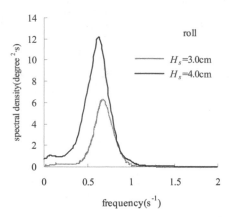

Fig. 7. Frequency spectra of the tunnel element motion responses for different significant wave heights ($d$=30cm, $T_p$=1.1s)

### 3.1.4 Influence of the peak frequency period on the tunnel element motions

Fig. 8 shows the results of the frequency spectra of the tunnel element motion responses in the test conditions $d$=30cm and $H_s$=3.0cm for different peak frequency periods of waves. It can be seen that the peak frequency period has an important influence on the motion responses of the tunnel element. The peak values of the frequency spectra of the motion responses increase markedly with the increase of the peak frequency period. Thus, the larger is the peak frequency period of waves, the larger are the motion responses of the tunnel element.

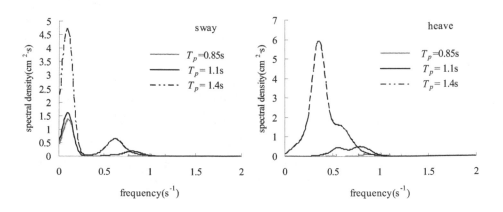

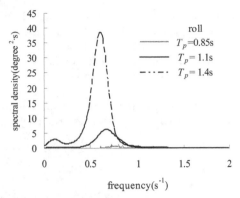

Fig. 8. Frequency spectra of the tunnel element motion responses for different peak frequency periods of waves ($d$=30cm, $H_s$=3.0cm)

Furthermore, in the figure it is shown that the frequency spectra of the tunnel element motion responses all have a peak at the frequency corresponding to the respective peak frequency period of waves, besides a peak corresponding to the low-frequency motion of the tunnel element. For the sway, the peak frequency corresponding to the low-frequency motion of the tunnel element is the same in the cases of different peak frequency periods. However, for the heave, the peak frequency corresponding to the tunnel element low-frequency motion varies with the peak frequency period. It increases as the peak frequency period increases. The reason may be that there occurs slack state in the cables during the movement of the tunnel element when the peak frequency period increases, for which there is no more the restraint from the motion of the tunnel element in the vertical direction from the cables at this time.

### 3.2 Cable tensions
### 3.2.1 Time series of cable tensions
As a typical case, Fig. 9 shows the time series of the cable tensions in the wave conditions $H_s$=4.0cm, $T_p$=1.1s and $d$=30cm within the time 160s. In the figure, C11 represents the front cable at the onshore side, C12 the back cable at the onshore side, C21 the front cable at the offshore side and C22 the back cable at the offshore side. It can be seen that the time series of tensions of the cables C11 and C12 at the onshore side are very similar, as well similar are those of the cables C21 and C22 at the offshore side. It shows that under the normal incident irregular wave actions the tunnel element does only two-dimensional motions. This can also be observed in the experiment from the movement of the tunnel element.

### 3.2.2 Cable tensions for the different immersing depth of the tunnel element
Fig. 10 shows the results of the frequency spectra of the cable tensions in the wave conditions $H_s$=4.0cm and $T_p$=1.1s for different immersing depths of the tunnel element. From the peak values of the frequency spectra curves and the areas under the frequency spectra, it is seen that the tensions acting on the cables are comparatively large in the case of comparatively small immersing depth, as is corresponding to the motion responses of the tunnel element. Furthermore, the peak values and the areas of the frequency spectra of the cable tensions at the offshore side are all larger than those of the cable tensions at the onshore side for different immersing depths. It indicates that the total force of the cables at the offshore side is larger

than that of the cables at the onshore side. It is also shown that in the figure there are at least two peaks in the curves of the frequency spectra of the cable tensions, which are respectively corresponding to the wave-frequency motions and low-frequency motions of the tunnel element. When the tunnel element is at the position of a relatively small immersing depth, the frequency spectra of the cable tensions have other small peaks besides the two peaks at the wave frequency and the low frequency. It illustrates that the case of the forces generating in the cables is more complicated for the comparatively strong motion responses of the tunnel element under the wave actions when the immersing depth is relatively small.

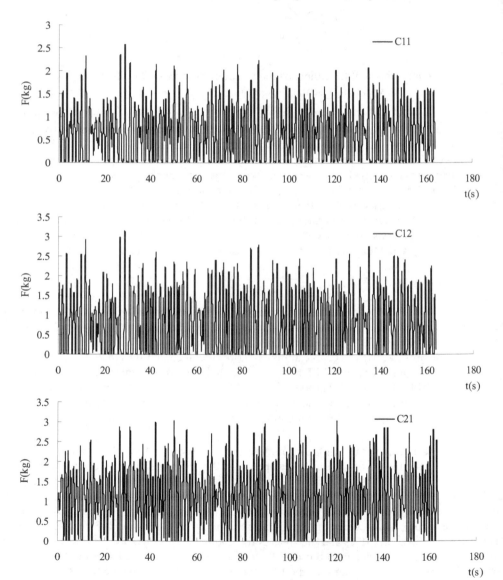

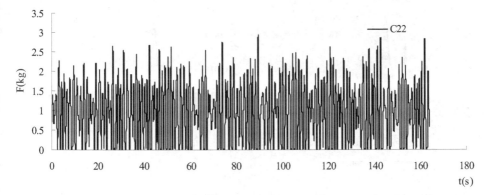

Fig. 9. Time series of tensions acting on the cables ($d$=30cm, $H_s$=4.0cm, $T_p$=1.1s, C11: front cable at the onshore side, C12: back cable at the onshore side, C21: front cable at the offshore side, C22: back cable at the offshore side)

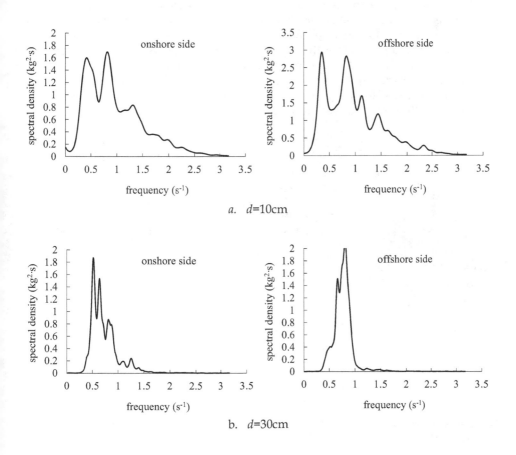

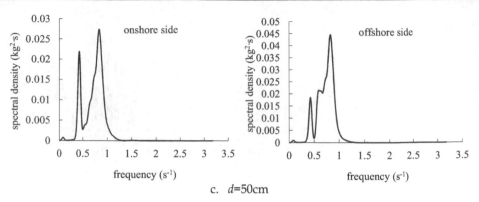

c.  *d*=50cm

Fig. 10. Frequency spectra of the cable tensions for different immersing depths ($H_s$=4.0cm, $T_p$=1.1s)

### 3.2.3 Influence of the significant wave height on the cable tensions

Fig. 11 gives the results of the frequency spectra of the cable tensions for different significant wave heights in the test conditions *d*=30cm and $T_p$=1.1s. It is shown that the area under the frequency spectrum of the cable tensions for the significant wave height $H_s$=4.0cm is larger than that for $H_s$=3.0cm. Therefore, the larger is the significant wave height, the larger are the cable tensions accordingly. This is corresponding to the case that the motion responses of the tunnel element are larger for the larger significant wave height. When the significant wave height increases, the wave effects on the tunnel element increase. Accordingly, the forces acting on the cables also become larger.

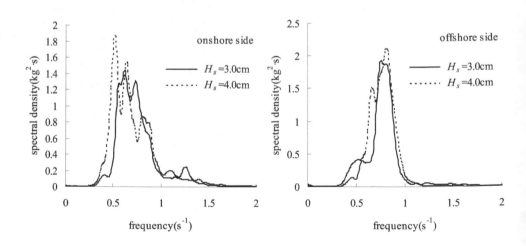

Fig. 11. Frequency spectra of the cable tensions for different significant wave heights (*d*=30cm, $T_p$=1.1s)

### 3.2.4 Influence of the peak frequency period on the cable tensions

The results of the frequency spectra of the cable tensions for different peak frequency periods of waves in the test conditions $d$=30cm and $H_s$=3.0cm are shown in Fig. 12. It is seen that the cable tensions are largely influenced by the peak frequency period. The peak values of the frequency spectra of the cable tensions increase rapidly as the peak frequency period increases. Corresponding to the case of the motion responses of the tunnel element for different peak frequency periods, the larger is the peak frequency period, the larger are also the cable tensions. For different peak frequency periods, the frequency spectra of the cable tensions all have a peak at the corresponding frequency. Besides, from the figure, it can be observed that the peaks of the frequency spectra at the lower frequency are obvious when the peak frequency period $T_p$=1.4s. This reflects that the low-frequency motions of the tunnel element become large with the increase of the peak frequency period of waves.

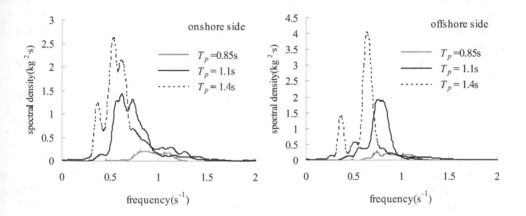

Fig. 12. Frequency spectra of the cable tensions for different peak frequency periods of waves ($d$=30cm, $H_s$=3.0cm)

### 3.3 Relation between the tunnel element motions and the cable tensions

The tunnel element moves under the irregular wave actions, and at the same time, the tunnel element is restrained by the cables in the motions. So the wave forces and cable tensions together result in the total effect of the motions of the tunnel element. On the other hand, the restraint of the cables from the movement of the tunnel element makes the cables bear forces. Hence, the motions of the tunnel element and the cable tensions are coupled. According to the discussion in the above context, in the case when the immersing depth is small and the significant wave height and the peak frequency period are large comparatively, the motion responses of the tunnel element are relatively large. And in the case of that, the variations of the cable tensions are accordingly more complicated.

Moreover, corresponding to the wave-frequency peak and low-frequency peak of the frequency spectra of the cable tensions, there occur the wave-frequency motions and low-frequency motions in the tunnel element. This also reflects directly the interrelation of the tunnel element motions and the cable tensions.

## 4. Conclusions

The motion dynamic characteristics of the tunnel element and the tensions acting on the controlling cables in the immersion of the tunnel element under irregular wave actions are experimentally investigated in this chapter. The irregular waves are considered normal incident and the influences of the immersing depth of the tunnel element, the significant wave height and the peak frequency period of waves on the tunnel element motions and the cable tensions are analyzed. Some conclusions are drawn as follows.

As the immersing depth is comparatively small, the motion responses of the tunnel element are relatively large. Besides the wave-frequency motions, the tunnel element has also the low-frequency motions that result from the actions of cables. For the sway of the tunnel element, for different immersing depth the low-frequency motion is always larger than the wave-frequency motion. While for the heave, with the increase of the immersing depth, the motion turns gradually from that the low-frequency motion is dominant into that the wave-frequency motion is dominant.

For the large significant wave height, the motion responses of the tunnel element are accordingly large. The peak values of the frequency spectra of the motion responses increase rapidly with the increase of the peak frequency period of waves. Especially, for the heave motion of the tunnel element, the peak frequency of the response spectrum corresponding to the low-frequency motion increases with the increasing peak frequency period.

The total force of the cables at the offshore side is larger than that of the cables at the onshore side of the tunnel element. Corresponding to the motion responses of the tunnel element, the cable tensions are relatively large and their variations are more complicated in the case as the immersing depth is small and the significant wave height and the peak frequency period are large comparatively. The changing laws of the tunnel element motions and the cable tensions reflect the interrelation of them.

In this chapter, the immersion of the tunnel element is done from the fixed trestle in the experiment, by ignoring the movements of the barges on the water surface. Actually, when the movements of the barges are relatively large, they have influences on the motions of the tunnel element. The influences of the movements of the barges on the tunnel element motions will be considered in the further researches. The numerical investigation will also be carried out on the motion dynamics of the tunnel element in the immersion under irregular wave actions.

## 5. Acknowledgment

This work was partly supported by the Scientific Research Foundation of Third Institute of Oceanography, SOA (Grant No. 201003), and partly by the National Natural Science Foundation of China (Grant No. 51009032).

# 6. References

Anastasopoulos, I., Gerolymos, N., Drosos, V., Kourkoulis, R., Georgarakos, T. & Gazetas, G. (2007). Nonlinear Response of Deep Immersed Tunnel to Strong Seismic Shaking, *Journal of Geotechnical and Geoenviron-mental Engineering*, Vol. 133, No 9, (September 2007), pp. 1067-1090, ISSN 1090-0241

Aono, T., Sumida, K., Fujiwara, R., Ukai, A., Yamamura K. & Nakaya, Y. (2003). Rapid Stabilization of the Immersed Tunnel Element, *Proceedings of the Coastal Structures 2003 Conference*, pp. 394-404, ISBN 978-0-7844-0733-2, Portland, Oregon, USA, August 26-30, 2003

Chen, S. Z. (2002). *Design and Construction of Immersed Tunnel*, Science Press, ISBN 7-03-010112-X, Beijing, China. (in Chinese)

Chen, Z. J., Wang, Y. X., Wang, G. Y. & Hou, Y. (2009a). Frequency responses of immersing tunnel element under wave actions, *Journal of Marine Science and Application*, 2009, Vol. 8, pp. 18-26, (March 2009), ISSN 1671-9433

Chen, Z. J., Wang, Y. X., Wang, G. Y. & Hou, Y. (2009b). Time-domain responses of immersing tunnel element under wave actions, *Journal of Hydrodynamics, Ser. B*, Vol. 21, No. 6, (December 2009), pp. 739-749, ISSN 1001-6058

Chen, Z. J., Wang, Y. X., Wang, G. Y. & Hou, Y. (2009c). Experimental Investigation on Immersion of Tunnel Element, *28th International Conference on Ocean, Offshore and Arctic Engineering*, pp. 1-8, ISBN 978-0-7918-4344-4, Honolulu, Hawaii, USA, May 31–June 5, 2009

Ding, J. H., Jin, X. L., Guo, Y. Z. & Li., G. G. (2006). Numerical Simulation for Large-scale Seismic Response Analysis of Immersed Tunnel, *Engineering Structures*, Vol. 28, No. 10, (January 2006), pp. 1367-1377, ISSN 0141-0296

Gursoy, A., Van Milligen, P. C., Saveur, J. & Grantz, W. C. (1993). Immersed and Floating Tunnels, *Tunnelling and Underground Space Technology*, Vol.8, No.2, (December 1993), pp. 119-139, ISSN 0886-7798

Hakkaart, C. J. A. (1996). Transport of Tunnel Elements from Baltimore to Boston, over the Atlantic Ocean, *Tunnelling and Underground Space Technology*, Vol. 11, No. 4, (October 1996), pp. 479-483, ISSN 0886-7798

Ingerslev, L. C. F. (2005). Considerations and Strategies behind the Design and Construction Requirements of the Istanbul Strait Immersed Tunnel, *Tunnelling and Underground Space Technology*, Vol. 20, (October 2005), pp. 604-608, ISSN 0886-7798

Kasper, T., Steenfelt, J. S., Pedersen, L. M., Jackson P. G. & Heijmans, R. W. M. G. (2008). Stability of an Immersed Tunnel in Offshore Conditions under Deep Water Wave Impact. *Coastal Engineering*, Vol. 55, No. 9, (August 2008), pp. 753-760, ISSN 3783-3839

Zhan, D. X. & Wang, X. Q. (2001a). Experiments of hydrodynamics and stability of immersed tube tunnel on transportation and immersing. *Journal of Hydrodynamics, Ser. B*, Vol. 13, No. 2, (June 2001), pp. 121-126, ISSN 1001-6058

Zhan, D. X., Zhang, L. W., Zhao, C. B., Wu, J. P. & Zhang, S. X. (2001b) Numerical simulation and visualization of immersed tube tunnel maneuvering and immersing, *Journal of Wuhan University of Technology (Transportation Science and Engineering)*, Vol. 25, No. 1, (March 2001), pp. 16-20, ISSN 1006-2823 (in Chinese)

Zhao, Z. G. (2007). Discussion on Several Techniques of Immersed Tunnel Construction. *Modern Tunnelling Technology*, Vol. 44, No.4, (August 2007), pp. 5-8, ISSN1009-6582 (in Chinese)

# 11

# Formation and Evolution of Wetland and Landform in the Yangtze River Estuary Over the Past 50 Years Based on Digitized Sea Maps and Multi-Temporal Satellite Images

Xie Xiaoping
*School of Geography and Tourism,*
*Qufu Normal University, Qufu*
*China*

## 1. Introduction

The Yangtze River originates in the Qinghai-Tibet Plateau and extends more than 6300 km eastward to the East China Sea, a tectonic subsidence belt (Li & Wang, 1991). It is one of the largest rivers in the world, in terms of suspended sediment load, water discharge, length, and drainage area. The Yangtze River Estuary is located in the east China. There are three main islands including Chongming Island, Changxing Island, and Hengsha Island as well as several shoals in the Yangtze River Estuary (Fig. 1). These islands once are shoals emerged from the water and merged to the north bank or coalesced together. In the Yangtze River Estuary, most of the sediments from the drainage basin are suspended. The spatial and temporal variations of the suspended sediment concentration in the estuarine field survey indicate that the sediment is suspended, transported, and deposited under riverine and marine processes, such as river flow, waves, tidal currents, and local topography (Cao et al., 1989; Chen, 2001; Gao, 1998; Li et al., 1995, Huang and Chen, 1995; Xu et al., 2002; Pan and Sun, 1996). In longitudinal section, these islands and shoals stand out on the link between the -10 m isobathic line (the zero elevation means the 1956 Yellow Sea Water Surface in Qingdao Tidal Station, Qingdao, Shandong Province, China) from the upper reach section to the lower reach section, it is a convex geomorphic unit in the Yangtze River Estuary (Fig.2 A-A´ and Fig. 3), in transverse section, these shoals and islands sit in between the channels and distributaries (Fig.2 B-B´ and Fig. 4). In order to analyze the formation and evolution of the wetland and landform of the Yangtze River Estuary, related sea maps from 1945 to 2001 and satellite images from 1975 to 2001 are collected and analyzed. Water and sediment discharge from 1950 to 2003 at the Datong Hydrologic Station 640 km upstream from the estuary mouth are also collected. Datong Hydrologic Station is the most downstream hydrologic station on the free-flowing Yangtze River, where the tidal influence can affect flows hundreds of kilometers upstream. All related sea maps are digitized using Mapinfo7.0, and the sediment volume deposited in this area is calculated from a series of processes dealt in Surfer7.0. The relation between formation and evolution of the wetland and landform of the Yangtze Estuary over the past 50 years were analyzed via Geographical Information System technology and a Digital Elevation Model.

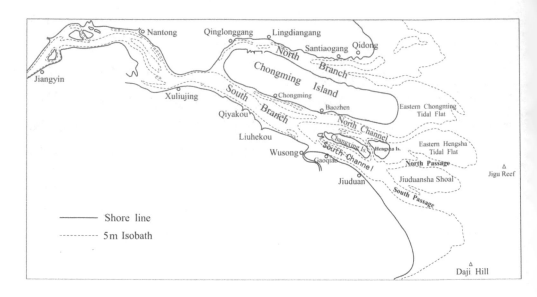

Fig. 1. The sketch map of the Yangtze River Estuary

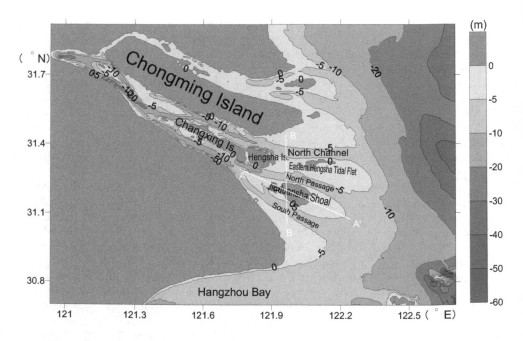

Fig. 2. Location of the Jiuduansha Shoal in Yangtze River Estuary

Formation and Evolution of Wetland and Landform in the Yangtze River Estuary Over the Past 50 Years Based on Digitized Sea Maps and Multi-Temporal Satellite Images

217

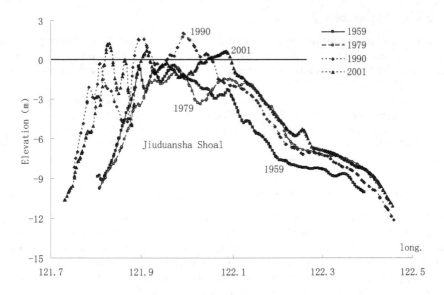

Fig. 3. Longitudinal section at 121°35'E, 31°16'N-122°25'E, 31°5'N (shown on Fig. 2 as A-A') of the Jiuduansha Shoal from 1959 to 2001

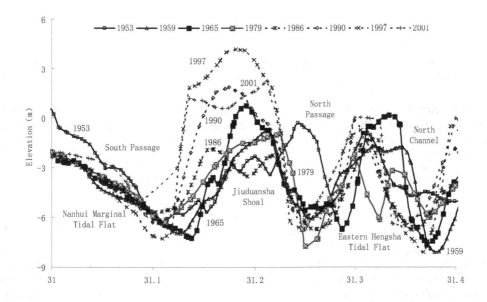

Fig. 4. Sketch map of the cross section at 122°E (shown on Fig. 2 as B-B') of the Yangtze River Estuary from 1953 to 2001

## 2. Data and methodology

In order to analyze the formation and evolution of the Yangtze River Estuary in past 50 years, related sea maps from 1945 to 2001 and satellite images from 1975 to 2001 are collected and analyzed. Landsat MSS (multi-spectral scanner) data acquired on 1975 and 1979, Landsat TM (Thematic Mapper) and Landsat ETM+ (Enhanced Thematic Mapper Plus) from 1990 to 2001, ASTER (Advanced Spaceborne Thermal Emission and Reflection Radiometer) data from2002 to 2005 were collected and analysed. All these remote sensing data were corrected geometrically. Image processing of these satellites remote sensing data were used ENVI4.6 and Erdas9.0. And formation and evolution of the landform over the past 50 years are analyzed in detail.

## 3. Formation and evolution of wetland and landform

### 3.1 Formation mechanism of the Yangtze River Estuary

The Yangtze River Estuary is nearly 90 km wide at the mouth from the Southern cape to the Northern cape. Coriolis force and centrifugal force are strong enough to cause a horizontal separation of the flow, forming an ebb tide dominated channel and a flood tide dominated channel, respectively. Because of the river bed friction, tidal currents and wave power decreased during the tidal currents flow into the mouth and wave form began to change, and the flood tidal range in the northern part is larger than that in the southern part of the same cross section, while in the ebb tide period, the longitudinal water surface gradient and the transverse water surface slope increase (Zhang and Wang, 1987). The transverse water surface slope caused by curve bend circulation is $J_B$,

$$J_B = \left(1 + 5.75\frac{g}{C^2}\right)\frac{V_{cp}^2}{gr} \tag{1}$$

where $C$ is the Chézy roughness coefficient, $V_{cp}$ is the vertical mean velocity, $r$ is the river bend radius of curvature, and $g$ is the acceleration of gravity. For example, when $V_{cp} = 2$ m/s，$r = 10,000$ m，$C = 90$ m$^{1/2}$/s，and $g = 9.81$ m/s$^2$, then $J_B = 4.1 \times 10^{-5}$.

Another factor that might affect transverse water surface slope in the Yangtze River Estuary is the Coriolis force. The transverse water surface slope caused by the Coriolis force was studied by Zou (1990), in this case the transverse water surface slope is $J_C$,

$$J_C = \frac{2\omega V_{cp}\sin\varphi}{g} \tag{2}$$

where $\omega$ is the rotational angular velocity of the earth, $\omega = 7.27 \times 10^{-5}$(s$^{-1}$); $\varphi$ is the stream section latitude, $\varphi$ is 32°. If $V_{cp}$ is equal to 2.0 m/s in the calculation like in curve bend circulation, and $g$ is 9.81 m/s$^2$, then $J_C = 1.57 \times 10^{-5}$.

Comparing $J_C$ and $J_B$, shows that for similar condition, the slope caused by the Coriolis force is smaller than that caused by curve bend circulation. However, due to the long term action of the Coriolis force, the thalweg of the ebb current and river flow is directed to the right bank and formed the Ebb Channel, while the thalweg of the flood current is directed to the left and formed the Flood Channel, the main tide direction is nearly 305° progressing from the East China Sea toward the river mouth area while the ebb tide current direction is nearly

90°-115°. The ebb tide current is not in a direction opposite to the flood tide direction; there is a 10°-35° angle between the extension line of the flood and ebb tidal currents because of the Coriolis force(Shen et al., 1995). Ebb tidal current is obviously diverted to the south, while the flood current is diverted to the north. Thus, between the flood and ebb tidal currents in the river mouth area there is a slack water region where sediment rapidly deposited to form shoals, and eventually coalesced to form estuarine islands (Chen et al., 1979). This is the evolutionary history of the three larger islands (Chongming Island, Changxing Island and Hengsha Island, respectively) in the estuary. These islands form three orders of bifurcation and four outlets in the Yangtze River Estuary. The first order of the bifurcation is the North Branch and the South Branch separated by Chongming Island. The South Branch is further divided into the North Channel and the South Channel by Changxing Island and Hengsha Island. The South Channel is further divided into the North Passage and the South Passage by the Jiuduansha Shoal (Fig. 1). Therefore, the Yangtze River Estuary has North Branch, North Channel, North Passage and South Passage four outlets through which the water and sediment from the Yangtze River discharge into the East China Sea.

From 1950 to 2003, the annual water discharge at the Datong Hydrologic Station did not substantially change. The total annual discharge is about $9481 \times 10^8$ cubic meters per year and the sediment load is about $3.52 \times 10^8$ tons/yr. The sediment discharge during the flood season (from May to October) constituted 87.2% of the annual sediment load before the 1990s, but decreased in the 1990s (Fig. 5). Most of the suspended sediment are silt and clay, which are transported to the East China Sea where they are carried away from the delta by the longshore currents. Part of the suspended load is deposited in mouth bars and a subaqueous delta area to form the tidal flats and mouth bars in the Yangtze Estuary. A broad mouth bar system and tidal flats were formed. The runoff and the sediment discharge

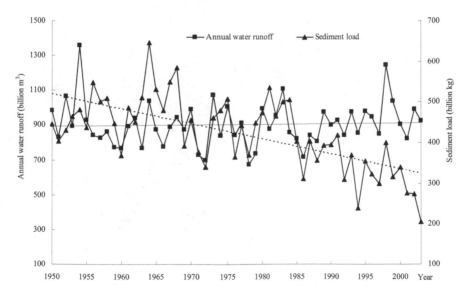

Fig. 5. Water and Sediment discharge from 1950 to 2003 at Datong Hydrologic Station.

during the flood season vary between 71.7 and 87.2 % of the annual total value based on data from the Datong Hydrologic Station. According to previous research (Gong and Yun, 2002; Niu et al., 2005), at discharges greater than 60,000 m³/s at the Datong Hydrologic Station, the estuarine riverbed has obviously changed due to erosion and deposition; when the flood water discharge greater than 70,000 m³/s, can form new branches on the river and cluster ditches because of the floodplain flows, these changes affect the estuary and new navigation channel development. In 1954 (from June 18th to October 2nd), the average water discharge at the Datong Hydrologic Station was about 60,000 m³/s, and the highest discharge was about 92,600 m³/s. Water discharge greater than 60,000 m³/s, increase the water surface gradient and the sediment carrying capacity in the estuary (Yang et al., 1999).

Estuarine sedimentation and landform features have been observed and studied in various settings around the world, including the Thames Estuary, Cobequid Bay, and the Bay of Fundy (Dalrymple and Rhodes, 1995; Knight, 1980; Dalrymple et al., 1990), as well as Chesapeake Bay (Ludwick, 1974) and Moreton Bay (Harris et al., 1992). These studies found that tidal bars in all these estuarine settings are important sedimentary features. Because estuaries are areas where freshwater and seawater mix, the systems react very sensitively to small changes in geomorphology of the estuary, and the results can reveal the changes of the estuarine environment.

According to the evolution history of the Yangtze River Estuary (Wang et al., 1981; Li et al., 1983; Qin and Zhao, 1987; Qin et al., 1996; Chen et al., 1985, 1991; Chen and Stanley 1993, 1995; Stanley and Chen, 1993; Hori, K. et al., 2001a, 2001b, 2002; Saito, Y. et al., 2001), the main delta was formed by the step-like seaward migration of the river mouth bars from Zhenjiang and Yangzhou area, the apex of the delta, to the present river mouth (Fig. 6).

The newer generation island is Jiuduansha Shoal, it was once the southern part of the Tongsha Tidal Flat. In 1945, under the processes of ebb and flood tidal currents, one pair of a

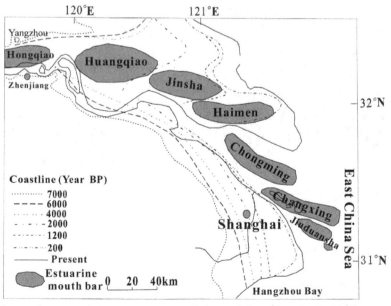

Fig. 6. Evolution history of the Yangtze River Estuary (after Chen et al., 2000)

Formation and Evolution of Wetland and Landform in the Yangtze River Estuary Over the Past 50 Years Based
on Digitized Sea Maps and Multi-Temporal Satellite Images

221

flood channel and an ebb channel developed on the southern part of the Tongsha Tidal Flat,
but the Jiuduansha Shoal had not formed as an isolated shoal (Fig.7).

In 1954, the ebb channel and flood channel on the Tongsha Tidal Flat linked up, the linked
ebb and flood channel formed the North Passage under the Flood from the drainage basin.
While the -2 m isobath line linked up the ebb channel and the flood channel, the Jiuduansha
Shoal was isolated, and the Jiuduansha Shoal formed as a new island in the Yangtze River
Estuary (Fig.8).

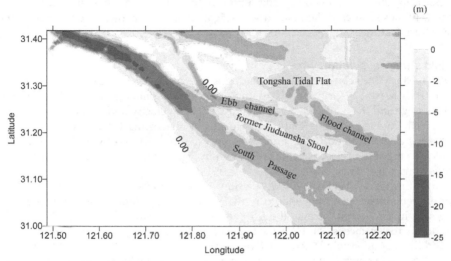

Fig. 7. Former Jiuduansha Shoal in 1945

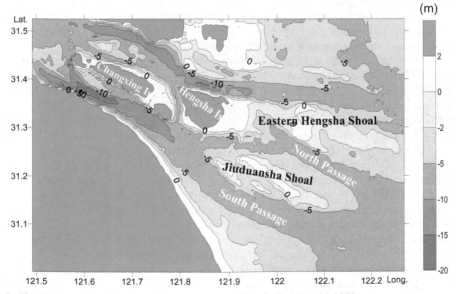

Fig. 8. The Jiuduansha Shoal and the North and South Passage in 1959.

The formation and landform evolution of the Yangtze River Estuary are related to the water and the sediment coming from the drainage basin and human activities, and also related to the riverine and marine processes. The Yangtze River Estuary is an irregular semidiurnal tidal estuary, there is a clearly different tidal range in a day, especially, the daily mean higher high tide is 1.47 m higher than lower high tide (Shen and Pan, 1988). in a tidal cycle, a flow diversion period exists, and this period differs throughout the year because of the different flood and dry seasons, and different spring and neap cycles. The channel bed changes easily and frequently under the actions of the runoff and the tidal current, while the human activities such as reclamation and navigation channel construction is also influence the landform features.

## 3.2 Field survey evaluation

In order to study the relation between the deposition and erosion of the tidal flat during the flood and dry season at spring and neap tides, field survey data for the middle section of the North Passage and the South Passage are analyzed. The velocity and sediment concentration in the North Passage and the South Passage during the spring tidal cycle obtained in the field survey use OBS 5 and DCDP and water and sediment samples which measured in the laboratory, part of the related results are shown in Fig. 9 and Fig. 10, and a summary of the collected data is listed in Table 1.

Data from this field survey show that the flow velocity and sediment concentration in the dry and flood seasons at spring and neap tidal cycles are different. In the dry season during spring tide in the South Passage, the flow velocity at the water surface (H is the relative water depth, the surface is 0H, 1H is the bottom) in the ebb tide period is higher than that in the flood tide period (Table 1). At a relative depth of 0.4H, the ebb tide velocity is lower than that the flood tide current. At 0.8H relative depth from the water surface, the flow velocity of the ebb tide is lower than that the flood tide current. In the neap tidal cycle in the South Passage, the ebb tide and river flow velocity at relative depth of 0H and 0.4H depth are higher than that flood tide velocity respectively, but the flood velocity at relative depth of 0.8H is higher than that ebb and river flow velocity.

In the dry season during spring tide in the North Passage, the velocity of ebb tide and river flow at relative depth of 0H is little lower than that flood velocity, but at relative depth of 0.4H and 0.8H are little higher than that flood velocity, respectively. While during the neap tide period, the ebb and river flow velocity at relative depth of 0H, 0.4H, and 0.8H are higher than that flood tide velocity respectively.

In the flood season during the spring and neap tidal cycle in the South and North Passage, the ebb tide and river flow velocity at relative depth of 0H, 0.4H, and 0.8H depth are correspondingly higher than that flood velocity, respectively.

In most cases, the mean sediment concentration during ebb tide period in the South and North Passage in the dry and flood season during the spring tidal cycle at relative depth of 0H, 0.4H, and 0.8H are higher than that flood tide period, respectively. But in some cases, the sediment concentration at relative depth of 0H and 0.4H are different because of the different riverine mechanics during the spring and neap tidal cycle.

Through the comparison of the velocity of ebb tide and river flow with flood tide velocity during the spring and neap tidal cycle in flood and dry season, in most cases, the ebb tide and river flow velocity at water surface is higher than the flood tide velocity, while at relative depth of 0.4H and 0.8H, in some cases, the flood tide velocity is higher than that the ebb tide and river flow velocity. That is during the flood tide period, flood tide current start

Formation and Evolution of Wetland and Landform in the Yangtze River Estuary Over the Past 50 Years Based on Digitized Sea Maps and Multi-Temporal Satellite Images

223

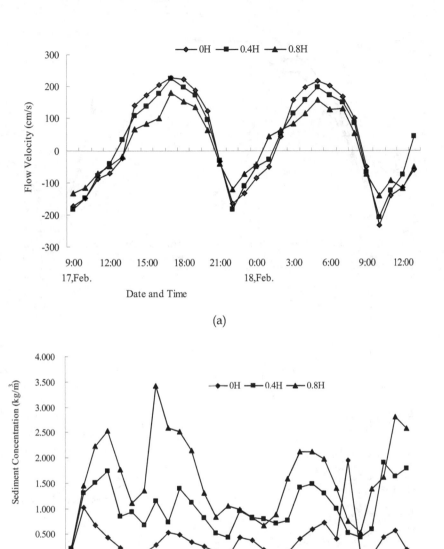

(a)

(b)

Fig. 9. Variation of the (a) flow velocity and (b) sediment concentration in the dry season at spring tide in the North Passage, where 0H means measured at the surface, 0.4H and 0.8H means measured at the 0.4 and 0.8 fractions of depth (H) from the surface, respectively.

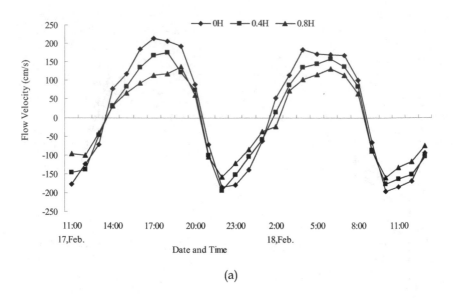

(a)

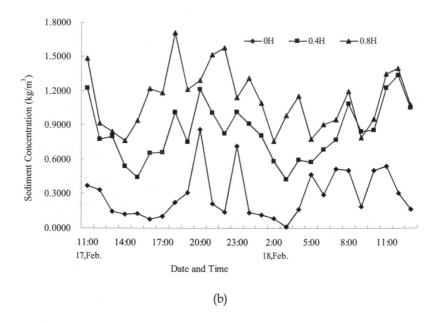

(b)

Fig. 10. Variation of the (a) flow velocity and (b) sediment concentration in the dry season at spring tide in the South Passage, where 0H means measured at the surface, 0.4H and 0.8H means measured at the 0.4 and 0.8 fractions of depth (H) from the surface, respectively.

**In the flood season**

**North Passage**

| spring tide (15-16, July) | | | | | | neap tide (20-21, July) | | | | | |
|---|---|---|---|---|---|---|---|---|---|---|---|
| Max Flood Velocity (cm/s) | 0H | 122 | Sediment Concentration (kg/m³) | 0H | 0.21 | Max Flood Velocity (cm/s) | 0H | 74 | Sediment Concentration (kg/m³) | 0H | 0.11 |
| | 0.4H | 113 | | 0.4H | 0.34 | | 0.4H | 86 | | 0.4H | 0.22 |
| | 0.8H | 111 | | 0.8H | 0.78 | | 0.8H | 76 | | 0.8H | 0.48 |
| Max Ebb Velocity (cm/s) | 0H | 195 | Sediment Concentration (kg/m³) | 0H | 0.87 | Max Ebb Velocity (cm/s) | 0H | 139 | Sediment Concentration (kg/m³) | 0H | 0.51 |
| | 0.4H | 193 | | 0.4H | 0.16 | | 0.4H | 129 | | 0.4H | 0.55 |
| | 0.8H | 145 | | 0.8H | 1.53 | | 0.8H | 111 | | 0.8H | 0.56 |

**South Passage**

| spring tide (15-16, July) | | | | | | neap tide (20-21, July) | | | | | |
|---|---|---|---|---|---|---|---|---|---|---|---|
| Max Flood Velocity (cm/s) | 0H | 198 | Sediment Concentration (kg/m³) | 0H | 0.40 | Max Flood Velocity (cm/s) | 0H | 79 | Sediment Concentration (kg/m³) | 0H | 0.22 |
| | 0.4H | 191 | | 0.4H | 1.16 | | 0.4H | 77 | | 0.4H | 0.22 |
| | 0.8H | 159 | | 0.8H | 1.82 | | 0.8H | 61 | | 0.8H | 0.38 |
| Max Ebb Velocity (cm/s) | 0H | 246 | Sediment Concentration (kg/m³) | 0H | 0.88 | Max Ebb Velocity (cm/s) | 0H | 180 | Sediment Concentration (kg/m³) | 0H | 0.29 |
| | 0.4H | 232 | | 0.4H | 1.97 | | 0.4H | 164 | | 0.4H | 0.28 |
| | 0.8H | 184 | | 0.8H | 1.97 | | 0.8H | 116 | | 0.8H | 1.09 |

**In the dry season**

**North Passage**

| spring tide (17-18, Feb.) | | | | | | neap tide (23-24, Feb.) | | | | | |
|---|---|---|---|---|---|---|---|---|---|---|---|
| Max Flood Velocity (cm/s) | 0H | 233 | Sediment Concentration (kg/m³) | 0H | 0.07 | Max Flood Velocity (cm/s) | 0H | 147 | Sediment Concentration (kg/m³) | 0H | 1.02 |
| | 0.4H | 207 | | 0.4H | 0.59 | | 0.4H | 125 | | 0.4H | 0.46 |
| | 0.8H | 141 | | 0.8H | 1.37 | | 0.8H | 88 | | 0.8H | 2.30 |
| Max Ebb Velocity (cm/s) | 0H | 228 | Sediment Concentration (kg/m³) | 0H | 0.53 | Max Ebb Velocity (cm/s) | 0H | 220 | Sediment Concentration (kg/m³) | 0H | 0.14 |
| | 0.4H | 225 | | 0.4H | 0.74 | | 0.4H | 190 | | 0.4H | 0.70 |
| | 0.8H | 180 | | 0.8H | 2.59 | | 0.8H | 117 | | 0.8H | 1.23 |

**South Passage**

| spring tide (17-18, Feb.) | | | | | | neap tide (23-24, Feb.) | | | | | |
|---|---|---|---|---|---|---|---|---|---|---|---|
| Max Flood Velocity (cm/s) | 0H | 197 | Sediment Concentration (kg/m³) | 0H | 0.51 | Max Flood Velocity (cm/s) | 0H | 130 | Sediment Concentration (kg/m³) | 0H | 0.70 |
| | 0.4H | 194 | | 0.4H | 0.83 | | 0.4H | 121 | | 0.4H | 0.88 |
| | 0.8H | 160 | | 0.8H | 0.95 | | 0.8H | 101 | | 0.8H | 1.53 |
| Max Ebb Velocity (cm/s) | 0H | 213 | Sediment Concentration (kg/m³) | 0H | 0.10 | Max Ebb Velocity (cm/s) | 0H | 188 | Sediment Concentration (kg/m³) | 0H | 0.11 |
| | 0.4H | 175 | | 0.4H | 1.01 | | 0.4H | 143 | | 0.4H | 0.36 |
| | 0.8H | 137 | | 0.8H | 1.21 | | 0.8H | 93 | | 0.8H | 1.09 |

Table 1.

from the bottom firstly, and during the ebb tide period, the ebb tide current start from the surface firstly. These different riverine and marine mechanics may cause the sediment deposited during the flood season because of the more longer slack water period, during the dry season, because of the lower water level and less water discharge from the drainage basin, Jiuduansha Shoal will be eroded. During the neap tidal cycle, the flow velocity is lower than that during the spring tidal cycle, and the sediment concentration is lower than that the spring tidal cycle, that means the more sediment deposited on the tidal flat and estuarine river channel. During spring tidal cycle, because of the higher flow velocity and stronger tidal current, some of the deposited sediment in the channel and tidal flats maybe eroded and maximum turbidity formed.

### 3.3 Evolution of the Eastern Chongming Tidal Flat, Jiuduansha Shoal and Nanhui Marginal Tidal Flat

The Yangtze River transported a quantity of sediment into the estuarine region, and deposited sediment in estuary formed the tidal flats and shoals. About 2/3 of land area are expanded because of reclamation of the tidal flat, in nearly 50 years, about 800km² been reclaimed. From 1978 on, about 338.4km² tidal flat been reclaimed, especially in Eastern Chongming Tidal Flat and Nanhui Marginal Tidal Flat, and the reclamation still continued at present.

Eastern Chongming Tidal Flat, Jiuduansha Shoal and Nanhui Marginal Tidal Flat are the three very important wetlands in Yangtze Estuary. The Eastern Chongming Tidal Flat is an important wetland in the Yangtze River Estuary, from 1975 to 2005, the reclaimed area of the upper tidal flat is about 82 km², that means about all tidal flat over 0m isobathic line had been reclaimed under 1992, 1998 and 2001 levees construction (Fig. 11-14). Nanhui Marginal Tidal Flat is the main tidal flat in the south bank of the Yangtze River Estuary, but continued reclamation in past 30 years, about 140km² tidal flat had been reclaimed, and the tidal flat has lost the ecological significance because of the human actions (Fig. 15-16). After the formation of the Jiuduansha Shoal because of the Flood in 1954, the area, volume, and elevation of the Jiuduansha Shoal increased respectively. Figure 11 and figure 12 to 13 show a comparison of the Jiuduansha Shoal during 1975 to 2005. In 1975, the Jiuduansha shoal still formed by three shoals named Shangsha, Zhongsha and Xiasha respectively, there is only 9.5km² over 1m isobathic line (Yellow Sea Level) (Fig.17), under the riverine and marine processes, and human actions in the Yangtze Estuary, and Zhongsha and Xiasha coalesced in 2001 (Fig.18), and then three shoals of Jiuduansha Shoal coalesced in 2005 (Fig.19).

Fig. 11. Main tidal flat in Yangtze Estuary in 1975

Formation and Evolution of Wetland and Landform in the Yangtze River Estuary Over the Past 50 Years Based
on Digitized Sea Maps and Multi-Temporal Satellite Images
227

Fig. 12. Eastern Chongming Tidal Flat in 1990

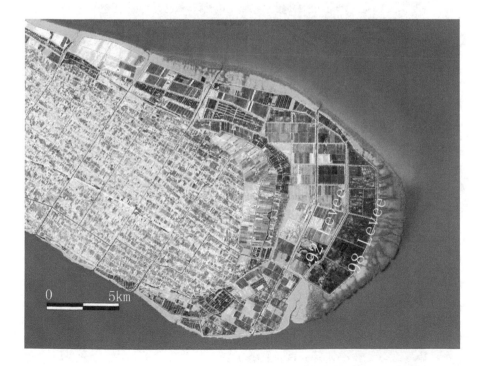

Fig. 13. Eastern Chongming Tidal Flat in 2001

Fig. 14. Eastern Chongming Tidal Flat in 2005

Fig. 15. Nanhui Marginal Tidal Flat in 2001

Formation and Evolution of Wetland and Landform in the Yangtze River Estuary Over the Past 50 Years Based on Digitized Sea Maps and Multi-Temporal Satellite Images

229

Fig. 16. Nanhui Marginal Tidal Flat in 2005

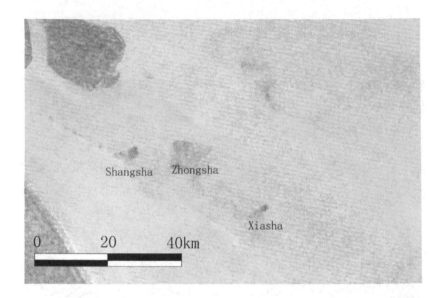

Fig. 17. Jiuduansha shoal in 1975

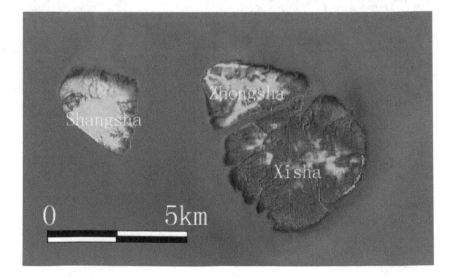

Fig. 18. Jiuduansha shoal in 2001

Formation and Evolution of Wetland and Landform in the Yangtze River Estuary Over the Past 50 Years Based on Digitized Sea Maps and Multi-Temporal Satellite Images

231

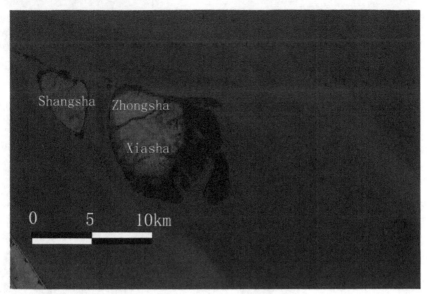

Fig. 19. Jiuduansha shoal in 2005

## 4. Conclusions

1.  Eastern Chongming Tidal Flat is increased consistently in area and altitude. After the construction of 1992 and 1998 levee and 2001 dike, the higher tidal flat has been reclaimed, but due to the deposition of the sediment from the drainage basin, the higher tidal flat, inter-tidal flat and lower tidal flat are increased continuously.
2.  The Jiuduansha Shoal formed in 1954 because of the historically large Flood in the Yangtze River Basin, the Flood caused the ebb channel and flood channel merge, and the Jiuduansha Shoal isolated from the Southern Tongsha Tidal Flat. Because of the Siltation on the Jiuduansha Shoal, the area and altitude of the Jiuduansha shoal increased consistently.
3.  Nanhui Marginal Tidal Flat once an important tidal flat in southern bank of the Yangtze River Estuary, it is had lost the ecological situation because of the reclamation to 0m isobathic line.

## 5. Acknowledgements

This work was supported by the NSFC (41072164) , National Key Basic Research and Development Program (Grant No. 2003CB415206) and MHREG (MRE201002).

## 6. References

Cao, Peikui, Hu, Fangxi, Gu, Guochuan, and Zhou, Yueqin, 1989, Relationship between suspended sediments from the Changjiang Estuary and the evolution of the embayed muddy coast of Zhejiang Province, Acta Oceanologica Sinica, 8(2), 273-283.

Chen, Shenliang, 2001, Seasonal, neap-spring variation of sediment concentration in the joint area between Yangtze Estuary and Hangzhou Bay, Science in China (Series B), Vol. 44 Supp, 57-62.

Chen, J., Zhu, H., Dong, Y., Sun, J., 1985. Development of the Changjiang Estuary and its submerged delta. Continental Shelf Research. 4: 47-56

Chen, J.Y., Yun, C.X., Xu, H.G., Dong, Y.F., 1979. The evolutional model of the Changjiang river mouth since 2000 years. Oceanol. Sin., 1 (1): 103-111 (in Chinese with English abstract)

Chen, Z., Stanley, D.J., 1993. Changjiang delta, eastern China: 2. Late Quaternary subsidence and deformation. Marine Geology 112, 13– 21

Chen, Z., Stanley, D.J., 1995. Quaternary subsidence and river channel migration in the Changjiang delta plain, eastern China. Journal of Coastal Research 11, 927–945

Chen, Z., Xu, Sh., Yan, Q., 1991. Sedimentary facies of Holocene subaqueous Changjiang river delta. Oceanologia & limnologia Sinica. 22 (1) 29-37 (in Chinese with English abstract)

Dalrymple, R.W., Knight, R.J., Zaitlin, B.A., Middleton, G.V., 1990. Dynamics and facies model of a macrotidal sandbar complex. Cobequid Bay–Salmon River Estuary (Bay of Fundy). Sedimentology 37, 577–612.

Dalrymple, R.W., Rhodes, R.N., 1995. Estuarine dunes and bars. In: Perillo, G.M.E. (Ed.), Geomorphology and Sedimentology of Estuaries. Developments in Sedimentology 53, Elsevier, Amsterdam, pp. 359–422

Gao, Jin, 1998, The evolutional rule of Changjiang River mouth and hydrodynamic effect, Acta Geographica Sinica, 53(3), 264-269 (in Chinese with English summary).

Gong, Cailan and Yun, Caixing, 2002, Floods rebuilding the riverbed of the Changjiang Estuary, The Ocean Engineering, 20(3), 94-97, (in Chinese with English summary).

Harris, P.T., Pattiaratchi, C.B., Cole, A.R., Keene, J.B., 1992. Evolution of subtidal sandbanks in Moreton Bay, eastern Australia. Mar. Geol. 103, 225–247

Huang, Weikai and Chen, Jiyu, 1995, Prediction of topography changes in Changjiang River Estuary bar, Oceanologia & Limnologia Sinica, 26(4), 343-349 (in Chinese with English abstract).

Kazuaki Hori, Yoshiki Saito, Quanhong Zhao, Pinxian Wang, 2002. Architecture and evolution of the tide-dominated Changjiang (Changjiang) River delta, China. Sedimentary Geology. 146: 249-264

Kazuaki Hori, Yoshiki Saito, Quanhong Zhao, Xinrong Cheng, Pinxian Wang, Yoshio Sato, Congxian Li, 2001a. sedimentary facies of the tide-dominated paleo-Changjiang (Changjiang) estuary during the last transgression. Marine Geology. 177, 331-351

Kazuaki Hori, Yoshiki Saito, Quanhong Zhao, Xinrong Cheng, Pinxian Wang, Yoshio Sato, Congxian Li, 2001b. Sedimentary facies and Holocene progradation rates of the Changjiang (Changjiang) delta, China. Geomorphology. 41.233-248

Knight, R.J., 1980. Linear sand bar development and tidal current flow in Cobequid Bay, Bay of Fundy, Nova Scotia. In: McCann, S.B. (Ed.), The Coastline of Canada. Geol. Surv. Can. Pap., 80-10, 123–152.

Li, Jiufa, Shen, Huanting, and Xu, Haigen, 1995, The bedload movement in the Changjiang River Estuary, Oceanologia & Limnologia Sinica, 26(2), 138-145 (in Chinese with English abstract).

Li, C., Wang, P., 1991. Stratigraphy of the Late Quaternary barrier–lagoon depositional system along the coast of China. Sedimentary Geology 72, 189–200.

Li, C.X., Li, P., Chen, X.R., 1983. Effects of marine factors on the Changjiang River channel below Zhengjiang. Acta Geographica Sinica 38 (2): 128−140 (in Chinese, with English abstract).

Ludwick, J.C., 1974. Tidal currents and zig-zag sand shoals in a wide estuary entrance. Geol. Soc. Am. Bull. 85, 717–726.

Niu, Xinqiang, Xu, Jianyi, and Li, Yuzhong, 2005, Preliminary analysis for influence of flow and sediment variation on underwater bar growth in the estuary of the Yangtze River, Yangtze River, 36(8), 31-33 (in Chinese with English Summary).

Pan, Ding'an and Sun, Jiemin, 1996, The sediment dynamics in the Changjiang River Estuary mouth bar area, Oceanologia & Limnologia Sinica, 27(2), 279-286 (in Chinese with English abstract).

Qin, Y., Zhao S., 1987. Sedimentary structure and environment evolution of submerged delta of Changjiang river since late Pleistocene. Acta Sedimentologica Sinica 5 (105-112) (in Chinese with English abstract).

Qin, Y., Zhao, Y., Chen, L., Zhao, S., (Eds.), 1996. Geology of the East China Sea. Science Press. Beijing. 357pp.

Shen, Huanting and Pan, Ding'an, 1988, The characteristics of tidal current and its effects on the channel of the Changjiang Estuary, In: Chen, Jiyu, Shen, Huanting, and Yun, Caixing, eds. Processes of Dynamics and Geomorphology of the Changjiang Estuary, Shanghai Scientific and Technical Publishers, 80-90 (in Chinese).

Shen, H.T., Li, J.F., Jin, Y.H., 1995. Evolution and regulation of flood channels in estuaries. Oceanol. Limnol. Sin. 26 (1): 82-89 (in Chinese with English abstract)

Stanley, D.J., Chen, Z., 1993. Changjiang delta, eastern China: 1. Geometry and subsidence of Holocene depocenter. Marine Geology 112, 1– 11

Wang, J., Guo, X., Xu, S., Li, C., 1981. Evolution of the Holocene Changjiang delta. Acta Geologica Sinica 55, 67-81 (in Chinese with English abstract)

Xu, Fumin, Yan, Yixin, and Mao, Lihua, 2002, Analysis of hydrodynamic mechanics on the change of the lower section of the Jiuduansha sandbank in the Yangtze River Estuary, Advances in Water Science, 13(2), 166-171 (in Chinese with English abstract).

Yang, Shilun, Yao, Yanming, and He, Songlin, 1999, Coastal profile shape and erosion-accretion changes of the sediment island in the Changjiang River Estuary, Oceanologia & Limnologia Sinica, 30(6), 764-769 (in Chinese with English abstract).

Yoshiki Saito, Zuosheng Yang, Kazuaki Hori. 2001. The Huanghe (Yellow River) and Changjiang (Yangtze River) deltas: a review on their characteristics, evolution and sediment discharge during the Holocene. 41: 219-231

Zhang, Hongwu and Wang, Jiayin, 1987, River bend hydraulics, Yellow River Institute of Hydraulic Research, 65-68 (in Chinese).

Zou, Desen, 1990, The hydrodynamic and PLT in the mouth region of Yangtze River and its training, Journal of Sediment Research, 3, 27-34 (in Chinese with English summary).

# Part 4

## Multiphase Phenomena:
## Air-Water Flows and Sediments

# 12

# Sediment Gravity Flows:
# Study Based on Experimental Simulations

Rafael Manica

*Instituto de Pesquisas Hidráulicas - Universidade Federal do Rio Grande do Sul*

*Brazil*

## 1. Introduction

Gravity (or density currents) currents are a general class of flows (also known as stratified flows) in which flow takes place because of relatively small differences in density between two flows (Middleton, 1993). Gravity currents that are driven by gravity acting on dispersed sediment in the flow were called *sediment gravity flows* (Middleton & Hampton, 1973). Sediment gravity flows may occur in both subaerial (e.g. avalanches, pyroclastic flows and so on) and subaqueous ambients (e.g. bottom currents, turbidity currents, debris flow – see Simpson, 1997) and may flow above, below or inside the ambient fluid. The distinction regarding sediment gravity flows and open-channel flows is due to the order of magnitude of the density difference between the fluids. Sediment gravity flow are generally of the same order of magnitude, whilst open-channel flow the difference in density between the flow (e.g. rivers) and the ambient air is much higher than that.

The interest in these types of flows are mainly due to four factors: (i) phenomenon comprehension highlighting the origin, transport and deposition processes; (ii) their great magnitude and unpredictability (potential environmental hazards); (iii) the lack of monitoring these events in nature and; (iv) because of their economic significance, since some deposits generated by such currents are prospective reserves of hydrocarbon.

Despite the great progress addressing theoretical and analytical evaluation of these phenomena, particularly on the origin, transport and deposition of this class of flow, even today, they are not completely comprehended. Generally, the complexity of the phenomenon can be expressed by: (i) interaction between the flow and the bed morphology; (ii) the quantity and the composition of sediment transported and (iii) the complex mixing processes. As a consequence, the origin and the hydrodynamics properties of these flows are less understood than open-channel flows (Baas et al., 2004). Simple definitions, such as volumetric concentrations of sediments, its composition and size distribution of solid particles in the mixture as well as the sediment-support mechanisms are difficult to measure in nature which is also an indicative of such complexity.

Kneller & Buckee (2000) commented that difficulties in understanding the dynamics of suspended sediment are extremely complex by virtue of turbulence. In that case, the phenomenon is: non-linear; non-uniform (variation in space) and unsteady (variation in time). If the flow contains large loads of sediments and/or cohesive sediments in suspension this complexity increases even more. Besides the variation of density with time and space (open boundary conditions), the mechanical properties (rheology) of the suspensions involved (thixotropy, viscosity and gravitational forces) must be taken into account as well as the sediment-support mechanism and the influence of shear stress on the upper layer (Kuenen, 1950). Because of such uncertainty and complexity, many terms, concepts, models and particular descriptions (over than 30) have being introduced and applied to interpret these classes of flows and deposits along the years (e.g. Gani, 2004; Lowe, 1982; Middleton & Hampton, 1973).

Sediment gravity flows can be divided into five broad categories according to Parsons et al., (2010). Each flow type has a range of concentrations, Reynolds numbers, duration, grain size and rheology behaviour, enclosing a general overview of the flows transformation along time and space (Fischer, 1983). Two types of flows have been regularly studied along the last 60 years: turbidity currents and debris flows. Both represent the contrast of the sediment gravity flows categories (not considering mass flows, like slides and slumps - see also Middleton & Hampton, 1973). Succinctly, the main properties attributed and well accepted in the literature to turbidity currents are: diluted (low-density), Newtonian behaviour, turbulent regime, and Bouma sequence type deposit (Bouma, 1962) usually called turbidites. On the other side, debris flows are characterized by great influence of non-cohesive material, non-Newtonian behaviour, matrix strength, bipartite and chaotic (ungraded) deposits.

The interest of many fields of academy and industry do not only concern the comprehension of those two particular types of flows. In fact, all classes of sedimentary gravity currents are motivating researchers to face the problem from different approaches and methods, for instance: studies based on outcrops analogy (generally by sedimentologists and correlated areas); numerical and analytical modelling (which is improving through time) and, finally experimental simulation which has been a powerful tool of visualization and measurement of flow dynamics properties as well as of generated deposit.

The scope of this chapter is to outline the experimental study on sediment gravity flows in order to characterize and comprehend this phenomenon regarding their rheological behaviour, hydrodynamics and depositional properties. The simulations covered a wide range of concentration and/or different amount of cohesive sediments in the mixture. The properties of the flow and deposit were evaluated, classified and compared to literature background. The chapter is structured in five sections; first, a general description of sediment gravity flows will be presented followed by the experimental approach applied. Then, the rheology tests it will be reported and finally, the careful evaluation of the experimental results in terms of time-space and vertical profiles will be described in order to extrapolate the results to natural sediment gravity flows.

## 1.1 Sediment gravity flow anatomy

In nature, subaqueous sediment gravity flow behaves like a river system, i.e., originating (source zone), flowing (transfer zone) and decelerating up to the point where all suspended sediment settled down (depositional zone). In general, the initiation of sediment gravity flows is strongly related to two processes of sediment remobilization in the natural field: Firstly, by the occurrence of catastrophic events such as earthquakes, sedimentary failures, storms and volcanic eruptions which cause high instabilities and remobilize large amount of sediments instantaneously (Normark & Piper, 1991); Secondly, by continuous river supply in which the river discharge is connected into water body (usually reservoirs, lakes and oceans) generating plumes and/or hyperpycnal flows due the density difference (positive or negative). After the process starts, the mixture of suspended sediment (concentration, size and composition) is transported ahead by the flow (transfer zone). Concomitantly, dynamical and depositional processes occur along time and space, causing flow transformations, such as: sediment transport, erosion and/or deposition, mixing, entrainment (Elisson & Turner, 1959) and so on (Fig. 1).

The sediment gravity flows which maintain buoyancy flux throughout movement are called *conservative* (i.e. do not interact with their boundary). Otherwise, flows are called *non-conservative sediment gravity flows* (i.e. open boundary interaction such as erosion and deposition).

Generally, gravity currents are divided geometrically into three distinct parts: head, body and, tail.

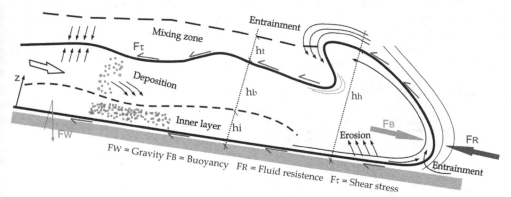

Fig. 1. Schematic of a sediment gravity flow (description of all terms is provided in the list of nomenclature).

The *head* or front of the current is roughly shaped as a semielipse. In most cases, the head is thicker than the body and tail, because of the resistance imposed by the ambient fluid (fluid resistance) to its advance. The head plays an important role on flow dynamics because is characterized by strong three-dimensionality effects and intense mixing (Simpson, 1997). The most advanced point of the front is called *nose* and it is located slightly above the bottom surface, as a result of the no-slip condition at the bottom as well as the resistance (shear) at upper surface (Britter & Simpson, 1978). In the head, two types of instabilities are the main responsible for mixing with the ambient fluid (*entrainment*). The first type of instability is a complex pattern of *lobes and clefts* caused by second order gravitational instabilities at front surface (Kneller et al., 1999; Simpson, 1972). The second type of instability is a series of billows associate to Kelvin-Helmholtz instabilities (Britter & Simpson, 1978), which takes place just behind the head and produced by viscous shear at the head and body (upper surface). This zone behind head creates a large-scale turbulence mixing and also divides the head from the body (symbolically called: *neck* of the flow).

Generally, the velocity of the *body* is greater than the head velocity by 30% or 40% (Baas et al., 2004; Kneller & Buckee, 2000). One reason for this is the presence of a large billow behind the head which cause a locally diluted zone (entrainment of ambient fluid). Thus, in order to the flow maintain its constant rate of advance, the current increases the velocity of the body to compensate the deficit of density created (Middleton, 1993). The body is divided into two zones: near the bottom zone, where the density is higher; and above this, a suspended/mixing zone, where the mixing with the fluid ambient occurs. The interface between these layers (bipartite flow) point out a discontinuity in the body (water-column stratification) that is reflected by an abrupt gradient of velocity, concentration and viscosity (Postma et al., 1988).

The third part of sediment gravity flow is characterized by a deceleration zone and final dilution stage of the current, normally called *tail*.

In terms of dynamics properties of the flow, sediment gravity flows differ significantly from open-channel flows (e.g. rivers) regarding their velocity profile. In that case of sediment gravity flow, the main difference is due to the fact that is not possible to ignore the shear effects in the upper surface of the current (*see* Fig. 2 a, b). Then, the sediment gravity flows velocity profile has null values at the upper and bottom surfaces and values grow towards to the middle (balance of drag forces acting on those surfaces), creating a front point (maximum value) usually at 0.2 to 0.3 times the height of the current. Depending on the concentration and composition of sediments in suspension, both velocity and concentration profiles may present completely different shape (Fig. 2 c) as the inner dynamic of the flow became more complex (e.g. matrix strength, cohesive forces).

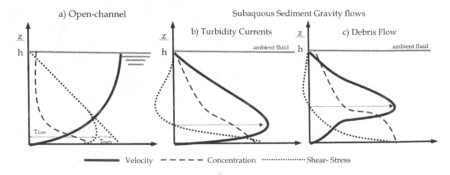

Fig. 2. Vertical profiles of velocity, concentration and shear stress for: a) open-channel flows; b) turbidity current and; c) debris Flow.

The two most known classes of sedimentary gravity flows (described earlier) have differences regarding their internal dynamics. The dynamics of turbidity currents is complex due to the processes of erosion and deposition. Because of this, the three-dimensional representation of this phenomenon through analytical equations is not simple, which leads to simplification (e.g. shallow water flows – Parson et al., 2010; Parker et al., 1986). In the same way, the debris flows are extremely complex too, as the existence of yield strength caused by the high density and the presence of clay implies in shear-like flow and plug-like flows as illustrated in Fig. 3.

Generally, the hydrodynamic of a sediment gravity flow is closely associated to sediment-transport capacity (total amount of sediment transported by the flow) and competence (ability of the flow to carry particular grain size) as well as to the sediment-support mechanism, whose the main role is to keep the sediments in suspension for a long period of time (and distance). For each class of flow may occur different mechanisms of sediment-support, as it depends on grain-size and composition, concentration of sediments and the rheological properties of the mixture.

For turbidity currents, the main sediment-support mechanisms are vertical component of turbulence and buoyancy. However, for flows of high concentration (high-density) several sediment-support mechanisms may occur simultaneously, such as: *hindered settling*, in which grains deposition is inhibited because the number of particles increases in an certain zone, creating a slower-moving mixture than would normally be expected (effect of population of grains); *dispersive pressure:* in which the grains are held in suspension by their interaction forces (collision) and; *matrix strength:* a mixture of interstitial fluid and fine sediment (cohesive), which has a finite yield strength that supports coarse grains (Lowe, 1979; Middleton & Hampton, 1973).

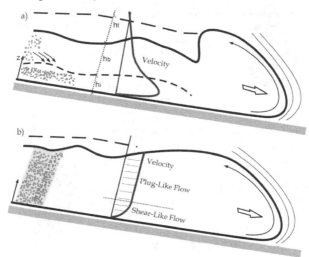

Fig. 3. The difference between the internal dynamics of the turbidity current (a) and debris flow (b).

The effect of high concentration on the dynamics of sediment gravity flows is expressed by changes in the mixture and flow properties such as: density of the fluid; increase of the potential energy and momentum of the flow and; viscosity of the mixture (rheological behaviour). Also, the settling velocity of particles is strongly influenced by the increase in fluid concentration mainly because: the fall of the particles induces an upward movement of water; the buoyancy of the particle increases due to high-density fluid, and by the interaction between particles (effect of population - *hindered settling*). The transport capacity of the flow tends to increase with high sediment concentration; however, these changes also depend on the composition of sediment present in suspension.

In contrast, the presence of cohesive sediment implies a different scenario in which the flocs of cohesive particles will settle down during the flow, creating a clay/mud near-bed layer with high content of water inside. Despite the fact the turbulence can be produced in this clay/mud layer (due to shear flow), there is also a significant increase in viscous forces (non-Newtonian behaviour), which could reduced the flow ability to transport great amounts of sediment downstream.

## 2. Apparatus and experimental simulations

In order to understand the hydrodynamic of natural sediment gravity, an experimental study was performed with different types of sediments, such as: non-cohesive particles represented by very fine sand and silt sized glass beads, and cohesive particles represented by kaolin clay. Both sediments have density approximately of 2600 kg/m³. In total, 21 experiments (Fig. 4) were carried out with eight values of bulk volumetric concentration (2.5%, 5%, 10%, 15%, 20%, 25%, 30% and 35%). In addition, for each value of concentration were used three different proportions of clay in the mixture from 0% (pure non-cohesive flows) passing to 50% (mixed) and finally, 100% (pure cohesive flows).

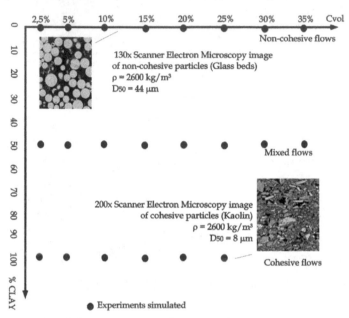

Fig. 4. Initial properties of the mixtures simulated and the particles properties.

The experiments were performed in a 2D Perspex tank (4.50 m long x 0.20 m wide x 0.50 m height). A 120 litres mixture was prepared in a mixing box (full capacity of 165 litres) connected at the upstream part of the tank through a removable lock-gate (0.21 m wide and 0.70 m high). An electric-mechanical mixer was installed within that box to assure the full mixing of sediment mixture. The tank also had a dispersion zone (approximately 1.00 m length) in which the water (and flow) were drained after the experiment.

In all sets of experiments were used lock-exchange methodology characterized by the instantaneously release of the mixture (lock-gate opening) reproducing a catastrophic event on nature. As soon as the mixture entered into the channel, the dense flow was generated.

In order to measure the flow properties during the experiments, a group of equipments was installed within the tank. Four UHCM's (*Ultrasonic High-Concentration Meter*) were set along the vertical profile (at 1.0; 3.2; 6.4 and 10 cm from the bottom) to acquire time-series concentration data, whilst ten UVP's (*Ultrasonic Doppler Velocity Profiler*) of 2 MHz transducers were set along vertical profile (15 cm) to register time-series of velocity data. Both equipments were located at 340 cm from the gate. With both velocity and concentration data, the hydrodynamic properties were established for all flows such as: time series of velocity and concentration, mean vertical profiles, non-dimensional parameters for the head, body and tail zones.

Additionally, all flows were recorded with a digital video-camera placed on the side of the tank in order to evaluate the time series of geometric features of the current (see Fig. 1), such as: *the current height ($h_t$)*; *thickness of the body ($h_b$)* defined as the height of the body not considering the mixing zone at the upper surface and; *thickness of the internal layer ($h_i$)*, which considers the interface layer created by the presence of a more concentrated zone near the bottom. The depositional properties (e.g. deposition rate) were also evaluated through the video images.

After the experiment, the ambient fluid was slowly drained and the final deposit properties (e.g. thickness, grain-size and mass balance) were measured (and/or sampled).

## 3. Rheology of mixtures

The rheology is the study of deformation and flow of matter and is a property of the fluid that expresses its behaviour under an applied shear stress. Through the rheological characterization of mixtures (water and sediment), it is possible to establish the relationship between shear stress and strain rate (shear rate), and consequently the coefficient of dynamic viscosity (and/or apparent) as well as the constitutive equations in terms of volumetric concentration and presence of clay.

In natural flows, the non-conservative condition of the sediment gravity flows, i.e. erosion and deposition during the movement, modifies the mechanisms of transport and deposition of particles within the flow (e.g. local concentration, size and composition of grains in suspension), which impact also their rheological behaviour.

Based on this, a rheological characterization of mixtures was carried out aiming to establish such property of the mixtures and verify its behaviour for different initial conditions. To do that, it was used a Rheometer device with two types of spindle (cone plate and parallel plate). For the tests, the mixtures were prepared following the same proportions of sediment used in the experimental work and also considering the same temperature ($\sim 19°C$). The rheogram - output data of the Rheometer consisting in the ratio of shear stress and strain rate - was compared to typical rheological models found in literature. The simplest rheological model of imposed stress ($\tau_x$) related to strain rate ($\delta u/\delta z$) is the *Newtonian* model (due to the definition of Newton's law of viscosity) and it can be expressed for two-dimensional flow in the x – z plane as:

$$\tau_x = \mu \frac{\partial u}{\partial z} \tag{1}$$

The equation (1) shows a linear relationship between the imposed shear stress and strain rate (gradient of deformation). As a consequence, the viscosity of the fluid or mixture (*coefficient of dynamic viscosity - $\mu$*) is constant for all values of shear rate. Any deviation from linearity between the stress-strain curve converts the rheological property to non-Newtonian behaviours, which can be generally divided into four more groups: *plastics* in which there is no deformation of the flow until the critical initial stress (yield strength - $\tau_0$) is overcome; *dilatant and pseudoplastic*, in which the deformation (strain rate) is expressed by a power law type (if coefficient of power law n > 1 then the fluid is *dilatant* otherwise (n < 1) is *pseudoplastic*) and; *Herschel-Bulkley* in which the fluids has a plastic behaviour (yield strength - $\tau_0$) followed by a power law behaviour. The *Herschel-Bulkley* model can be expressed for two-dimensional flow in the x – z plane as:

$$\tau_{ap} = \tau_0 + K \left( \frac{\partial u}{\partial z} \right)^n \tag{2}$$

To non-Newtonian mixtures, the determination of viscosity (curve slope at the rheogram) is no longer direct, implying that for each value of gradient of deformation (strain rate) applied, there will be a different coefficient of dynamic viscosity. When this occurs, the viscosity is called *apparent viscosity of the fluid* ($\mu_{ap}$) rather than the dynamic viscosity.

From the results obtained with the rheometry tests, it was defined two distinct groups for the mixtures simulated in terms of different values of concentration and clay content: the Newtonian group of mixtures and the Herschel-Bulkley plastic group of mixtures (Fig. 5).

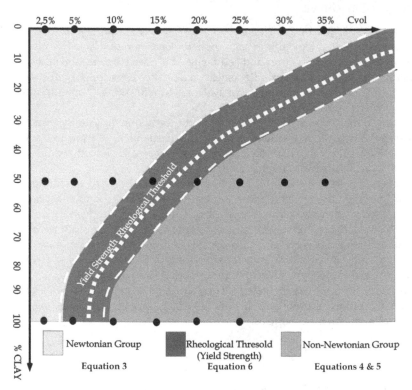

Fig. 5. Rheological characterization of the mixtures simulated and the constitutive equations in terms of volumetric concentration and presence of clay for each group.

For the group of Newtonian mixtures (above threshold line) it was possible to establish an empirical relationship (linear) between the values of dynamic viscosity with the volumetric concentration and clay presence, which allows properly assess the effect of viscosity on the hydrodynamic parameters for this group of mixtures (eq. 3). The coefficient values were similar to those found in literature for non-cohesive grain mixtures (e.g. Coussot, 1997; Einstein, 1906). The rheological characterization was carried out to the volumetric concentration of 35% only. Extrapolation to higher values must be handled carefully (see Coussot, 1997; Wan & Wang, 1994).

$$\frac{\mu}{\mu_0} = 1 + C_{vol} \left( 2.24 + 0.44 \ \%Clay \right) \tag{3}$$

The threshold line represents the transition from Newtonian to non-Newtonian behaviour (plastic) and can be represented by the occurrence of yield strength. Clearly, there is not a unique value representing this change of rheological behaviour. A transition interval must be considered (dashed line around the threshold) to more accurate analysis. In addition, different composition of clay may move the position of the curve, for instance; the threshold of montmorillonite shows similar shape. However this curve of yield strength (high values for this particular type of clay) is moved into the top-left of the diagram.

For the group of Herschel-Bulkley plastic mixtures (high concentration and more presence of clay - below the threshold line) the constitutive equations were empirically determined (eq. 4, 5) correlating the apparent viscosity, the clay content in the mixture, the bulk concentration of the mixtures and, the gradient of deformation (strain rate) for this group of mixtures.

$$\frac{\mu_{ap}}{\mu_0} = \left[ 1.39 \ e^{(31 \ C_{vol})} \left( \frac{\partial u}{\partial z} \right)^{(0.24 - 1.8 \cdot C_{vol})} \right] \cdot C_{clay} \tag{4}$$

where

$$C_{clay} = 0.0016 \ e^{(8.7 \cdot \%Clay)} \ \frac{\partial u}{\partial z}^{(0.59 - \%Clay)} \tag{5}$$

It was also established an empirical relationship to yield strength in terms of the volumetric concentration and the presence of clay in the mixture.

$$\tau_i = 0.00104 \ e^{(2790 \ \%Clay \ C_{vol})} \tag{6}$$

## 4. Experimental results

The rheological characterization (rheometry) has classified the mixtures into two distinct groups as it was illustrated in Fig. 5. Based on that approach, all data and results obtained through experimental work were compared in order to establish groups with similar properties. A total of 15 parameters divided into seven categories were used to fully characterize and distinguish each group: geometry, rheology, analysis of mean vertical profiles, time-series of data, internal dynamics of the flow, depositional features and, non-dimensional parameters as seen in Fig. 6.

After applying this method of analysis, it was possible to identify six regions (or groups) of similar sediment gravity flows generated experimentally. Each one has typical properties and characteristics in terms of rheology, geometry, hydrodynamic and depositional processes along time and space. Moreover, the relationship with initial properties (concentration and clay content) demonstrates the cause-consequence of the experiments (from source to deposit) and the entire dynamic involved. The Fig. 7 illustrates this diagram-phase with delimited boundaries amongst the regions.

Each region properties will be completely described below from non-cohesive dominated flows (regions I, II and III) to cohesive dominated flows (regions IV, V and VI). The averaged vertical profiles will be discussed apart (item 4.6).

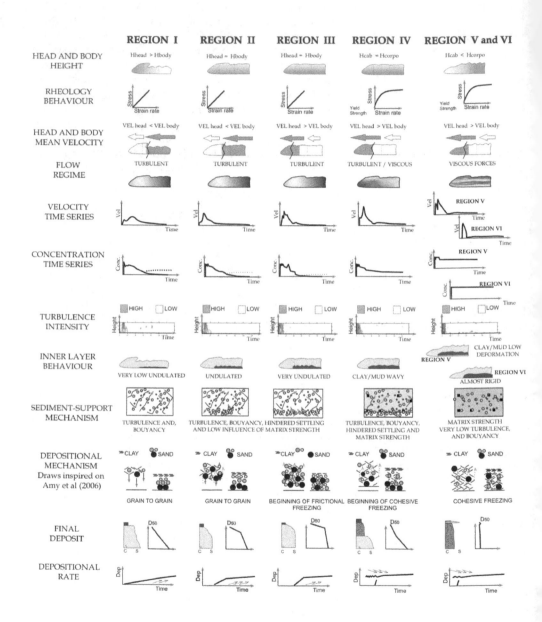

Fig. 6. Results obtained through the experimental simulations

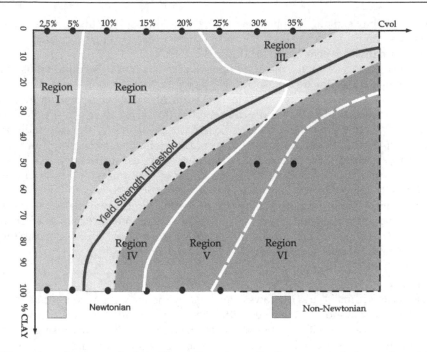

Fig. 7. Six regions (or groups) of similar sediment gravity flows generated experimentally

## 4.1 Region I - Turbidity currents like sediment gravity flows

Sediment gravity flows generated considering the properties of the region I (Newtonian, low-volumetric concentration (< 5%) regardless of the amount of clay) reproduces a classic behaviour of turbidity currents widely discussed in the literature (Kneller & Buckee, 2000; Middleton, 1966; Simpson, 1997). The current accelerates (waxing flow – Kneller, 1995) due to the buoyancy flux with clearly defined head at the front. The thickness of the head is greater than the body, indicating the flow undergoes a large resistance of the ambient fluid and also from gravitational forces acting over the body. As consequence, a large billow (shear vortex) takes place behind the head (high mixing zone).

The body presents the peak of velocity and after this point the flow starts to decelerate gradually (waning flows - Kneller, 1995). Concomitantly, the concentration of sediment within the flow follows the velocity behaviour. In the head, sediments are held in suspension by virtue of the high-turbulence intensity (no depositional zone) and then, the suspended sediments start to settle (fall out) with the decrease in velocity. The current becomes diluted and finally there is only the sedimentation of finer particles by decantation (very long time for cohesive particles because of low settling velocity).

In these currents the main mechanism of grain support is turbulence (inertial forces) with high Reynolds numbers along the entire current (despite the low concentration of mixtures) except for the final stages of the flow (tail). The evaluation of the turbulent intensity (*root mean square* - RMS) shows that turbulence occurs mainly in the head and particularly in the vortex generated behind the head whilst in the body, turbulence occurs around shear layer (mixing zone). Along the vertical profile there was absence of high RMS values near the bottom, which may explain the initiation of the deposition just after the passage of the front. For the flows of this region, the presence of cohesive sediments at low concentrations (< 5%) implies in no significant changes on the flow behaviour.

During the flow movement, the main mechanism of deposition was by individual particles (grain-to-grain) falling out from suspension by gravity (decelerating flow). Consequently, the dissipation of turbulence caused the lost of sediment-transport capacity of the flow and the grains segregated naturally, i.e. the coarse grains (high setting velocity) were deposited first followed by fine grains and then by colloidal particles (after the stop indeed). As a result, the deposits generated normal gradation (decreasing mean grain size towards to the top - fining upward). For the flows containing clay in suspension, the deposit is characterized by a non-cohesive layer of grains near the bottom with a layer of clay (as a resulted of settling) at the top. The contact between the non-cohesive and cohesive grains is very sharp, clearly indicating different stages of deposition. Despite the fact that clay may form flocs, due to the cohesion of their particles, there was no evidence of the formation of large flocs. The depositional rate for this class of flows was linear (deposit thickness increased at constant rate) starting just after the passage of the head.

## 4.2 Region II

The Newtonian sediment gravity flows originated by the increase in concentration and the presence of clay (around 50%) showed differences in the properties of the flow dynamics and deposition. Both velocity and buoyancy flux increased in the flow causing a decreased in the head height. As a result, the average velocities of the body and head were almost identical, showing that buoyancy forces present in the head are in balance with gravitational forces. Yet, the head of the current is slightly higher then the body and is characterized by intense mixing zone. The main difference comparing with region I can be noticed after the peak of velocity in the body, since the flow rapidly decelerates reaching low values of velocity until completely stopped (tail). The quick deceleration is related to formation of an inner layer of grains more concentrated near the bottom. For a short time the flow becomes stratified (bipartite) changing the velocity and concentration profile instantly and implying in different mechanisms of deposition.

The sediment-support mechanism of the non-cohesive flows (low content of clay) is driven by the turbulence of the flow (in the head), Kelvin-Helmholtz instabilities behind the head and along the mixture layer at the upper and the bottom surfaces. The high values of turbulence intensity were measured throughout vertical profile explaining a period of no deposition at early stages of the flow. Also, the concentrated near-bed layer (mainly non-cohesive) is characterized by high turbulence intensity and its internal undulations are closely related to instabilities at the upper surface.

The flows generated from experiments adjacent to rheological threshold in Fig. (7), the increase of amount of clay and/or concentration caused a decrease in turbulence intensity. Also, not only the turbulence plays a key role on these flows but also the influence of the matrix strength and cohesive interaction of the grains started to become relatively significant. This fact is reflected on the behaviour of near-bed layer (mainly cohesive) which is characterized by undulations and deformations, although not as considerable as those presented by pure non-cohesive flows.

The mechanism of deposition in such flows differs from region I. Besides the grain-to-grain sedimentation caused by dissipation of the turbulence intensity (typical behaviour of flows next to boundary between regions I and II), other depositional processes start to play in the flows regarding the amount of clay in the mixture.

In non-cohesive flows, the sediment in suspension settled down creating a concentrated near-bed layer that was constantly fed by sediments from the top (fall out). As consequence, the space between grains became more restricted causing rapid deposition of sediments (high depositional rate in the first stages of the flow where there was insufficient time for the natural segregation of the grains). Hence, the deposit generated *partially graded beds*, i.e., massive (coarse size) deposits at the bottom followed by fining upwards particles on the top (final stages of flow with low depositional rate).

Despite the presence of clay in the mixtures gives the impression to modify the mechanism of deposition of these currents, again, for this group of experiments the deposits show a clear division between the non-cohesive grains (at the bottom) and cohesive grains (at the top).

## 4.3 Region III

The region III corresponds to Newtonian flows with high-concentration and low presence of clay (up to maximum of 20%). The hydrodynamics followed the processes described before (region II), considering the higher values of velocity (amongst all regions) and also the flux of buoyancy, which does not allow the grains settled down in the early stages of the flow. The magnitude of forces acting over the head (mainly buoyancy) and over the body (mainly gravitational) was similar reducing the head height. Once more the flow generates a very wavy concentrated layer close to the bottom, creating a bipartite flow which caused sudden deposition (high-depositional rate) of large amount of sediments. Then, the diluted current flows over the bed previously deposited (low-depositional rate).

The support mechanism of grain in these flows is basically turbulence generated at the head (high values of RMS) as well as the upper and lower surfaces. However, additional sediment-support mechanisms as hindered settling and dispersive pressure may occur within the concentrated near-bed layer (mainly non-cohesive). On the other side, the mechanism of deposition for these flows represents an evolution of the processes described in region II. Since the suspended load of sediment becomes progressively concentrated towards the bottom, the continuous supply of the grain from the top (fall out) compress the inner layer reducing space for grains to move. At this point, there is a rapid deposition of grains. This process may be a first signal of frictional freezing, where non-cohesive grains settle quickly (collapse) without segregating grains by size. As a result, deposit is partially graded; being massive graded near the bottom and normally graded (fining upward) on the top.

## 4.4 Region IV

The flows classified as region IV are non-Newtonian, which consequently leads to changes in hydrodynamic properties, such as sediment-support mechanism and depositional processes, mainly because of the yield strength.

In this class of flows dominated by cohesive particles, the hydrodynamic processes are closely related to the region II. The head of the current is the local of high velocity, turbulent intensity and mixing, whilst the viscous forces play a significant role on the body causing deceleration and then, the early stage of deposition. It was also verified the formation of a concentrated layer (mainly dominated by clay) at the bottom. The presence of this deformable clay/mud near-bed layer is followed by a constant value of inner concentration.

The sediment-support mechanism is influenced by the content of clay once the turbulence is damped within the current (being only verified in the head of the flow). The cohesive matrix begins to act internally changing the hydrodynamic behaviour of the current. The buoyancy of the interstitial fluid (water and clay) and pore-pressure also contribute to keep the grains in suspension inside the clay/mud near-bed. This behaviour differs from Newtonian non-cohesive flows (regions II and III). In region IV, the concentration has not yet reached the gelling concentration for cohesive mixtures (Winterwerp, 2002).

During the flow, it was possible clearly identify the shear-like flow near the bed and plug-like flow above that, which is dominated by viscous forces acting on the flow. However, the flow can not be classified as completely laminar, since spots of turbulence (high intensity) can be generated within this layer. Also, in the plug-like flow, fluid shear stress is lower than yield strength of the mixture, generating an instantaneously mass deposit (cohesive freezing). As it occurs suddenly, there is no segregation (selection) of the grains. On the other side, the shear stress at the bed is higher enough to allow the settled of non-cohesive sediments. As a result, the final deposit is divided into three distinct depositional layers: low-content clay (~ 5%) bottom layer (shear-like flow); an intermediate ungraded matrix of sand and clay/mud layer (plug-like flow) and; a clay dominant layer on the top (tail and settling deposition).

### 4.5 Region V and Region VI - Debris flow like sediment gravity flow

Regions V and VI have very similar behaviour with high concentration and high amount of cohesive material (Herschel-Bulkley rheological model). This region represents the other extreme of sediment gravity flows evolution and their transformations.

The hydrodynamic of the current was influenced by the clay content presenting a strong waxing flow-phase (high-turbulence intensity only at the head) and abrupt deceleration, after the arrival of deformable clay/mud near-bed layer (for Region V) and practically not undulating/deformable (for region VI). The plug-like flow in the body induced cohesive freezing, in which a large amount of sediments are deposited in few seconds (high-depositional rate). In this region, the content of clay in the mixture at high concentrations is influenced by the gelling concentration. According to the literature, this occurs at concentrations of clay between 80 and 180 g/l, equivalent to a solid volume fraction of 0.03 and 0.07 (Whitehouse et al., 2000; Winterwerp, 2001, 2002). The mixtures simulated in the regions V and VI correspond to this range of values. Therefore, the cohesive forces acting on these deposits are transmitted to all mass deposited and not only to each single particle causing a thick ungraded chaotic deposit.

The sediment-support mechanism is highly influenced by the increased of apparent viscosity of the mixture and matrix strength which is induced by electrostatic interactions of clay particles. Thus, turbulence is damped throughout the flow, with local spots of high-turbulence intensity close to the bottom (high values), as well as at the interface between the deposit generated by clay/mud near-bed layer and the remaining flow (body and tail). This final stage of the flow generates a normally graded deposit (coarse-tail grading on the top) associated to the mechanism of deposition described in the region I (turbidity currents like flows).

### 4.6 Mean vertical profiles

Based on the experimental results, Fig. (8) illustrated the idealized pattern for each region concerning the average velocity, concentration and sediment flux vertical profiles.

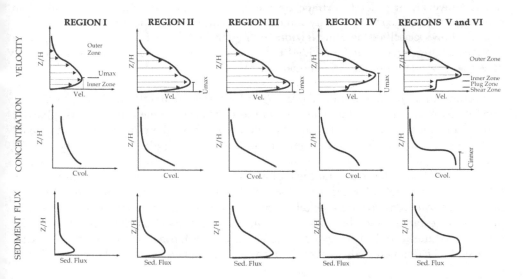

Fig. 8. Mean vertical profiles of velocity, concentration and sediment flux for the six regions of sediment gravy flows.

### 4.6.1 Velocity profiles

Concerning the flows classified as Newtonian (regions I, II and III), the velocity profile presented the classical behaviour of turbidity current (see description section 1.1) with a maximum velocity point located at some distance from the bottom and two distinct zones: an inner zone near the wall and an outer zone up to the top surface (Fig 8, top-left). Applying the model developed by Michon et al., (1955) and modified by Altinakar, (1988) it was possible to establish analytical equations for non-dimensional velocity profiles in terms of initial concentration of the flow.

The model consists in a relationship between a non-dimensional velocity and geometry parameters and also separates the velocity profile in two zones (the threshold is height of maximum velocity - $h_m$). The equations below present the results of applied methodology for the inner zone ($z < h_m$) including the parameters fitted for this group of experiments.

$$\frac{u}{U_{max}} = \left(\frac{z}{h_{max}}\right)^{0.4} \tag{7}$$

And for the outer zone ($z > h_m$) is,

$$\frac{u}{U_{max}} = e^{\left[-2.7\left(\frac{z-h_m}{h_t-h_m}\right)^{1.9}\right]} \tag{8}$$

Those equations can be applied to a wider range of currents with different behaviours as the first approximation of the non-dimensional velocity profile for Newtonian sediment gravity flows. However, in order to extrapolate the results to natural fields, it must take into consideration the maximum velocity value and its location within the current.

For the flows classified as non-Newtonian (regions IV, V and V) the velocity profile changes drastically and can be divided in four zones (Fig. 8 top-right): the *shear-like flow zone* (near the bottom), strongly influenced by viscous sublayer; the *plug-like flow zone:* occurs when the value of the shear stress is lower than yield strength; and the *other two zones* from the remaining diluted current (similar to Newtonian flows described above). The first two zones involve the evaluation of the shear stress at the wall (viscous sublayer) and the thickness of the plug. For the last two zones above the plug-like flow the model of Michon et al., (1955) can be adjusted adding the plug-like flow velocity and its thickness.

The velocity profiles measured for the high-density currents were similar to those cited by (McCave & Jones, 1988; Postma et al., 1988, Talling et al., 2007). To express these profiles in terms of equations require a detailed analysis of stress distribution along the vertical profiles (to establish the shear zone and plug zone) as well as the estimative of the thickness of the near–bed layer (inner flow). In nature those parameters are not easily estimated. The detailed evaluation of these parameters can be found in Manica, (2009).

### 4.6.2 Concentration profile

The mean concentration profile measured in the experiments show the transition between the six regions of sediment gravity flows (Fig. 8). For flows classified as Newtonian (regions I, II and III) the profile is practically more invariable along the vertical (region I) with a slight increase (creating an inflexion point) at the concentration values near the bottom. The curve is similar to an exponential trend (regions II and III), corresponding typical profiles of open-channel flows (e.g. rivers). An empirical exponential law can be fit in such type of curves considering non-dimensional parameters defined as: local concentration divided by concentration measured at 5% of the total height of the flow (concentration of reference – $C_r$); and the distance from the bottom divided by total height of the current ($z/h_t$). The equation fitted for the experimental results is expressed by

$$\frac{C(z)}{C_r} = 1.22 \cdot e^{\left(-4.0\frac{z}{h_t}\right)} \tag{9}$$

Considering the non-Newtonian sedimentary gravity flows (regions IV, V and VI), the vertical profile of concentrations is strongly influenced by the clay/mud inner layer, which generates high-levels of concentration and, practically stratified the profile into two regions (threshold is the inner layer thickness – see Fig. 1). In terms of analytical adjustment of these peculiar curves, the definition of this threshold point is crucial, once it can be presumed for $z < h_i$ that concentration assumes the value of concentration of reference. Above the clay/mud inner layer, equation (9) can be applied.

The methodology presented here to obtain the non-dimensional concentration profiles (Fig. 9) was straightforward in order to simplify at maximum the input parameters. Methodologies found in literature such as (Graf & Altinakar, 1998, Parker et al., 1987) were tested and applied showing very similar results.

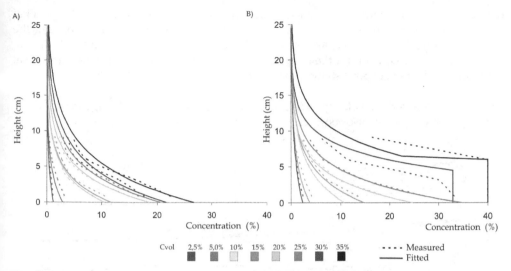

Fig. 9. Concentration profiles measured and fitted curves for the two groups of sediment gravity flows simulated: a) regions I, II and III; b) regions IV, V and VI.

### 4.6.3 Reduced Flux of Sediment

The evaluation of the reduced flux of sediments gives the idea of the mass conservation during the flow, since velocity, concentration and, initial properties of the flow (reduced gravity) are taken into account (eq. 10). Through the evaluation of this parameter, it is possible to check which zone within the flow the sediments are being transported as seen in the experimentally-derived profiles in Fig. 8.

$$S_{Flux} = g \cdot \left( \frac{\rho_m - \rho_a}{\rho_a} \right) \cdot C_{vol} \cdot h_{mean} \cdot u_{mean} \tag{10}$$

The differences among all classes of sedimentary gravity flows simulated were evidenced, particularly, the influence of the cohesive particles (non-Newtonian regions), which implying in a great amount of sediments at the bottom of the current.

## 5. Spatial evolution of the sediment gravity flows

The limitations of the simulations in terms of the length of the tank do not allow a complete study on spatial variability of the sediment gravity flows from their origin to the final deposit. Nonetheless, the full characterization of the main parameters involved in the flow such as time series, vertical profiles, rheology, deposition and so on, may be applied in order to extrapolate the results to natural ambient. Based on that, a detailed spatial analysis was accomplished and the flow evolution for each region will be described below.

Concerning the flows from regions I and II, both concentration and presence of clay increased sediment capacity of transport of the flow, indicating the current could flow further. At the early stages (near the source) the gravity flow is more concentrated with high buoyancy flux and high-turbulence. As the flow propagates downstream, the hydrodynamic processes (e.g. entrainment of ambient water at the upper surface) and depositional processes (e.g. deposition of sediment over time and space) take place, transforming the inner properties of the flow. As a result, the current become more diluted due to deceleration of the flow, losing their capacity of transport (grains settled down) and then, tend to stop. The final deposit shows coarse grains in the proximal areas, due to deposition by gravity (high-settling velocity) and a gradually grain size thinning towards to downstream (low-settling velocity). The Fig. 10 illustrates a model of propagation for flows from region I to region II, also considering their transition points.

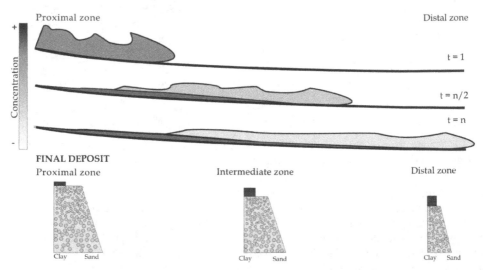

Fig. 10. Spatial evolution scheme of the sediment gravity flow for regions I and II.

In the flows characterized by high concentration and high content of clay (regions II and III) the hydrodynamic properties of the flow change during the run by virtue of the presence of concentrated near-bed layer. The suspended sediment rapidly settled down after the formation of this concentrated layer, causing a reduction in buoyancy flux. As consequence, the remaining diluted current is not able to travel further. This process occurs mainly in the proximal and intermediate zones, where the final deposit is, basically, massive graded. After this zone, the deposits were mainly generated by settling of the grains (gravity) up to distal zone (Fig. 11).

The spatial evolution of the deposit for non-Newtonian mixtures (regions IV, V and VI) showed a distinct behaviour. The increase of flow sediment capacity of transport by reason of high concentration and high presence of clay was counter-balanced by viscous forces, which dominated the flow dynamics and consequently, the generation of the clay/mud near-bed layer. Thus, the bulk of sediment from regions IV, V and VI was not able to travel long distances. Within the plug-like flow, the shear stress of the flow was not enough to prevail over the yield strength of the mixture. As a result, the deposit showed a great quantity of sediment in the proximal zone, whilst only a remaining diluted current flows (with more fine particles) moving to distal zones. The Fig. 12 illustrates this idealized model. The main difference between the idealized transition models of evolution to non-Newtonian sediment gravity flow regards the dynamic of the clay/mud near-bed layer and the final deposit. From region IV to V, a concentrated inner layer presents a high deformation and undulation over time, with a shear–like flow near the bottom and a plug-like flow above (generating the three layer deposit commented on section 4.4). On the other side, from region V to VI (Fig. 13), the near-bed layer is practically a solid mass of mixture flowing downstream, generating a thick clay/muddy deposit at proximal zone.

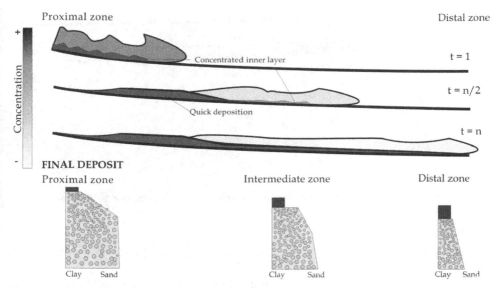

Fig. 11. Spatial evolution scheme of the sediment gravity flow for regions II and III.

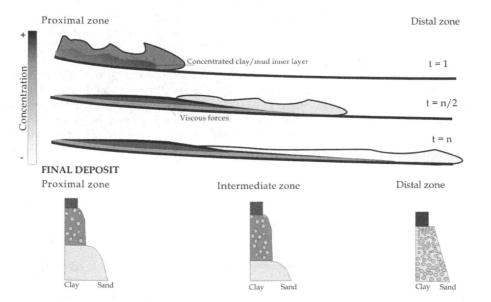

Fig. 12. Spatial evolution scheme of the sediment gravity flow for regions VI and V

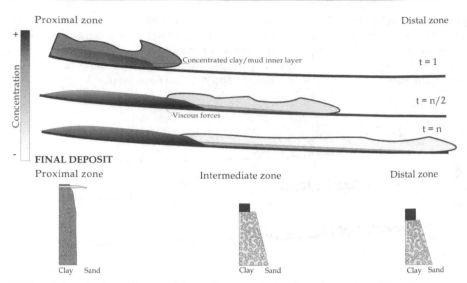

Fig. 13. Spatial evolution scheme of the sediment gravity flow for regions V and VI

## 6. Conclusion

This chapter presented an experimental study of sediment gravity flows in which six types of flows were distinguished based on a comparison of hydrodynamic, depositional and rheological properties. A phase diagram was created, showing the boundaries between these flow types in terms of rheological behaviour, bulk volumetric concentration and clay concentration. The main characteristics of the flow types are summarized below:

Type I: Low density flow; Newtonian; grains supported by upward component of turbulence; no hindered settling; segregation of grains and normally graded beds; Type II: Newtonian; grains supported by turbulence; turbulent flow with gently undulating high-concentration near-bed layer; partial hindered settling and partial size segregation forming partially graded beds; Type III: Newtonian; fully turbulent flow with strongly undulating high-concentration near-bed layer; hindered settling resulting in rapid deposition and generation of partially graded beds; Type IV: non-Newtonian (plastic); viscous flow; formation of plug and shear flow (mud layer close the bottom); viscous forces cause freezing of the flow and forming graded beds of muddy sand and; Types V and VI: non-Newtonian (plastic); viscous flow with thick mud layer; grain support by matrix strength; weakly undulating internal mud layer (type VI show no undulations); cohesive freezing forms an ungraded muddy sand with coarse-tail grading on top.

The six types of flow/deposits classified represent the transition between the two most known types of sedimentary gravity flows: from turbidity currents (low-concentration, low clay and Newtonian behaviour) to debris flow (high-concentration and high clay content and non-Newtonian behaviour). The experimental study allows the comparison and extrapolation of the results obtained from physical model to natural environments. However it must be considered the experimental simplifications adopted. Apart from that, the rheological properties of mixtures and some hydrodynamic (e.g. cohesion effects) and depositional (e.g. settling velocity) properties are scale-independent and can be applied for further interpretation.

The experiments simulated a single catastrophic event and do not consider a continuous sediment supply from rivers (for instance, plumes and hyperpycnal flows among others) which can change some properties of the flow along time and space. Moreover, the limit of maximum value of volumetric concentration was 35% by volume. In this case, regions III (see Amy et al., 2006) and region VI (see e.g. Hampton, 1972; Ilstad et al., 2004; Marr et al., 2001; Mohrig et al., 1999; Mohrig & Marr, 2003) were left with an open boundary to further experiments and perhaps the creation of a complementary experimental-derived classification of sediment gravity flows.

## 7. Nomenclature

| | |
|---|---|
| %Clay | clay content in the mixture (%) |
| $C_r$ | concentration of reference (%) |
| $C_{vol}$ | volumetric concentration (%) |
| $C(z)$ | volumetric concentration at the point z (%), |
| $\partial u/\partial z$ | strain rate (1/s) |
| g | acceleration of gravity  $(m/s^2)$ |
| $h_b$ | body height  (m), |
| $h_h$ | head height  (m), |
| $h_i$ | inner layer thickness (m), |
| $h_m$ | height of the point of maximum velocity (m), |
| $h_{mean}$ | mean current height (m), |
| $h_t$ or H | overall height of the current (cm), |
| K | consistence coefficient |
| n | power law coefficient |
| $S_{flux}$ | sediment flux $(m^3/s^3)$ |
| u | current velocity (m/s), |
| $U_{max}$ | maximum current velocity (m/s), |
| $u_{mean}$ | mean current velocity (m/s), |
| Z or z | distance to bottom (cm) |

*Greek letters*

| | |
|---|---|
| $\mu_o$ | coefficient of dynamic viscosity of pure water (Pa.s) |
| $\mu$ | dynamic viscosity coefficient (Pa.s) |
| $\mu_{ap}$ | apparent viscosity coefficient (Pa.s) |
| $\rho_a$ | density of ambient fluid $(kg/m^3)$ |
| $\rho_m$ | density of mixture $(kg/m^3)$ |
| $\tau_i$ | yield strength - critical shear stress (Pa) |
| $\tau_0$ | shear stress at the bottom (Pa) |
| $\tau_{lam}$ | laminar component of shear stress (Pa) |
| $\tau_{turb}$ | turbulent component of shear stress (Pa) |
| $\tau_x$ | shear stress (Pa) |

## 8. Acknowledgement

A grateful thanks to: CNPq – Brazilian National Council for Scientific and Technological Development - to support my PhD "sandwich" program at University of Leeds; the head of

NECOD – Density Currents Research Center, IPH/UFRGS - Professor Rogério D. Maestri; to Professor Ana Luiza de O. Borges and professor Jaco H. Baas for their support and also to my colleagues Eduardo Puhl and Richard E. Ducker.

## 9. References

Amy, L. A., Talling, P. J., Edmonds, V. O., Sumner, E. J. & Leseur, A. (2006) An experimental investigation on sand-mud suspension settling behaviour and implications for bimodal mud content of submarine flow deposits. *Sedimentology* Vol.53, pp. 1411–1434, ISSN 0037-0746

Altinakar, M. S. (1988). Weakly depositing turbidity currents on small slopes. PhD thesys à *Ecol. Pol. Fed. Lausanne.*

Baas, J.H.; Kesteren, W.V.; Postma, G. (2004). Deposits of depletive high-density turbidity currents: a flume analogue of bed geometry, structure and texture. *Sedimentology,* v51, pp. 1053-1088 ISSN 13653091 |

Bouma, A. H. (1962). *Sedimentology of some flysch deposits: a graphic approach to facies interpretation.* Amsterdam. Netherlands

Britter, R. E.; Simpson, E. J. (1978). Experiments on the dynamics of a gravity current head. *Journal of Fluid Mechanics,* Vol. 88, pp. 223-240. ISSN: 0022-1120.

Coussot, P. (1997). *Mudflow Rheology and Dynamics.* Taylor & Francis. ISBN 905410693X.

Einstein, A. (1906) Investigations on the theory of the Brownian movement (from the *annalen der physik,* Vol. 4), No. 19, pp. 289-306

Elisson, T.H.; Turner, J. S. (1959). Turbulent entrainment in stratified flows. *Journal of Fluid Mechanics,* Vol. 6, pp. 423-448 ISSN: 00221120

Fischer, R. V. (1983). Flow transformations in sediment gravity flows. *Geology,* Vol. 11, pp. 273-274, ISSN 0091-7613

Gani, M. R. (2004). From Turbid to Lucid: A straightforward approach to sediment gravity flows and their deposits. *The Sedimentary Record.* September, 2004, Vol. 2, No. 3, pp. 4-8.

Graf, W. e Altinakar, M. S. (1998). Turbidity currents, In: *Fluvial Hydraulics: Flow and Transport processes in channels of simple geometry.* W. H. Graff, (Ed.) 468-516, Jonh Wiley e Sons, ISBN: 978-0-471-97714-8 New York- US.

Hampton, M.A. (1972). The role of subaqueous debris flow in generating turbidity currents. *Journal of Sedimentary Petrology,* Vol. 42, No 4, pp. 775-793 ISSN 0016-7606

Ilstad, T., Elverhøi, A., Issler, D., & Marr, J. (2004). Subaqueous debris flow behaviour and its dependence on the sand/clay ratio: a laboratory study using particle tracking. *Marine Geology,* Vol. 213, pp. 415–438, ISSN 0025-3227

Kneller, B. (1995). Beyond the turbidite paradigm: physical models for deposition of turbidites and their implications for reservoir prediction. In: *Characterization of deep marine clastic systems.* A.J. Hartley (Ed). Pp. 31-49, Geological Society London. ISBN: 1897799357. London, UK.

Kneller, B.C., Bennett, S.J. & McCaffrey, W.D. (1999) Velocity structure, turbulence and fluid stresses in experimental gravity currents. *Journal of Geophys. Research. Oceans.,* Vol. 94, pp. 5281-5291.

Kneller, B. & Buckee, C. (2000). The structure and fluid mechanics of turbidity currents: a review of some recent studies and their geological implications. *Sedimentology,* Vol.47, Suppl. 1, pp. 62-94, ISSN 0091-7613.

Kuenen, P. H. (1950). Turbidity currents of high density. In: *Proceedings of 18th International Geological Congress (1948)*, London, pp. 44-52.

Lowe, D. R. (1979). Sediment gravity flows: Their classification and some problems of application to natural flows and deposits. *SEPM (Special Publication n. 27), pp 75-82.*

Lowe, D. R. (1982). Sediment gravity flows: II. Depositional models with special reference to the deposits of high-density turbidity currents. *Journal of Sedimentary Petrology*, Vol. 52, No. 1, pp. 279-297.

Normark, W.R. & Piper, D. J. W. (1991). Initiation processes and flow evolution of turbidity currents: implications for the depositional record. *SEPM (Special Publication n. 46).* pp. 207-230.

Manica, R. (2009). *Correntes de turbidez de alta densidade: condicionantes hidráulicos e deposicionais.* PhD thesys, Instituto de Pesquisas Hidráulicas, Universidade Federal do Rio Grande do Sul, Porto Alegre - Brazil.

Marr, J.G., Harff, P.A., Shanmugan, G. & Parker, G. (2001) Experiments on subaqueous sandy gravity flow: the role of clay and water content in flow dynamics and depositional structures. *GSA Bulletin*, Vol. 113, No. 11, pp. 1377-1386.

McCave, I, N & Jones, P.N (1988). Deposition of ungraded muds from high-density non turbulent turbidity currents. *Nature*, Vol. 333, pp 250-252, ISSN 0016-7649

Michon, X.; Goddet, J. & Bonnefille, R. (1955). *Etude Theorique et experimentale des courants de densite.* 2 vol. Lab. Nat. d'Hydralique Chatou, França.

Middleton, G. V. (1966). Experiments on density and turbidity currents I. Motion of the head. *Canadian Journal of Earth Sciences*, Ottawa, Vol. 3, pp. 523-546.

Middleton, G. V. (1993). Sediment deposition from turbidity currents. *Annual Review of Earth Planet Science*, Vol. 21, pp. 89-114.

Middleton, G. V. & Hampton, M. A. (1973). Sediment gravity flows: mechanics of flow and deposition. In *Turbidites and Deep Water Sedimentation* G. V. Middleton and A. H. Bouma (eds.). Anaheim, California, SEPM. Short Course Notes, 38p

Mohrig, D.; Elverhoi, A.; & Parker, G. (1999). Experiments on the relative mobility of muddy subaqueous and subaerial debris flows, and their capacity to remobilize antecedent deposits. *Marine Geology.* Vol. 154, pp. 117-129.

Mohrig, D. & Marr, J.G. (2003). Constraining the efficiency of turbidity current from submarine debris flow and slides using laboratory experiments. *Marine and Petroleum Geology*, Vol. 20, pp. 883-899

Parker, G., Fukushima, Y., & Pantin, H.M., (1986). Self accelerating turbidity currents: *Journal of Fluid Mechanics*, Vol. 171, pp. 145-181.

Parker, G., Garcia, M., Fukushima, Y. & YU, W. (1987) Experiments on turbidity currents over a erodible bed. *Journal of Hydraulic Research*, Vol. 25, pp. 123-147.

Parsons, J. D; Friedrichs C. T.; Traykovsky P. A.; Mohrig, D.; Imran, J.; Syvitsky, P. M.; Parker, G.; Puig, P.; Buttles, J. L. & Garcia, M. H. (2010). The mechanics of sediment gravity flows. In *Continental Margin Sedimentation: From Sediment Transport to Sequence Stratigraphy.* C. A. Nittrouer, J. A. Austin, M. E. Field, J. H. Kravitz, J. P. M. Syvitski, P. L. Wiberg, John Wyley &Sons, ISBN 9781405169349, New York, US.

Postma, G.; Nemec, W. & Kleinspehn, K.L. (1988). Large floating clasts in turbidites: a mechanism for their emplacement. *Sedimentary Geology*, Vol. 58, pp. 47-61.

Simpson, E. J. (1972). Effects of the lower boundary on the head of a gravity current. *Journal of Fluid Mechanics*, Vol. 53, pp. 759-768.

Simpson, E. J. (1997). *Gravity currents in the enviroment and the laboratory.* 2.ed. Cambridge University, ISBN 0521664012, UK

Talling, P. J.; Wynn, R. B.; Masson; D. G., Frenz, M.; Cronin, B. T.; Schiebel,R; Akhmetzhanov, A. M., Dallmeier-Tiessen, S., Benetti, S.; Weaver, P. P. E.; Georgiopoulou, A.; Zuhlsdorff, C. & Amy, L. A. (2007). Onset of submarine debris flow deposition far from original giant landslide. *Nature.* Vol. 450. pp. 541-544

Wan, Z.; &Wang, Z., (1994),. *Hyperconcentrated Flow.* IAHR Monograph Series, A. A. Balkema. Rotterdam. 290p

Whitehouse, R., Soulsby, R., Roberts, W. & Mitchener, H. (2000). *Dynamics of Estuarine Muds.* London, UK.

Winterwerp, J. C. (2001). Stratification effects by cohesive and non-cohesive sediment, *J. Geophys. Research*, Vol. 106, No. 22, pp. 559 -574.

Winterwerp, J.C. (2002). On the flocculation and settling velocity of estuarine mud. *Cont. Shelf Research.* Vol. 22, pp. 1339–1360.

# 13

# Stepped Spillways: Theoretical, Experimental and Numerical Studies

André Luiz Andrade Simões, Harry Edmar Schulz,
Raquel Jahara Lobosco and Rodrigo de Melo Porto
*University of São Paulo*
*Brazil*

## 1. Introduction

Flows on stepped spillways have been widely studied in various research institutions motivated by the attractive low costs related to the dam construction using roller-compacted concrete and the high energy dissipations that are produced by such structures. This is a very rich field of study for researchers of Fluid Mechanics and Hydraulics, because of the complex flow characteristics, including turbulence, gas exchange derived from the two-phase flow (air/water), cavitation, among other aspects. The most common type of flow in spillways is known as skimming flow and consists of: (1) main flow (with preferential direction imposed by the slope of the channel), (2) secondary flows of large eddies formed between steps and (3) biphasic flow, due to the mixture of air and water. The details of the three mentioned standards may vary depending on the size of the steps, the geometric conditions of entry into the canal, the channel length in the steps region and the flow rates. The second type of flow that was highlighted in the literature is called nappe flow. It occurs for specific conditions such as lower flows (relative to skimming flow) and long steps in relation to their height. In the region between these two "extreme" flows, a "transition flow" between nappe and skimming flows is also defined. Depending on the details that are relevant for each study, each of the three abovementioned types of flow may be still subdivided in more sub-types, which are mentioned but not detailed in the present chapter. Figure 1 is a sketch of the general appearance of the three mentioned flow regimes.

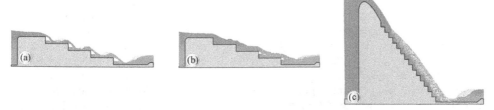

Fig. 1. Flow patterns on stepped chutes: (a) Nappe-flow, (b) transition flow and (c) skimming flow.

The introductory considerations made in the first paragraph shows that complexities arise when quantifying such flows, and that specific or general contributions, involving different points of view, are of great importance for the advances in this field. This chapter aims to provide a brief general review of the subject and some results of experimental, numerical and theoretical studies generated at the School of Engineering of Sao Carlos - University of São Paulo, Brazil.

## 2. A brief introduction and review of stepped chutes and spillways

In this section we present some key themes, chosen accordingly to the studies described in the next sections. Additional sources, useful to complement the text, are cited along the explanations.

### 2.1 Flow regimes
It is interesting to observe that flows along stepped chutes have also interested a relevant person in the human history like Leonardo da Vinci. Figure 2a shows a well-known da Vinci's sketch (a mirror image), in which a nappe-flow is represented, with its successive falls. We cannot affirm that the sketching of such flow had scientific or aesthetic purposes, but it is curious that it attracted da Vinci's attention. Considering the same geometry outlined by the artist, if we increase the flow rate the "successive falls pattern" changes to a flow having a main channel in the longitudinal direction and secondary currents in the "cavities" formed by the steps, that is, the skimming flow mentioned in the introduction. Figure 2b shows a drawing from the book *Hydraulica* of Johann Bernoulli, which illustrates the formation of large eddies due to the passage of the flow along step-formed discontinuities.

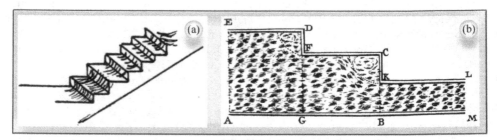

Fig. 2. Historical drawings related to the fields of turbulent flows in channels and stepped spillways: (a) Sketch attributed to Leonardo da Vinci (Richter, 1883, p.236) (mirror image), (b) Sketch presented in the book of Johann Bernoulli (Bernoulli, 1743, p.368).

The studies of Horner (1969), Rajaratnam (1990), Diez-Cascon et al. (1991), among others, presented the abovementioned patterns as two flow "regimes" for stepped chutes. For specific "intermediate conditions" that do not fit these two regimes, the transition flow was then defined (Ohtsu & Yasuda, 1997). Chanson (2002) exposed an interesting sub-division of the three regimes. The nappe flow regime is divided into three sub-types, characterized by the formation or absence of hydraulic jumps on the bed of the stairs. The skimming flow regime is sub-divided considering the geometry of the steps and the flow conditions that lead to different configurations of the flow fields near the steps. Even the transition flow regime may be divided into sub-types, as can be found in the study of Carosi & Chanson (2006).

Ohtsu et al. (2004) studied stepped spillways with inclined floors, presenting experimental results for angles of inclination of the chute between 5.7 and 55° For angles between 19 and 55° it was observed that the profile of the free surface in the region of uniform flow is independent of the ratio between the step height (s) and the critical depth ($h_c$), that is, $s/h_c$, and that the free surface slope practically equals the slope of the pseudo-bottom. This sub-system was named "Profile Type A". For angles between 5.7 and 19, the unobstructed flow slide is not always parallel to the pseudo-bottom, and the Profile Type A is formed only for small values of $s/h_c$. For large values of $s/h_c$, the authors explain that the profile of the free surface is replaced by varying depths along a step. The skimming flow becomes, in part, parallel to the floor, and this sub-system was named "Profile Type B".

Researchers like Essery & Horner (1978), Sorensen (1985), Rajaratnam (1990) performed experimental and theoretical studies and presented ways to identify nappe flows and skimming flows. Using results of recent studies, Simões (2011) presented the graph of Figure 3a, which contains curves relating the dimensionless $s/h_c$ and $s/l$ proposed by different authors. Figure 3b represents a global view of Figure 3a, and shows that the different propositions of the literature may be grouped around two main curves (or lines), dividing the graph in four main areas (gray and white areas in Fig 3a). The boundaries between these four areas are presented as smooth transition regions (light brown in Fig 3b), corresponding to the region which covers the positions of the curves proposed by the different authors.

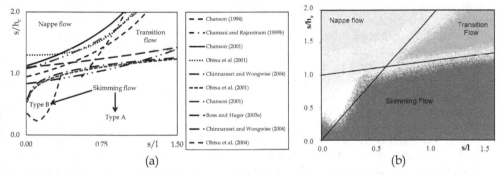

Fig. 3. Criteria for determining the types of flow: (a) curves of different authors (cited in the legend) and (b) analysis of the four main areas (white and gray) and the boundary regions (light brown) between the main areas (The lines are: $s/h_c=2s/l$; $s/h_c = 0.233s/l+1$).

## 2.2 Skimming flow
### 2.2.1 Energy dissipation
The energy dissipation of flows along stepped spillways is one of the most important characteristics of these structures. For this reason, several researchers have endeavored to provide equations and charts to allow predictions of the energy dissipation and the residual energy at the toe of stepped spillways and channels. Different studies were performed in different institutions around the world, representing the flows and the related phenomena from different points of view, for example, using the Darcy-Weisbach or the Manning equations, furnishing algebraic equations fitted to experimental data, presenting experimental points by means of graphs, or simulating results using different numerical schemes.

## Darcy-Weisbach resistance function ("friction factor")

The Darcy-Weisbach resistance function has been widely adopted in studies of stepped spillways. It can be obtained following arguments based on physical arguments or based on a combination of experimental information and theoretical principles. In the first case, dimensional analysis is used together with empirical knowledge about the energy evolution along the flow. In the second case, the principle of conservation of momentum is used together with experimental information about the averaged shear stress on solid surfaces. Of course, the result is the same following both points of view. The dimensional analysis is interesting, because it shows that the "resistance factor" is a function of several nondimensional parameters. The most widespread resistance factor equation, probably due to its strong predictive characteristic, is that deduced for flows in circular pipes. For this flows, the resistance factor is expressed as a function of only two nondimensional parameters: the relative roughness and the Reynolds number. When applying the same analysis for stepped channels, the resistance factor is expressed as dependent on more nondimensional parameters, as illustrated by eq. 1:

$$f = \Phi_1\left( Re, Fr, \alpha, \frac{k}{L_c}, \frac{\varepsilon_p}{L_c}, \frac{\varepsilon_e}{L_c}, \frac{\varepsilon_m}{L_c}, \frac{s}{L_c}, \frac{1}{L_c}, \frac{L_c}{B}, C \right) \tag{1}$$

f is the resistance factor. Because the obtained equation is identical to the Darcy-Weisbach equation, the name is preserved. The other variables are: Re = Reynolds number, Fr = Froude number, $\alpha$ = atg(s/l), k = scos$\alpha$, $L_c$ = characteristic length, $\varepsilon$ = sand roughness (the subscripts "p", "e "and "m" correspond to the floor of the step, to the vertical step face and the side walls, respectively), s = step height, l = step length, B = width of the channel, C = void fraction.

Many equations for f have been proposed for stepped channels since 1990. Due to the practical difficulties in measuring the position of the free surface accurately and to the increasing of the two-phase region, the values of the resistance factor presented in the literature vary in the range of about 0.05 to 5! There are different causes for this range, which details are useful to understand it. It is known that, by measuring the depth of the mixture and using this result in the calculation of f, the obtained value is higher than that calculated without the volume of air. This is perhaps one of the main reasons for the highest values. On the other hand, considering the lower values (the range from 0.08 to 0.2, for example), they may be also affected by the difficulty encountered when measuring depths in multiphase flows. Even the depths of the single-phase region are not easy to measure, because high-frequency oscillations prevent the precise definition of the position of the free surface, or its average value. Let us consider the following analysis, for which the Darcy-Weisbach equation was rewritten to represent wide channels

$$f = \frac{8gh^3 I_f}{q^2} \tag{2}$$

in which: g = acceleration of the gravity, h = flow depth, $I_f$ = slope of the energy line, q = unit discharge. The derivative of equation (2), with respect to f and h, results $\frac{\partial f}{\partial q} = -\frac{16gh^3 I_f}{q^3}$

and $\frac{\partial f}{\partial h} = \frac{24gh^2 I_f}{q^2}$, respectively, which are used to obtain equation 3.

This equation expresses the propagation of the uncertainty of f, for which it was assumed that the errors are statistically independent and that the function f = f (q, h) varies smoothly with respect to the error propagation.

$$\frac{\Delta f}{f} = \sqrt{4\left(\frac{\Delta q}{q}\right)^2 + 9\left(\frac{\Delta h}{h}\right)^2} \tag{3}$$

Assuming $I_f = 1\pm 0$ (that is, no uncertainty for $I_f$), h = 0.05 ± 0.001 m and q = 0.25 ± 0.005 m²/s, the relative uncertainty of the resistance factor is around 7.2%. The real difficulty in defining the position of the free surface imposes higher relative uncertainties. So, for $\Delta h = 3$ mm, we have $\Delta f/f = 18.4\%$ and for $\Delta h = 5$ mm, the result is $\Delta f/f = 30.3\%$. These $\Delta h$ values are possible in laboratory measurements.

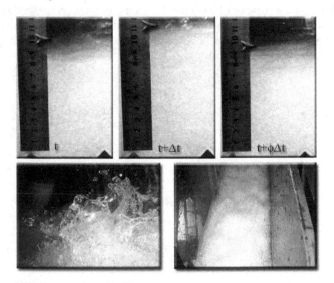

Fig. 4. Behavior of the free surface ($\phi > 1$)

Figure 4 contains sequential images of a multiphase flow, obtained by Simões (2011). They illustrate a single oscillation of the mean position of the surface with amplitude close to 15 mm. The first three pictures were taken under ambient lighting conditions, generating images similar to the perception of the human eye. The last two photographs were obtained with a high speed camera, showing that the shape of the surface is highly irregular, with portions of fluid forming a typical macroscopic interface under turbulent motion. It is evident that the method used to measure the depth of such flows may lead to incorrect results if these aspects are not well defined and the measurement equipment is not adequate.

Figure 4 shows that it is difficult to define the position of the free surface. Simões et al. (2011) used an ultrasonic sensor, a high frequency measurement instrument for data acquisition, during a fairly long measurement time, and presented results of the evolution of the two-phase flow that show a clear oscillating pattern, also allowing to observe a transition length between the "full water" and "full mixture" regions of the flows along stepped spillways. Details on similar aspects for smooth spillways were presented by

Wilhelms & Gulliver (2005), while reviews of equations and values for the resistance factor were presented by Chanson (2002), Frizell (2006), Simões (2008), and Simões et al. (2010).

**Energy dissipation**

The energy dissipated in flows along stepped spillways can be defined as the difference between the energy available near the crest and the energy at the far end of the channel, denoted by $\Delta H$ throughout this chapter. Selecting a control volume that involves the flow of water between the crest (section 0) and a downstream section (section 1), the energy equation can be written as follows:

$$z_0 + \frac{p_0}{\gamma} + \alpha_0 \frac{V_0^2}{2g} = z_1 + \frac{p_1}{\gamma} + \alpha_1 \frac{V_1^2}{2g} + \Delta H \tag{4}$$

According to the characteristics of flow and the channel geometry, the flows across these sections can consist of air/water mixtures. Assuming hydrostatic pressure distributions, such that $p_0/\gamma = h_0$ and $p_1/\gamma = h_1\cos\alpha$ (Chow, 1959), the previous equation can be rewritten as:

$$\Delta H = \overbrace{z_0 - z_1}^{H_{dam}} + h_0 + \alpha_0 \frac{q^2}{2gh_0^2} - \left(h_1\cos\alpha + \alpha_1\frac{q^2}{2gh_1^2}\right) =$$

$$= \left(H_{dam} + h_0 + \alpha_0\frac{h_c^3}{2h_0^2}\right)\left[1 - \left(h_1\cos\alpha + \alpha_1\frac{q^2}{2gh_1^2}\right) \bigg/ \left(H_{dam} + h_0 + \alpha_0\frac{h_c^3}{2h_0^2}\right)\right]$$

Denoting $H_{dam} + h_0 + \alpha_0\dfrac{h_c^3}{2h_0^2}$ by $H_{max}$, the previous equation is replaced by:

$$\frac{\Delta H}{H_{max}} = 1 - \left(\frac{h_1}{h_c}\right)\left(\frac{\cos\alpha + \alpha_1\dfrac{h_c^3}{2h_1^3}}{\dfrac{H_{dam}}{h_c} + \dfrac{h_0}{h_c} + \alpha_0\dfrac{h_c^2}{2h_0^2}}\right) \tag{5}$$

Taking into account the width of the channel, and using the Darcy-Weisbach equation for a rectangular channel in conjunction with equation 5, the following result is obtained:

$$\frac{\Delta H}{H_{max}} = 1 - \left\{\frac{\left[\dfrac{8I_f}{(1 + 2h_1/B)f}\right]^{-1/3}\cos\alpha + \dfrac{\alpha_1}{2}\left[\dfrac{8I_f}{(1 + 2h_1/B)f}\right]^{2/3}}{\dfrac{H_{dam}}{h_c} + \dfrac{h_0}{h_c} + \alpha_0\dfrac{h_c^2}{2h_0^2}}\right\} \tag{6}$$

Rajaratnam (1990), Stephenson (1991), Hager (1995), Chanson (1993), Povh (2000), Boes & Hager (2003a), Ohtsu et al. (2004), among others, presented conceptual and empirical equations to calculate the dissipated energy. In most of the cases, the conceptual models can be obtained as simplified forms of equation 6, which is considered a basic equation for flows in spillways.

## 2.2.2 Two phase flow

The flows along smooth spillways have some characteristics that coincide with those presented by flows along stepped channels. The initial region of the flow is composed only by water ("full water region" 1 in Figure 5a), with a free surface apparently smooth. The position where the thickness of the boundary layer coincides with the depth of flow defines the starting point of the superficial aeration, or inception point (see Figure 5). In this position the effects of the bed on the flow can be seen at the surface, distorting it intensively. Downstream, a field of void fraction $C(x_i, t)$ is generated, which depth along $x_1$ (longitudinal coordinate) increases from the surface to the bottom, as illustrated in Figure 5.

The flow in smooth channels indicates that the region (1) is generally monophasic, the same occurring in stepped spillways. However, channels having short side entrances like those used for drainage systems, typically operate with aerated flows along all their extension, from the beginning of the flow until its end. Downstream of the inception point a two-dimensional profile of the mean void fraction C is formed, denoted by $\overline{C}^*$. From a given position $x_1$ the so called "equilibrium" is established for the void fraction, which implies that $\overline{C}^* = \overline{C}^*(x_1)$. Different studies, like those of Straub & Anderson (1958), Keller et al. (1974), Cain & Wood (1981) and Wood et al. (1983) showed results consistent with the above descriptions, for flows in smooth spillways. Figure 5b shows the classical sketch for the evolution of two-phase flows, as presented by Keller et al. (1974). Wilhelms & Gulliver (2005) introduced the concepts of entrained air and entrapped air, which correspond respectively to the air flow really incorporated by the water flow and carried away in the form of bubbles, and to the air surrounded by the twisted shape of the free surface, and not incorporated by the water.

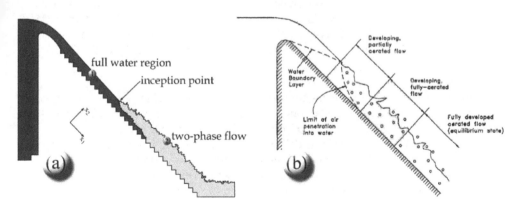

Fig. 5. Skimming flow and possible classifications of the different regions
Sources: (a) Simões (2011), (b) Keller *et al.* (1974)

One of the first studies describing coincident aspects between flows along smooth and stepped channels was presented by Sorensen (1985), containing an illustration indicating the inception point of the aeration and describing the free surface as smooth upstream of this point (Fig. 6a). Peyras et al. (1992) also studied the flow in stepped channels formed by gabions, showing the inception point, as described by Sorensen (1985) (see Figure 6b).

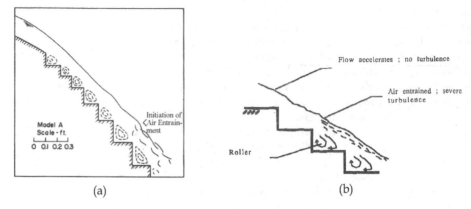

(a)                                                  (b)

Fig. 6. Illustration of the flow
Reference: (a) Sorensen (1985, p.1467) and (b) Peyras *et al.* (1992, p.712).

The sketch of Figure 6b emphasizes the existence of rolls downstream from the inception position of the aeration. Further experimental studies, such as Chamani & Rajaratnam (1999a, p.363) and Ohtsu et al. (2001, p.522), showed that the incorporated air flow distributes along the depth of the flow and reaches the cavity below the pseudo-bottom, where large eddies are maintained by the main flow.

The mentioned studies of multiphase flows in spillways (among others) thus generated predictions for: (1) the position of the inception point of aeration, (2) profiles of void fractions (3) averages void fractions over the spillways, (4) characteristics of the bubbles. As mentioned, frequently the conclusions obtained for smooth spillways were used as basis for studies in stepped spillways. See, for example, Bauer (1954), Straub & Anderson (1958), Keller & Rastogi (1977), Cain & Wood (1981), Wood (1984), Tozzi (1992), Chanson (1996), Boes (2000), Chanson (2002), Boes & Hager (2003b) and Wilhelms & Gulliver (2005).

### 2.2.3 Other topics

In addition to the general aspects mentioned above, a list of specific items is also presented here. The first item, cavitation, is among them, being one of major relevance for spillway flows. It is known that the air/water mixture does not damage the spillway for void fractions of about 5% to 8% (Peterka, 1953). For this reason, many studies were performed aiming to know the void fraction near the solid boundary and to optimize the absorption of air by the water. Additionally, the risk of cavitation was analyzed based on instant pressures observed in physical models. Some specific topics are show below:

1. Cavitation;
2. Channels with large steps;
3. Stepped chutes with gabions;
4. Characteristics of hydraulic jumps downstream of stepped spillways;
5. Plunging flow;
6. Recommendations for the design of the height of the side walls;
7. Geometry of the crest with varying heights of steps;
8. Aerators for stepped spillways;
9. Baffle at the far end of the stepped chute;

10. Use of spaced steps;
11. Inclined step and end sills;
12. Side walls converging;
13. Use of precast steps;
14. Length of stilling basins.

As can be seen, stepped chutes are a matter of intense studies, related to the complex phenomena that take place in the flows along such structures.

## 3. Experimental study

### 3.1 General information

The experimental results presented in this chapter were obtained in the Laboratory of Environmental Hydraulics of the School of Engineering at São Carlos (University of Sao Paulo). The experiments were performed in a channel with the following characteristics: (1) Width: B = 0.20 m, (2) Length = 5.0 m, 3.5 m was used, (3) Angle between the pseudo bottom and the horizontal: $\alpha$ = 45°; (4) Dimensions of the steps s = l = 0.05 m (s = step height l = length of the floor), and (5) Pressurized intake, controlled by a sluice gate. The water supply was accomplished using a motor/pump unit (Fig. 7) that allowed a maximum flow rate of 300 L/s. The flow rate measurements were performed using a thin-wall rectangular weir located in the outlet channel, and an electromagnetic flow meter positioned in the inlet tubes (Fig. 7b), used for confirmation of the values of the water discharge.

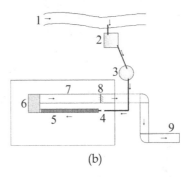

|   (a)   |   (b)   |

Fig. 7. a) Motor/pump system.; b) Schematic drawing of the hydraulic circuit: (1) river, (2) engine room, (3) reservoir, (4) electromagnetic flowmeter, (5) stepped chute, (6) energy sink, (7) outlet channel; (8) weir, (9) final outlet channel.

The position of the free surface was measured using acoustic sensors (ultrasonic sensors), as previously done by Lueker et al. (2008). They were used to measure the position of the free surface of the flows tested in a physical model of the auxiliary spillway of the Folsom Dam, performed at the St. Anthony Falls Laboratory, University of Minnesota. A second study that employed acoustic probes was Murzyn & Chanson (2009), however, for measuring the position of the free surface in hydraulic jumps.

In the present study, the acoustic sensor was fixed on a support attached to a vehicle capable of traveling along the channel, as shown in the sketch of Figure 8. For most experiments, along the initial single phase stretch, the measurements were taken at sections distant 5 cm from each other. After the first 60 cm, the measurement sections were spaced 10 cm from each other. The sensor was adjusted to obtain 6000 samples (or points) using a frequency of 50 Hz at each longitudinal position. These 6000 points were used to perform the statistical calculations necessary to locate the surface and the drops that formed above the surface. A second acoustic sensor was used to measure the position of the free surface upstream of the thin wall weir, in order to calculate the average hydraulic load and the flow rates used in the experiments. The measured flow rates, and other experimental parameters of the different runs, are shown in Table 1.

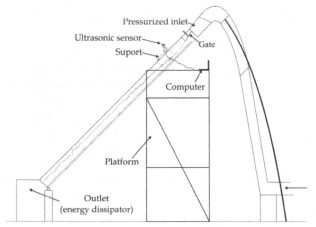

Fig. 8. Schematic of the arrangement used in the experiments

| Nº | Experiment name | Q [m³/s] | Profile | q [m²/s] | $h_c$ [m] | $s/h_c$ [-] | $h(0)$ [m] |
|---|---|---|---|---|---|---|---|
| 1 | Exp. 2 | 0.0505 | $S_2$ | 0.252 | 0.187 | 0.268 | 0.103 |
| 2 | Exp. 3 | 0.0458 | $S_2$ | 0.229 | 0.175 | 0.286 | 0.101 |
| 3 | Exp. 4 | 0.0725 | $S_2$ | 0.362 | 0.238 | 0.211 | 0.106 |
| 4 | Exp. 5 | 0.0477 | $S_2$ | 0.239 | 0.180 | 0.278 | 0.087 |
| 5 | Exp. 6 | 0.0833 | $S_3$ | 0.416 | 0.261 | 0.192 | 0.092 |
| 6 | Exp. 7 | 0.0504 | $S_2$ | 0.252 | 0.187 | 0.268 | 0.089 |
| 7 | Exp. 8 | 0.0073 | $S_2$ | 0.0366 | 0.051 | 0.971 | 0.027 |
| 8 | Exp. 9 | 0.0074 | $S_2$ | 0.0368 | 0.052 | 0.967 | 0.024 |
| 9 | Exp. 10 | 0.0319 | $S_2$ | 0.159 | 0.137 | 0.364 | 0.058 |
| 10 | Exp. 11 | 0.0501 | $S_3$ | 0.250 | 0.186 | 0.269 | 0.06 |
| 11 | Exp. 14 | 0.0608 | $S_2$ | 0.304 | 0.211 | 0.237 | 0.089 |
| 12 | Exp. 15 | 0.0561 | $S_2$ | 0.280 | 0.200 | 0.250 | 0.087 |
| 13 | Exp. 16 | 0.0265 | $S_2$ | 0.133 | 0.122 | 0.411 | 0.046 |
| 14 | Exp. 17 | 0.0487 | $S_2$ | 0.244 | 0.182 | 0.274 | 0.072 |
| 15 | Exp. 18 | 0.0431 | $S_2$ | 0.216 | 0.168 | 0.298 | 0.074 |
| 16 | Exp. 19 | 0.0274 | $S_2$ | 0.137 | 0.124 | 0.402 | 0.041 |
| 17 | Exp. 20 | 0.0360 | $S_2$ | 0.180 | 0.149 | 0.336 | 0.068 |
| 18 | Exp. 21 | 0.0397 | $S_2$ | 0.198 | 0.159 | 0.315 | 0.071 |

Table 1. General data related to experiments

As can be seen in Figure 4, the positioning of the free surface is complex due to its highly irregular structure, especially downstream from the inception point. One of the characteristics of measurements conduced with acoustic sensors is the detection of droplets ejected from the surface. These values are important for the evaluation of the highest position of the droplets and sprays, but have little influence to establish the mean profiles of the free surface. This is shown in Figure 9a, which contains the relative errors calculated considering the mean position obtained without the outliers (droplets). The corrections were made using standard criteria used for box plots. The maximum percentage of rejected samples (droplets) was 8.3% for experiment N° 5.

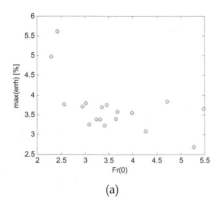 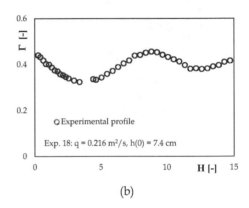

(a)                                                                              (b)

Fig. 9. (a) Maximum relative deviations corresponding to the eighteen experiments, in which: errh = $100 | | h^{(1)} - h^{(2)} | | / h^{(2)}$, $h^{(i)}$ = mean value obtained with the acoustic sensor, i = 1 (original sample), i = 2 (sample without outliers) and Fr(0) = Froude number at x = 0; (b) Mean experimental profile due to Exp.18. The deviations were used to obtain the maximum position of the droplets, but were ignored when obtaining the mean profile of the surface.

Figure 9b presents an example of a measured average profile obtained in this study. As can be seen, an $S_2$ profile is formed in the one-phase region. The inception point of the aeration is given by the position of the first minimum in the measured curve. It establishes the end of the $S_2$ curve and the beginning of the "transition length", as defined by Simões et al. (2011). As shown by the mentioned authors, the surface of the mixture presents a wavy shape, also used to define the end of the transition length, given by the first maximum of the surface profile.

## 3.2 Results
### 3.2.1 Starting position of the aeration (inception point)
As mentioned, the starting position of the aeration was set based on the minimum point that characterizes the far end of the $S_2$ profile. In some experiments, this minimum showed a certain degree of dispersion, so that the most probable position was chosen. To quantify the position of the inception point of the aeration, the variables involved in a first instance were $L_A/k$ and $F_r^*$, adjusting a power law between them, as already used by several authors. (e.g., Chanson, 2002; Sanagiotto, 2003) Equation 7 shows the best adjustment obtained for the present set of data, with a correlation coefficient of 0.91. Considering the four variables $L_A/k$, $h(0)/k$, Re(0), and $F_r^*$ (see figure 6a for the definitions of the variables), a second

equation is presented, as a sum of the powers of the variables. Equation 8 presents a correlation coefficient of 0.98, leading to a good superposition between data and adjusted curve, as can be seen in Fig.10b.

$$\frac{L_A}{k} = 1.61 F_r^{*1.06} \tag{7}$$

$$\frac{L_A}{k} = 699.97 F_r^{*-6.33} + 34.22 \left[\frac{h(0)}{k}\right]^{0.592} - 49.45 Re(0)^{-0.0379} \tag{8}$$

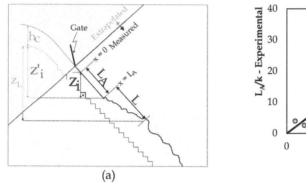

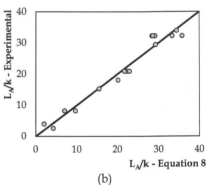

(a)                                                                 (b)

Fig. 10. Definition of variables related to the start of aeration (a) and comparison between measured data and calculated values using the adjusted equation 8.

Equations 7 and 8 show very distinct behaviors for the involved parameters. For example, the dependence of $L_A/k$ on $F_r^*$ shows increasing lengths for increasing $F_r^*$ when using equation 7, and decreasing lengths for increasing $F_r^*$ when using equation 8. Additionally, the influence of $h(0)/k$ appears as relevant, when considering the exponent 0.592. This parameter was used to verify the relevance of $F_r^*$ to quantify the inception point. Although the result points to a possible relevance of the geometry of the flow ($h(0)$), the adequate definition of this parameter for general flows is an open question. It is the depth of the flow at a fixed small distance from the sluice gate in this study, thus directly related to the geometry, but which correspondent to general flows, as already emphasized, must still be defined. In the present analysis, following restrictions apply: $2.09 \leq F_r^* \leq 20.70$, $0.69 \leq h(0)/k \leq 2.99$ and $1.15 \times 10^5 \leq Re(0) \leq 7.04 \times 10^5$.

Equation 7 can be rewritten using $z_i/s$ and $F$, in which $z_i = L_A \sin\alpha$, and $F$ is the Froude number defined by Boes & Hager (2003b) as $F = q/\sqrt{gs^3 \sin\alpha}$. In this case $F_r^* = F/(\cos^3\alpha)^{1/2}$. The correlation coefficient is the same obtained for equation 7, and the resulting equation, valid for the same conditions of the previous adjustments, is:

$$\frac{z_i}{s} = 1.397 F^{1.06} \tag{9}$$

The power laws proposed by Boes (2000) and Boes & Hager (2003b) were similar to equation 9, but having different coefficients. In order to compare the different proposals, equation 9 was modified by replacing $z_i$ by $z_i'$ (see Figure 10a). The energy equation was also used, for the region between the critical section (Section 1, represented by the subscript "c") and the initial section of the experiments (Section 2, at x = 0, represented by (0)). The Darcy-Weisbach equation was applied with average values for the hydraulic radius and the velocity. The resulting equation is similar to that proposed by Boes (2000, p.126), who also used the Coriolis coefficient, assumed unity in the present study. Equation (10) is the equation adopted in the present study:

$$z_c - z(0) = \frac{h(0)\cos\alpha + h_c^3/[2h(0)^2]-(3/2)h_c}{1-f\dfrac{h_c^3(h(0)+h_c)}{(h_c h(0))^2 16\sin\alpha}\dfrac{(B+h_c+h(0))}{B}} \tag{10}$$

The calculation of $z_i'$ requires the resistance factor, obtained here applying the methodology described by Simões et al. (2010). The obtained equation presented a correlation coefficient of 0.98 when compared to the measured data, having the form:

$$z_i'/s = 3.19F^{0.837} \tag{11}$$

where: $z_i' = z_i + z_c - z(0)$. This equation is valid for the same ranges of the variables shown for equations 7 and 8. Equation 11 furnishes lower values of $z_i/s$ when compared to the equations of Boes (2000, p.126) and Boes & Hager (2003b). Two reasons for the difference are mentioned here: (1) The method used to calculate the resistance factor, and (2) the definition of the position of the inception point. The mentioned authors defined the starting position of the aeration as the point where the void fraction at the pseudo-bottom is 1%, while the present definition corresponds to the final section of the $S_2$ profile. Using the equation of Boes & Hager (2003b), $z_i'/s = 5.9F^{0.8}$, it is possible to relate these two positions, as shown by equation 12. The position of the void fraction of 1% at the pseudo-bottom occurs approximately at 1.85 times the position of the final section of the $S_2$ profile, with both lengths having their origin at the crest of the spillway.

$$\frac{z_i')_{1\%}}{z_i')_{S_2}} = \frac{1.85}{F^{0.037}} \tag{12}$$

In equation 12, $z_i')_{1\%}$ corresponds to the length defined by the equation of Boes & Hager (2003b), and $z_i')_{S2}$ corresponds to the length defined by eq. 11. Transforming equation 11 considering the variables $L_{A*}/k$ and $F_r^*$, in which $L_{A*}$ correspond to $z_i'$, the following relation between the Froude numbers is obtained

$$F = \eta F_r^* \Rightarrow \frac{q}{\sqrt{gs^3\sin\alpha}} = \eta\frac{q}{\sqrt{gk^3\sin\alpha}} \Rightarrow \eta = \left(\frac{k}{s}\right)^{3/2} \Rightarrow F = \left(\frac{k}{s}\right)^{3/2} F_r^*$$

and remembering that $L_{A*}/k = (z_i'/s)/(\sin\alpha\cos\alpha)$, leads then to:

$$\frac{L_A^*}{k} = 4.13F_r^{*0.837} \tag{13}$$

The behavior of equation 13 in comparison with experimental data is illustrated in Figure 11, which contains experimental data found in the literature, as well as two additional predictive curves. Observe that, except for the first two points (obtained in the present study), the results are located close to the curve defined by Matos (1999). It is also interesting to note that the equations proposed by Matos (1999) and Sanagiotto (2003) are approximately parallel.

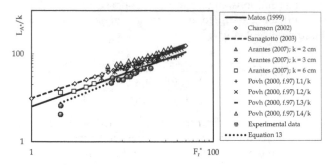

Fig. 11. Starting position of the aeration: a comparison between the experimental data of this research, the equation obtained in this work and data (experimental and numerical) of different authors.

### 3.2.2 Depths at the end of $S_2$

As in the previous case, power laws and sum of power laws were used to quantify the flow depth at the inception point, that is, in the final section of the $S_2$ profile. Equations 14 and 15 were then obtained, with correlation coefficients 0.97 and 0.98, respectively. Figure 12 contains a comparison with data from different sources.

$$\frac{h_A}{k} = 0.363 F_r^{*0.609} \tag{14}$$

$$\frac{h_A}{k} = 0.791 F_r^{*-6.98} + 1.285 \left[ \frac{h(0)}{k} \right]^{0.567} - 19.56 \, Re(0)^{-0.322} \tag{15}$$

Equations 14 and 15 are restricted to: $2.09 \leq F_r^* \leq 20.70$, $0.69 \leq h(0)/k \leq 2.99$ and $1.15 \times 10^5 \leq Re(0) \leq 7.04 \times 10^5$.

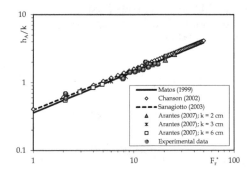

Fig. 12. Depth in the starting position of the aeration based on the final section of the $S_2$ profile: comparison with experimental and numerical data of different sources.

### 3.2.3 Transition to two-phase flow

The experiments showed that the averaged values of the depths form a free surface profile composed by a decreasing region ($S_2$) followed by a growing region that extends up to a maximum depth, from which a wavy shape is formed downstream, as illustrated by Figure 13a. The maximum value which limits the growing region is denoted by $h_2$. The length of the transition between the minimum ($h_A$) and the maximum ($h_2$) is named here "transition length", and is represented by L, a distance parallel to the pseudo bottom, as shown in Figure 13a. $h_A/k$ was related to $h_2/k$ using a power law (equation 16), showing a good superposition between experimental data and the adjusted equation, as shown in figure 13b, with a correlation coefficient of 0.99.

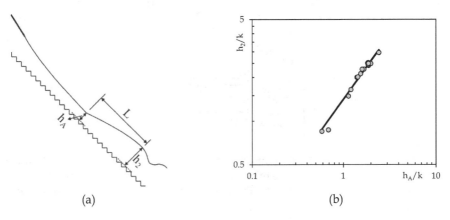

(a)                                                                                 (b)

Fig. 13. a) Definition of the depth $h_2$ and the transition length L, and b) correlation between the depth at the start of aeration ($h_A/k$) and the depth corresponding to the first wave crest ($h_2/k$), expressed by equation 16.

$$\frac{h_2}{k} = 1.408 \left( \frac{h_A}{k} \right)^{0.879} \tag{16}$$

As in previous cases, also the parameters $F_r^*$, $h(0)/k$ and $Re(0)$ were used to quantify $h_2/k$, for which equation 17 was obtained, with a correlation coefficient of 0.99. The ranges of validity of equations 16 and 17 are the same as for equations 7 and 8.

$$\frac{h_2}{k} = 0.319 F_r^{*0.553} + 0.529 \left[ \frac{h(0)}{k} \right]^{0.744} - 1.6 \times 10^4 \, Re(0)^{-2.1 \times 10^5} \tag{17}$$

The transition length between the last "full water" section (the last $S_2$ section) and the first "full mixture" section, or, in other words, the section at which the air reaches the pseudo-bottom, could be well characterized using the ultrasound sensor. From a practical point of view, this length is relevant because it involves a region of the spillway still unprotected, due to the absence of air near the bottom. Experimental verification of void fractions is still necessary to establish the void percentage attained at the pseudo-bottom in the mentioned section. An analysis is presented here considering the hypothesis that the "full mixture" section defined by the maximum of the measured depths corresponds to the 1% void fraction defined by Boes (2000) and Boes & Hager (2003b).

Combining the transition lengths with the values of $L_{A*}$ (or $z_i'$), the positions of the inception point considering this new origin are then obtained. This length was correlated with the dimensionless parameters $(z_i'+L\sin\alpha)/s = z_L/s$ and $F = q/(gs^3\sin\alpha)^{0.5}$. Equation 18 was then obtained, with a correlation coefficient of 0.95. Figure 14 illustrates the behavior of this adjustment in relation to the experimental data. The same figure also shows the curve obtained with the equation of Boes & Hager (2003b), showing that the two forms of analyses generate very similar results.

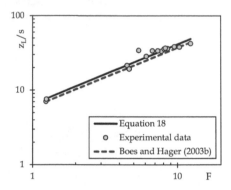

Fig. 14. Starting position of the aeration considering the present analysis (equation 18), corresponding to the position of the maximum of the measured depths, and the equation of Boes & Hager (2003b), $z_i'/s = 5.9F^{0.8}$ (in this case, $z_i'$ corresponds to a mean void fraction of 1% at the pseudo bottom).

$$\frac{z_L}{s} = 6.4F^{0.81} \tag{18}$$

The equations proposed for Boes (2000) and Boes & Hager (2003b) allow to relate $z_L/s$ with the position of the 1% void fraction on the pseudo bottom, leading to:

$$\frac{z_i')_{1\%}}{z_L} = 0.73F^{0.03} \quad z_i')_{1\%} \text{ from Boes (2000)} \tag{19}$$

$$\frac{z_i')_{1\%}}{z_L} = \frac{0.92}{F^{0.01}} \quad z_i')_{1\%} \text{ from Boes \& Hager (2003b)} \tag{20}$$

As can be seen, the results show different trends in relation to the Froude number. Such differences may be related to the values of the adjusted exponents, which are close, but not the same. Equation 18 is very similar to equation 11, with the Froude number in both equations having similar exponents, and the coefficient of equation 18 being 2 times bigger that the coefficient of equation 11. This result is close to the factor of 1.85 obtained with equation 12. Figure 15 shows that the results obtained with the present analysis are close to those obtained with the equation Boes & Hager (2003b), suggesting to use the maximum depth to locate the beginning of the bottom aeration, or, in other words, the position where there is a void fraction of 1% at the bottom. It is important to emphasize that the measurement of the position of the free surface is much simpler than the measurement of the concentrations at the pseudo-bottom, so that we suggest the present methodology to evaluate the position of the beginning of the "full mixed" region. Of course, the void fraction measurements at the bottom were important to allow the present comparison.

Of course, having a first confirmation, it is possible to obtain the same information involving the different axes used in spillway studies. For example, it is possible to $L_A^*/k$ with $F_r^*$, $L_A^*$ being the sum of $L_{A^*}$ with L. The result is equation 21, with a correlation coefficient of 0.95, and which behavior is illustrated in Figure 11, where it is compared with data from other sources. In general, there is a good agreement of equation 21 with most of the results of the cited studies. Special mention made be made for the data L4/k obtained by Povh (2000) (the L4 position corresponds to the fully-aerated section of the flow), the data of Chanson (2002), and Sanagiotto (2003).

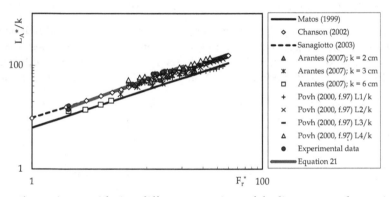

Fig. 15. Inception point considering different equations of the literature and equation 21: a comparison with data from different authors

$$\frac{L_A^*}{k} = 8.4F^{0.81} \tag{21}$$

Following the previous procedures followed in this section, equation 22 was also obtained, involving the geometrical information of the flow and the Reynolds number, presenting a correlation coefficient of 0.98.

$$\frac{L_A^*}{k} = 2397.09 F_r^{*-6.36} - 32.49\left[\frac{h(0)}{k}\right]^{-1.29} + 0.212\,Re(0)^{0.452} \tag{22}$$

The restrictions of this study are $2.09 \leq F_r^* \leq 20.70$, $0.69 \leq h(0)/k \leq 2.99$ and $1.15 \times 10^5 \leq Re(0) \leq 7.04 \times 10^5$.

### 3.2.4 Turbulence intensity and kinetic energy

The time derivatives of the position of the free surface were used to evaluate the turbulent intensity (w') and, assuming isotropy (as a first approximation), the turbulent kinetic energy ($k_e$), defined in equations (23) and (24):

$$w' = \sqrt{\overline{w^2}} \tag{23}$$

$$k_e = \frac{3}{2}w'^2 \tag{24}$$

Also a relative intensity and a dimensionless turbulent kinetic energy were defined, written in terms of the critical kinetic energy (all parameters per unit mass of fluid), which are represented by equations (25) and (26):

$$ir = \frac{w'}{V_c} \tag{25}$$

$$k_e^* = ir^2 \tag{26}$$

$V_c$ is given by $V_c = (gh_c)^{1/2}$; and $h_c = (q^2/g)^{1/3}$ (critical depth).

Figure 16 contains the results obtained in the present study for the relative intensities and dimensionless kinetic energy, both plotted as a function of the dimensionless position $z/z_i$, where $z$ = vertical axis with origin at $x=0$ and positive downwards. Four different regions may be defined for the obtained graphs: (1) Single-phase growing region, (2) Single-phase

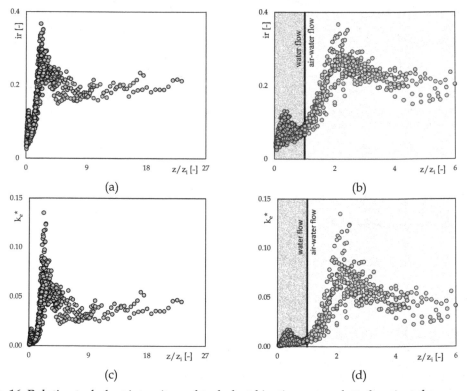

Fig. 16. Relative turbulent intensity and turbulent kinetic energy plotted against the dimensionless vertical position. The starting position of the aeration is defined as the final section of the $S_2$ profile. (a, b) turbulent intensity; (b, c) dimensionless kinetic energy.

decay region, which is limited downwards around the point $z/z_i=0.9$, (3) Two-phase growing region, limited by $\sim0.9<z/z_i<\sim2.11$, and (4) Two-phase decay region.

Considering the decay region limited by $2.5<z/z_i<14$, a power law of the type $k_e=a(z/z_i)^{-n}$ was adjusted, obtaining $n = 0.46$ with a correlation coefficient of 0.72. Using the terminology of the $k_e$-$\varepsilon$ model, for which the constant $C_{\varepsilon2}=(n+1)/n$ is defined, it implies in a $C_{\varepsilon2}=3.7$, which is about 1.7 times greater than the value of the standard model, $C_{\varepsilon2}=1.92$ (Rodi, 1993). This analysis was conducted to verify the possibility of obtaining statistical parameters linked to the kinetic energy, similar to those found in the literature of turbulence.

## 4. Numerical simulations

### 4.1 Introduction

Turbulence is a three dimensional and time-dependent phenomenon. If Direct Numerical Simulation (DNS) is planned to calculate turbulence, the Navier-Stokes and continuity equations must be used without any simplifications. Since there is no general analytical solution for these equations, a numerical solution which considers all the scales existing in turbulence must use a sufficiently refined mesh. According to the theory of Kolmogorov, it can be shown that the number of degrees of freedom, or points in a discretized space, is of the order of (Landau & Lifshitz, 1987, p.134):

$$\left(L_k / \eta\right)^3 = Re^{9/4} \qquad (27)$$

where: $L_k$ = characteristic dimension of the large-scales of the movement of the fluid, $\eta$ = Kolmogorov micro-scale of turbulence and Re = Reynolds number of the larger scales. Considering a usual Reynolds number, like Re = $10^5$, the mesh must have about $10^{11}$ elements. This number indicates that it is impossible to perform the wished DNS with the current computers. So, we must lower our level of expectations in relation to our results. A next "lower" level would be to simulate only the large scales (modeling the small scales) or the so called large-eddy simulation (LES). This alternative is still not commonly used in problems composed by a high Reynolds number and large dimensions. So, lowering still more our expectations, the next level would be the full modeling of turbulence, which corresponds to the procedures followed in this study. This chapter presents, thus, results obtained with the aid of turbulence models (all scales are modeled), which is the usual way followed to study flows around large structures and subjected to large Reynolds numbers.

### 4.2 Some previous studies

In recent years an increasing number of papers related to the use of CFD to simulate flows in hydraulic structures and in stepped spillways has been published. Some examples are Chen et al. (2002), Cheng et al. (2004), Inoue (2005), Arantes (2007), Carvalho & Martins (2009), Bombardelli et al. (2010), Lobosco & Schulz (2010) and Lobosco et al. (2011). Different aspects of turbulent flows were studied in these simulations, such as the development of boundary layers, the energy dissipation, flow aeration, scale effects, among others. The turbulence models k-$\varepsilon$ and RNG k-$\varepsilon$ were used in most of the mentioned studies, and Arantes (2007) also used the SSG Reynolds stress model (Speziale, Sarkar and Gatski, 1991). Some researchers have still adopted commercial softwares to perform their simulations, such as ANSYS CFX® and Fluent®. On the other hand, Lobosco & Schulz (2010) and Lobosco et al. (2011), for example, used a set of free softwares, among which the OpenFOAM® software. In this study we used the ANSYS CFX® software.

### 4.3 Results
### 4.3.1 Free surface comparisons

The experiments summarized in Table 1 were also simulated, in order to verify the possibilities of reproducing such flows using CFD. The Exp. 15 is the only one shown here, which main characteristics may be found in Table 1, and which was simulated considering the hypothesis of two-dimensional flow for the geometry sketched in Figure 17a. The inlet velocity was set as the mean measured velocity, with the value of 2.91 m/s, the outlet boundary condition was set to extrapolate the volume fractions of air and water. The analytical solution presented by Simões et al. (2010) was used to calculate the theoretical profile of the free surface for the single phase flow, for which  f = 0.041 resulted as the adjusted resistance factor. Figure 17b contains experimental data and numerical solutions calculated with different meshes and the following turbulence models: zero equation, k-ε, RNG k-ε and SSG. These results were obtained combining the non-homogeneous model and the free-surface model for the interfacial transfers. There is excellent agreement between the experimental points, the numerical results and theoretical curve for the one-phase region. It was found that the use of the mesh denoted by M2 (data indicated in the legend) led to results similar to those obtained with the mesh denoted by M1, two times more refined,

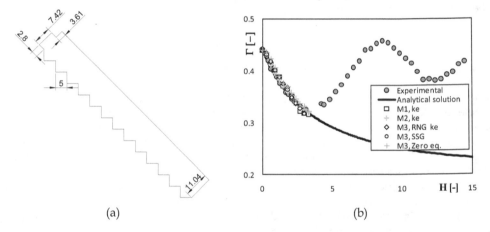

(a)                                                                                (b)

Fig. 17. Exp. 18 simulations: (a) Geometry and dimensions (in cm), (b) Experimental results, numerical solutions and analytical profile (ke=k-ε; M1 and M2 are unstructured meshes with ~0.5x10⁶ and ~0.25x10⁶ elements, respectively; M3 is a structured mesh with ~0.2x10⁶ elements).

when using the k-ε model. Further, the results obtained with the models RNG k-ε, SSG and without transport equations for turbulence (the three calculated using a third mesh denoted by M3) also superposed well the theoretical solution. In addition to the mentioned turbulence models, also the models k-ω, BSL and k-ε EARMS were tested. Only the k-ε EARMS model produced results with quality similar to those shown in Figure 17b. The k-ω and BSL models overestimated the depth of the flow, with maximum relative deviations from the experimental values near 8%. The same deviation was observed when using the mixture model in place of the free-surface model (for simulations using the k-ε model).

Persistent air cavities below the pseudo-bottom were observed in all simulated results, which is an "uncomfortable" characteristic of the simulations, because the experimental observations did not present such cavities (assuming, as usual, that the numerical solutions converged to the analytical solutions, which, on its turn, is viewed as a good model of the real flows). It must be emphasized that the predictions reproduce the one-phase flow, but that the two-phase flow presents undulating characteristics still not reproduced by numerical simulations. The undulating aspect is observed for different experimental conditions, as shown by Simões et al. (2011), and a complete quantification is still not available.

### 4.3.2 Simulations using prototype sizes

The inhomogeneous model and the k-ε turbulence model were selected to obtain the free-surface profiles using "numerical" scales compatible with prototype scales. The simulations were performed for steady state turbulence, with s = 0.60 m (27 simulations) and s = 2.4 m (four simulations with 1V:0.75H), considering two-dimensional domains, using high resolution numerical schemes and applying boundary conditions similar to those already described in the first example. The inlet condition is the same for all simulations (Figure 18a). The angles between the pseudobottom and the horizontal, and the dimensionless parameter $s/h_c$, chosen for the simulations were: 53.13° ($0.133 \leq s/h_c \leq 0.845$), 45° ($0.11 \leq s/h_c \leq 0.44$), 30.96° ($0.11 \leq s/h_c \leq 0.44$), and 11.31° ($0.133 \leq s/h_c \leq 0.44$). For each experiment a numerical value for the resistance factor was calculated, as described in item 4.3.1. Figure 18b, shows the analytical solution and the points obtained with the Reynolds Averaged

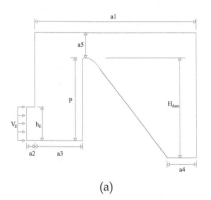

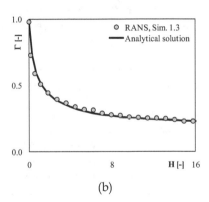

(a)                                                    (b)

Fig. 18. (a) Domain employed to perform the simulations for α=53.13 (colors = void fraction). The values of $a_i$ (i=1-5), P, $H_{dam}$, $h_E$ were chosen for each test (b) $S_2$ profile: numerical and analytical solution.

Navier-Stokes Equations (RANS) for multiphase flows. The axes shown in Figure 18b represent the following nondimensional parameters: $\Gamma = h/h_c$ and $H = z/h_c$. In this case, z has origin at the critical section, where $h = h_c$ (close to the crest of the spillway). All the results showed similar or superior quality to that presented in Figure 18.

Figure 19a shows the distribution of the friction factor values obtained numerically, for all simulations performed for the geometrical conditions described in Figure 18a. Figure 19b contains the distribution of f considering the experimental and the numerical data together. It was possible to adjust power laws between f and $k/h_c$ (k=scosα), in the form $f = a(k/h_c)^b$. The values of the adjusted "a" and "b", and the limits of validity of the adjusted equations, together with the geometrical information, are given in Table 2.

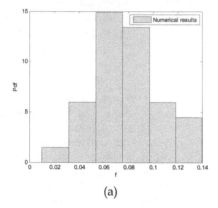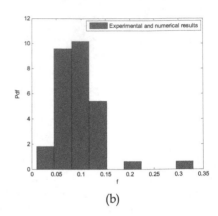

(a)                                             (b)

Fig. 19. Friction factor: (a) Probability distribution function for 11.31°≤α≤51.13° and numerical results; (b) Probability distribution function considering the numerical (11.31°≤α≤51.13°) and the experimental results together (α=45°). The total area covered by the bars is equal to 1.0 in both figures (R = correlation coefficient; Re = 4q/ν).

| α | a | b | $k/h_c$ | Re | R |
|---|---|---|---|---|---|
| [degrees] | [-] | [-] | [-] | [-] | [-] |
| 53.13 | 0.195 | 0.502 | 0.0798-0.485 | 8.0E6-1.2E8 | 0.97 |
| 45.00 | 0.146 | 0.355 | 0.0776-0.311 | 2.0E7-1.6E8 | 0.97 |
| 30.96 | 0.185 | 0.294 | 0.0942-0.377 | 2.0E7-1.6E8 | 0.96 |

Table 2. Coefficients for $f = a(k/h_c)^b$ and other details.

## 5. Conclusions

In this chapter different aspects of the flows over stepped spillways were described, considering analytical, numerical and experimental points of view. The results show characteristics not usually found in the literature, and point to the need of more studies in this field, considering the practical use of stepped chutes in hydraulic structures.

## 6. Acknowledgements

The authors thank CNPq(141078/2009-0), CAPES and FAPESP, Brazilian research support institutions, for the financial support of this study.

## 7. References

Arantes, E. J. (2007). Stepped spillways flow characterization using CFD tools. Dr Thesis, School of Engineering at São Carlos, University of São Paulo, São Carlos, Brazil [in Portuguese].

Bauer, W.J. (1954). Turbulent boundary layer on steep slopes. *Transactions*, ASCE, Vol. 119, Paper 2719, pp. 1212-1232.

Bernoulli, J. (1743). *Hydraulics*. Dover Publications, 2005.

Boes, R. (2000). Zweiphasenströmung und Energieumsetzung an Grosskaskaden. PhD Thesis – ETH, Zurich.

Boes, R.M. & Hager, W.H. (2003a). Hydraulic Design of Stepped Spillways. ASCE, Journal of Hydraulic Engineering. v.129, n.9, pp.671-679.

Boes, R.M. & Hager, W.H. (2003b). Two-Phase flow characteristics of stepped spillways. ASCE, *Journal of Hydraulic Engineering*. v.129, n.9, p.661-670.

Bombardelli, F.A.; Meireles, I. & Matos, J. (2010). Laboratory measurements and multi-block numerical simulations of the mean flow and turbulence in the non-aerated skimming flow region of steep stepped spillways. Environ. Fluid Mech., v.11, n.3, pp.263-288. Publisher: Springer Netherlands (DOI 10.1007/s10652-010-9188-6).

Cain, P. & Wood, I.R. (1981) Instrumentation for aerated flow on spillways. ASCE, *Journal of Hydraulic Engineering*, Vol. 107, No HY11.

Carosi, G. & Chanson, H. (2006). Air-water time and length scales in skimming flows on a stepped spillway. Application to the spray characterization. The University of Queensland, Brisbane, Austrália.

Carvalho, R.F. & Martins, R. (2009). Stepped Spillway with Hydraulic Jumps: Application of a Numerical Model to a Scale Model of a Conceptual Prototype. *Journal of Hydraulic Engineering*, ASCE 135(7):615–619.

Chamani, M.R. & Rajaratnam, N. (1999a). Characteristic of skimming flow over stepped spillways. ASCE, *Journal of Hydraulic Engineering*. v.125, n.4, p.361-368, April.

Chamani, M. R.; Rajaratnam, N. (1999b). Onset of skimming flow on stepped spillways. ASCE, *Journal of Hydraulic Engineering*. v.125, n.9, p.969-971.

Chanson, H. (1993). Stepped spillway flows and air entrainment. *Canadian Journal of Civil Engineering*. v.20, n.3, p.422-435, Jun.

Chanson, H. (1994). Hydraulics of nappe flow regime above stepped chutes and spillways. *Journal of Hydraulic Research*, v.32, n.3, p.445-460, Jan..

Chanson, H. (1996). *Air bubble entrainment in free-surface turbulent shear flows*. Academic Press, San Diego, California.

Chanson, H. (2002). *The hydraulics of stepped chutes and spillways*. A.A. Balkema Publishers, ISBN 9058093522, The Netherlands.

Chen, Q.; Dai, G. & Liu, H. (2002). Volume of fluid model for turbulence numerical simulation of stepped spillway overflow. *Journal of Hydraulic Engineering*, ASCE 128(7): pp.683–688.

Cheng X.; Luo L. & Zhao, W. (2004). Study of aeration in the water flow over stepped spillway. In: Proceedings of the World Water Congress 2004, ASCE, Salt Lake City, UT, USA.

Chow, V.T. (1959). Open channel hydraulics. New York: McGraw-Hill.

Diez-Cascon, J.; Blanco, J.L.; Revilla, J. & Garcia, R. (1991). Studies on the hydraulic behavior of stepped spillways. *Water Power & Dam Construction*, v.43, n.9, p.22-26, Sept..

Frizell, K.H. (2006). Research state-of-the-art and needs for hydraulic design of stepped spillways. *Water Resources Researches Laboratory*. Denver, Colorado, June.

Hager, W.H. (1995). *Cascades, drops and rough channels*. In.: Vischer, D.L.; Hager, W.H. (Ed.). Energy dissipators IAHR, Hydraulics Structures Design Manual. v.9, p.151-165, Rotterdam, Netherlands.

Essery, I.T.S. & Horner, M.W. (1978). The hydraulic design of stepped spillways. 2nd Ed. London: Construction Industry Research and Information Association, CIRIA Report No. 33.

Horner, M.W. (1969). An analysis of flow on cascades of steps. Ph.D. Thesis – Universidade de Birmingham, UK.

Inoue, F.K. (2005). Modelagem matemática em obras hidráulicas. MSc Thesis, Setor de Ciências Tecnológicas da Universidade Federal do Paraná, Curitiba.

Keller, R.J.; Lai, K.K. & Wood, I.R. (1974). Developing Region in Self-Aerated Flows. *Journal of Hydraulic Division*, ASCE, 100(HY4), pp. 553-568.

Keller, R.J. & Rastogi, A.K. (1977). Prediction of flow development on spillway. *Journal of Hydraulic Division*, ASCE, Vol. 101, No HY9, Proc. Paper 11581, Sept.

Landau, L.D. & Lifshitz, E.M. (1987). *Fluid Mechanics*. 2nd Ed. Volume 6 of Course of Theoretical Physics, ISBN 978-0750627672.

Lobosco, R.J. & Schulz, H.E. (2010). Análise Computacional do Escoamento em Estruturas de Vertedouros em Degraus. *Mecanica Computacional*, AAMC, Vol. XXIX, No 35, Buenos Aires, http://amcaonline.org.ar/ojs/index.php/mc/article/view/3252, pp.3593-3600.

Lobosco, R.J.; Schulz, H.E.; Brito, R.J.R. & Simões, A.L.A. (2011). Análise computacional da aeração em escoamentos bifásicos sobre vertedouros em degraus. 6° Congresso Luso-Moçambicano de Engenharia/3° Congresso de Engenharia de Moçambique.

Lueker, M.L.; Mohseni, O.; Gulliver, J.S.; Schulz, H.E. & Christopher, R.A. (2008). The physical model study of the Folsom Dam Auxiliary Spillway System, St. Anthony Falls Lab. Project Report 511. University of Minnesota, Minneapolis, MN.

Matos, J.S.G. (1999). Emulsionamento de ar e dissipação de energia do escoamento em descarregadores em degraus. Research Report, IST, Lisbon, Portugal.

Murzyn, F.& Chanson, H. (2009) Free-surface fluctuations in hydraulic jumps: Experimental observations. *Experimental Thermal and Fluid Science*, 33(2009), pp.1055-1064.

Ohtsu, I. & Yasuda, Y. (1997). Characteristics of flow conditions on stepped channels. In: Biennal Congress, 27, San Francisco, Anais… San Francisco: IAHR, pp. 583-588.

Ohtsu, I., Yasuda Y., Takahashi, M. (2001). Onset of skimming flow on stepped spillways – Discussion. *Journal of Hydraulic Engineering*. v. 127, p. 522-524. Chamani, M.R.; Rajaratnam, N. ASCE, Journal of Hydraulic Engineering. v. 125, n.9, p.969-971, Sep.

Ohtsu, I.; Yasuda, Y. & Takahashi, M. (2004). Flows characteristics of skimming flows in stepped channels. ASCE, *Journal of Hydraulic Engineering*. v.130, n.9, pp.860-869, Sept..

Peterka, A. J. (1953). The effect of entrained air on cavitation pitting. *Joint Meeting Paper*, IAHR/ASCE, Minneapolis, Minnesota, Aug..

Peyras, L.; Royet, P. & Degoutte, G. (1992). Flow and energy dissipation over stepped gabion weirs. *Journal of Hydraulic Engineering*, v.118, n.5, p.707-717, May..

Povh, P.H. (2000). Avaliação da energia residual a jusante de vertedouros em degraus com fluxos em regime skimming flow. [Evaluation of residual energy downstream of stepped spillways in skimming flows]. MSc Thesis. Department of Technology, Federal University of Paraná, Curitiba, Brazil [in Portuguese].

Rajaratnam, N. (1990). Skimming flow in stepped spillways. *Journal of Hydraulic Engineering*, v.116, n.4, p. 587-591, April.

Richter, J. P. (1883). *Scritti letterari di Leonardo da Vinci*. Sampson Low, Marston, Searle & Rivington, Londra. In due parti, p.1198 (volume 2, p.236). Available from <http:// www.archive.org/details/literaryworksofl01leonuoft (16/04/2008).

Rodi, W. (1993). *Turbulence Models and Their Application in Hydraulics*. IAHR Monographs, Taylor & Francis, 3ª Ed., ISBN-13: 978-9054101505.

Sanagiotto, D.G. (2003). Características do escoamento sobre vertedouros em degraus de declividade 1V:0.75H [Flow characteristics in stepped spillways of slope 1V:0.75H]. MSc Thesis. Institute of Hydraulic Research, Federal University of Rio Grande do Sul, Porto Alegre, Brazil, [in Portuguese].

Simões, A.L.A. (2008). Considerations on stepped spillway hydraulics: Nondimensional methodologies for preliminary design. MSc. Thesis. School of Engineering at São Carlos, University of São Paulo, Brazil, [in Portuguese].

Simões, A.L.A. (2011). Escoamentos em canais e vertedores com o fundo em degraus: desenvolvimentos experimentais, teóricos e numéricos. Relatório (Doutorado) – Escola de Engenharia de São Carlos – Universidade de São Paulo, 157 pp.

Simões, A.L.A.; Schulz, H.E. & Porto, R.M. (2010). Stepped and smooth spillways: resistance effects on stilling basin lengths. *Journal of Hydraulic Research*, Vol.48, No.3, pp.329-337.

Simões, A.L.A.; Schulz, H.E. & Porto, R.M. (2011). Transition length between water and air-water flows on stepped chutes. Computational Methods in Multiphase Flow VI, pp.95-105, doi:10.2495/MPF110081, Kos, Greece.

Sorensen, R.M. (1985). Stepped spillway hydraulic model investigation. *Journal of Hydraulic Engineering*, v.111, n.12, pp. 1461-1472. December.

Speziale, C.G.; Sarkar, S. & Gatski, T.B. (1991). Modeling the pressure-strain correlation of turbulence: an invariant dynamical systems approach. *Journal of Fluid Mechanics*, Vol. 227, pp. 245-272.

Stephenson, D. (1991). Energy dissipation down stepped spillways. *Water Power & Dam Construction*, v. 43, n. 9, p. 27-30, September.

Straub, L.G. & Anderson, A.G. (1958). Experiments on self-aerated flow in open channels. *Journal of Hydraulic Division*, ASCE Proc., v.87, n.HY7, pp. 1890-1-1890-35.

Tozzi, M.J. (1992). Caracterização/comportamento de escoamentos em vertedouros com paramento em degraus [Characterization of flow behavior in stepped spillways]. Dr Thesis. University of São Paulo, São Paulo, Brazil, [in Portuguese]. 302 pp. Dr Thesis – Universidade de São Paulo, São Paulo.

Wilhelms, S.C. & Gulliver, J.S. (2005). Bubbles and waves description of self-aerated spillway flow. *Journal of Hydraulic Research*, Vol. 43, No.5, pp. 522-531. 2005

Wood, I.R.; Ackers, P. & Loveless, J. (1983). General method for critical point on spillways. *Journal of Hydraulic Engineering*, Vol. 109, No. 27, pp. 308-312, 1985.

Wood, I.R. (1984). Air entrainment in righ speed flows. *Symposium on scale Effects in modelling hydraulic structures*, IAHR, Kobus, H. (Ed.), paper 4.1, Sep..

# Permissions

The contributors of this book come from diverse backgrounds, making this book a truly international effort. This book will bring forth new frontiers with its revolutionizing research information and detailed analysis of the nascent developments around the world.

We would like to thank Harry Edmar Schulz, André Luiz Andrade Simões and Raquel Jahara Lobosco, for lending their expertise to make the book truly unique. They have played a crucial role in the development of this book. Without their invaluable contribution this book wouldn't have been possible. They have made vital efforts to compile up to date information on the varied aspects of this subject to make this book a valuable addition to the collection of many professionals and students.

This book was conceptualized with the vision of imparting up-to-date information and advanced data in this field. To ensure the same, a matchless editorial board was set up. Every individual on the board went through rigorous rounds of assessment to prove their worth. After which they invested a large part of their time researching and compiling the most relevant data for our readers. Conferences and sessions were held from time to time between the editorial board and the contributing authors to present the data in the most comprehensible form. The editorial team has worked tirelessly to provide valuable and valid information to help people across the globe.

Every chapter published in this book has been scrutinized by our experts. Their significance has been extensively debated. The topics covered herein carry significant findings which will fuel the growth of the discipline. They may even be implemented as practical applications or may be referred to as a beginning point for another development. Chapters in this book were first published by InTech; hereby published with permission under the Creative Commons Attribution License or equivalent.

The editorial board has been involved in producing this book since its inception. They have spent rigorous hours researching and exploring the diverse topics which have resulted in the successful publishing of this book. They have passed on their knowledge of decades through this book. To expedite this challenging task, the publisher supported the team at every step. A small team of assistant editors was also appointed to further simplify the editing procedure and attain best results for the readers.

Our editorial team has been hand-picked from every corner of the world. Their multi-ethnicity adds dynamic inputs to the discussions which result in innovative outcomes. These outcomes are then further discussed with the researchers and contributors who give their valuable feedback and opinion regarding the same. The feedback is then collaborated with the researches and they are edited in a comprehensive manner to aid the understanding of the subject.

Apart from the editorial board, the designing team has also invested a significant amount of their time in understanding the subject and creating the most relevant covers. They scrutinized every image to scout for the most suitable representation of the subject and create an appropriate cover for the book.

The publishing team has been involved in this book since its early stages. They were actively engaged in every process, be it collecting the data, connecting with the contributors or procuring relevant information. The team has been an ardent support to the editorial, designing and production team. Their endless efforts to recruit the best for this project, has resulted in the accomplishment of this book. They are a veteran in the field of academics and their pool of knowledge is as vast as their experience in printing. Their expertise and guidance has proved useful at every step. Their uncompromising quality standards have made this book an exceptional effort. Their encouragement from time to time has been an inspiration for everyone.

The publisher and the editorial board hope that this book will prove to be a valuable piece of knowledge for researchers, students, practitioners and scholars across the globe.

# List of Contributors

António A.L.S. Duarte
University of Minho, Portugal

Shangming Li
Institute of Structural Mechanics, China Academy of Engineering Physics, Mianyang City, Sichuan Province, China

Luciana de Souza Cardoso
Universidade Federal do Rio Grande do Sul (UFRGS), Instituto de Biociências, Brazil

Rafael Siqueira Souza and David da Motta Marques
Universidade Federal do Rio Grande do Sul (UFRGS), Instituto de Pesquisas Hidráulicas (IPH), Brazil

Carlos Ruberto Fragoso Jr.
Universidade Federal de Alagoas (UFAL), Centro de Tecnologia, Brazil

Franklin Torres-Bejarano, Hermilo Ramirez and Clemente Rodríguez
Mexican Petroleum Institute, Mexico

Alan Cavalcanti da Cunha, Daímio Chaves Brito, Helenilza Ferreira Albuquerque Cunha and Eldo Santos
Federal University of Amapá - Environmental Science Department and Graduated Program in Ecological Sciences of Tropical Biodiversity, Brazil

Antonio C. Brasil Junior and Luis Aramis dos Reis Pinheiro
Universidade de Brasilia. Laboratory of Energy and Environment, Brazil

Alex V. Krusche
Environmental Analysis and Geoprocessing Laboratory CENA, Brazil

Renata Archetti and Maurizio Mancini
DICAM University of Bologna, Bologna, Italy

Steve Brenner
Department of Geography and Environment, Bar Ilan University, Israel

Jisheng Zhang and Chi Zhang
State Key Laboratory of Hydrology-Water Resources and Hydraulic Engineering, Hohai University, Nanjing, 210098, China

**XiuguangWu**
Zhejiang Institute of Hydraulics and Estuary, Hangzhou, 310020, China

**Yakun Guo**
School of Engineering, University of Aberdeen, Aberdeen, AB24 3UE, United Kingdom

**Ava Maxam and Dale Webber**
University of the West Indies, Jamaica

**Zhijie Chen, Binxin Zheng, Zhi Zeng and Jia He**
Open Lab of Ocean & Coast Environmental Geology, Third Institute of Oceanography, SOA, China

**Yongxue Wang and Weiguang Zuo**
State Key Laboratory of Coastal and Offshore Engineering, Dalian University of Technology, China

**Xie Xiaoping**
School of Geography and Tourism, Qufu Normal University, Qufu, China

**Rafael Manica**
Instituto de Pesquisas Hidráulicas - Universidade Federal do Rio Grande do Sul, Brazil

**André Luiz Andrade Simões, Harry Edmar Schulz, Raquel Jahara Lobosco and Rodrigo de Melo Porto**
University of São Paulo, Brazil